Alder, Kurt.
Electromagnetic excitation : theory of Coulomb excitation with heavy ions / Kurt Alder and Aage Winther. -- Amsterdam : North-Holland Pub. Co. ; New York : American Elsevier Pub. Co., 1975.
ix, 364 p. : ill. ; 23 cm.
Bibliography: p. 359-361.

ISBN 0-7204-0288-3 (North-Holland)
ISBN 0-444-10826-2 (American Elsevier)

(continued next card)
7/76

ELECTROMAGNETIC EXCITATION

THEORY OF COULOMB EXCITATION WITH HEAVY IONS

ELECTROMAGNETIC EXCITATION

THEORY OF COULOMB EXCITATION WITH HEAVY IONS

Kurt ALDER

Institute for Theoretical Physics, University of Basel, Switzerland

and

Aage WINTHER

The Niels Bohr Institute, University of Copenhagen, Denmark

1975

NORTH-HOLLAND PUBLISHING COMPANY, AMSTERDAM-OXFORD

AMERICAN ELSEVIER PUBLISHING COMPANY, INC.-NEW YORK

Library of Congress Catalog Card Number: 73–91445
North-Holland ISBN: 0 7204 0288 3
American Elsevier ISBN: 0 444 10826 2

95 graphs and illustrations, 21 tables

PUBLISHERS:
NORTH-HOLLAND PUBLISHING CO. - AMSTERDAM
NORTH-HOLLAND PUBLISHING COMPANY, LTD. - OXFORD

SOLE DISTRIBUTORS FOR THE U.S.A. AND CANADA:
AMERICAN ELSEVIER PUBLISHING COMPANY, INC.
52 VANDERBILT AVENUE, NEW YORK, N.Y. 10017

PRINTED IN THE NETHERLANDS

Preface

In 1964, having finished a preprint collection on Coulomb excitation, we decided to write a monograph on the same subject. There are several reasons why it has taken us 10 years to complete this project. It was clear from the beginning that several unsolved problems within the field had to be attacked and only slow progress was made during the short periods in which we had the opportunity to work together. The first years of our collaboration were a quiet period in this particular field and we had the pleasure of carefully working out details of the theoretical presentation. This situation was changed during recent years by the world-wide interest in the study of heavy ion reactions. We felt that it became urgent to finish the book so that the unpublished material which it contains would become available. The compromise which we had to make between completeness and a finite deadline is perhaps felt most seriously in the discussion of relativistic effects.

The topic of electromagnetic excitation of atomic nuclei is a well defined and closed subject. It is unique within the field of nuclear physics in that the interaction as well as the mechanism is well understood, and a model independent theory is possible and useful. In this situation it is natural that the presentation starts from the Schrödinger equation and the fundamental interactions, building up the detailed theory from which one may discuss the approximation methods that have to be used for the practical solutions under various circumstances. In order to facilitate the use of the text for reference purposes, we have tried to make each chapter and each section rather self-contained, and have, therefore, often preferred to repeat an argument or an equation rather than to quote it from distant parts of the book.

The text consists of ten chapters which are indicated by roman numerals (rather than arabic), and appendices which are placed at the end and are indicated by latin characters (rather than chinese). The chapters and appendices are subdivided in sections indicated by arabic numerals. The equations are numbered section by section and are referred to in the following way:

Equations within the same section are referred to simply by the number, e.g. eq. (2). Equations within the same chapter but different sections are referred to by section and equation number, e.g. eq. (3.2). Equations in other chapters are referred to in full, as e.g. eq. (IV.3.2). The figures and tables are numbered chapterwise and are, within the same chapter, referred to by the number only, e.g. fig. 3, but in full if the figure or table appears in another chapter, e.g. table V.4. The references are placed after the appendices and are indicated by the first three letters of the first author and the year of publication, e.g. [BOH 48].

During the preparation of the manuscript we enjoyed valuable assistance from many people. We want to acknowledge especially the help received from Dr. F. Roesel, who made most of the computations necessary for the preparation of the tables and figures and Mr. M. Hansen who has drawn all the figures. The assistance of Dr. J. de Boer was invaluable for the selection and preparation of the figures illustrating experimental results. Finally, we want to thank Drs. R. M. Steffen, J. de Boer, J. Newton and U. Smilansky for reading the manuscript.

Illhäusern,*** Mai 1974.

*** See [MIC 74].

Contents

CHAPTER I

Introduction

The possibility of exciting atomic nuclei by means of the electromagnetic field of impinging charged particles was realized already in the 1930s.* It was not before 1952, however, when the feasibility of exciting rotational states in deformed nuclei was pointed out by A. Bohr and B. Mottelson [MOT 52], that the process was experimentally confirmed [HUU 53, MCC 53], being in good agreement with the semiclassical theory of K. Ter-Martirosyan [TER 52]. In the years following the discovery of the process, which became known under the name Coulomb excitation, it was developed into an important tool for the investigation of low-lying nuclear states. The experimental results strongly supported the collective nuclear model and many new levels were discovered. At the same time, the theory was developed and refined so that it could give a detailed explanation of the experimental results lending in turn confidence also to the accuracy of the nuclear parameters which were extracted [ALD 56 and references contained therein].

In the early Coulomb excitation experiments, light ions were used as projectiles and the electromagnetic forces acting on the target were then so weak, that only a few nuclear states could be populated. The construction of accelerators for heavy ions opened up the possibilities of performing much more effective, but also more complex, Coulomb excitation experiments [STE 59]. Through the strong electromagnetic field from heavy projectiles the target nucleus may absorb several quanta and many nuclear states can be populated [ALD 60]. Such multiple Coulomb excitation processes offer a wide variety of experiments by which one may study the electromagnetic properties of nuclear states.

At the present time, several accelerators are planned and under construction which will make it possible to bring any nucleus up to the Coulomb barrier of even the heaviest elements. In the collision between such very heavy

* A detailed account on the early history is given in the monograph by L. C. Biedenharn [BIE 65].

ions one finds that Coulomb excitation cross sections are of the same order of magnitude or larger than elastic cross sections. This means that one cannot study the nuclear reactions between the heavy ions as e.g. transfer reactions, without considering the strong excitation of low-lying nuclear states in the target and the projectile which is caused by the electromagnetic interaction. In any quantitative analysis of heavy ion reactions one must take into account that target and projectile may be in excited states at the time when the nuclear reaction takes place.

In fact, many nuclear reactions between heavy ions can be described in terms of a semiclassical theory, which is quite analogous to the theory of Coulomb excitation [BRO 72, TRA 70]. In such a theory, nuclear reactions and Coulomb excitation are treated on equal footing through coupled equations, where the exchange of particles and the exchange of energy only differ through the recoil effects.

Inelastic scattering and transfer reactions occur also in collision between atoms. For a long time one has used, in atomic physics, semiclassical theories of these reactions. The process, which is most analogous to Coulomb excitation, is the excitation and ionization of atoms by impinging charged particles. These processes which give rise to a gradual slowing down of a projectile in matter were described in a completely classical manner by N. Bohr already in 1913 [BOH 13].

Another related subject is the electromagnetic excitation of nuclear states by high-energy electron or muon bombardment. Such reactions differ, however, strongly from the theory of Coulomb excitation with heavy ions since it is highly relativistic and dominated by the penetration of the particles into the nucleus [UEB 71].

In the present monograph we shall limit ourselves to a discussion of pure Coulomb excitation, i.e. of processes where the projectile energy is below the Coulomb barrier, so that no penetration into the region of nuclear forces occurs. Under these circumstances the excitation is caused solely by electromagnetic forces and the only nuclear properties which enter into the theory are the matrix elements of the electromagnetic multipole moments of the nuclei. Although this restriction is a practical one for the decision about which subjects to include, it does not impose very serious limitations on its applicability. Thus, at bombarding energies above the Coulomb barrier the cross section at forward scattering angles will still be dominated by partial waves of high orbital angular momenta which can be described in terms of the semiclassical theory. Furthermore it can be shown that even for lower orbital angular momenta the semiclassical description can be used as a basis for a quantal description, even in cases where the strict concept of the classical trajectory breaks down.

The theory of Coulomb excitation which is presented in the following emphasizes especially those aspects which are important for heavy ion experiments. The first order process has been discussed rather thoroughly in earlier review articles and monographs [BRE 59, BIE 65, ALD 56, STE 63, HUB 58, HEY 56], and reprints of early papers have been made available [ALD 66]. Below we shall give a qualitative survey of the theory which will serve as a guide to the contents of the following chapters.

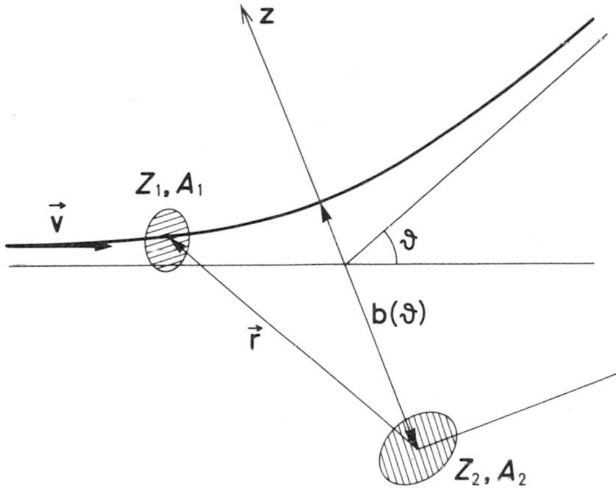

Fig. 1. Classical picture of the projectile orbit in the Coulomb field of the target nucleus. The hyperbolic orbit is shown in the relative frame of reference where the target is at rest. The charge and mass numbers of the projectile and target are indicated as well as the position vector $\boldsymbol{r}$ of the projectile. The deflection angle is denoted by ϑ and the velocity at large distances by $\boldsymbol{v}$. Furthermore the distance of closest approach $b(\vartheta)$ is indicated together with the coordinate system which is often used in this monograph.

In a collision between two heavy ions the electromagnetic interaction depends on the electromagnetic multipole moments of both nuclei. The great simplification in the treatment of Coulomb excitation below the barrier arises from the fact that only the Coulomb field (the monopole-monopole interaction) can ensure that the projectile does not penetrate into the nucleus. This means that the wave length $\lambda\!\!\!^-$ of the projectile must be much smaller than the distance of closest approach b in a head-on collision, i.e. the parameter

$$\eta = b/2\lambda\!\!\!^- = Z_1 Z_2 e^2/\hbar v \tag{1}$$

must be large compared to unity:

$$\eta \gg 1. \tag{2}$$

In eq. (1) Z_1 and Z_2 are the charge numbers of projectile and target nucleus, respectively, while v is thc relative velocity at large distances.

The inequality (2) is at the same time the condition for the applicability of classical physics for a description of the relative motion of the nuclei. Thus, if the condition is fulfilled one may describe the scattering in terms of wave-packets of dimensions which are small compared to the dimensions of the classical hyperbolic orbit and the wavepacket will follow the hyperbola quite accurately.

The hyperbola (see fig. 1) is completely specified by the charge numbers, the energy and the scattering angle ϑ. The differential scattering cross section is thus given by the classical Rutherford formula

$$(\mathrm{d}\sigma/\mathrm{d}\Omega)_{\mathrm{R}} = \tfrac{1}{4}a^2 \sin^{-4}(\tfrac{1}{2}\vartheta). \tag{3}$$

The quantity a is half the distance of closest approach, b, in a head-on collision, i.e.

$$a = \tfrac{1}{2}b = Z_1 Z_2 e^2/m_0 v^2, \tag{4}$$

where m_0 is the reduced mass of projectile and target. During the collision one or both nuclei may be excited. A detailed discussion of the mutual multipole-multipole interaction and of the question of mutual excitations will be given in chapter II. Let us assume here for simplicity that only the target nucleus is excited and that the probability of finding this nucleus in the state $|n\rangle$ after the collision is P_n. The differential cross section for the inelastic scattering to the state $|n\rangle$ is then given by

$$(\mathrm{d}\sigma/\mathrm{d}\Omega)_n = (\mathrm{d}\sigma/\mathrm{d}\Omega)_{\mathrm{R}} P_n. \tag{5}$$

In the classical picture the excitation is caused by the time-dependent electromagnetic field which sweeps over the target nucleus as the projectile moves along the hyperbolic orbit according to the classical equations of motion. In this picture one finds the excitation probabilities by solving the time-dependent Schrödinger equation for the target nucleus

$$\mathrm{i}\hbar \frac{\partial}{\partial t} |\psi(t)\rangle = (H_0 + V(\boldsymbol{r}(t)))\, |\psi(t)\rangle, \tag{6}$$

where H_0 is the Hamiltonian of the free nucleus, while $V(\boldsymbol{r}(t))$ is the electro-magnetic interaction which is indirectly a function of time through the relative position vector $\boldsymbol{r}(t)$. This wave equation for the target state vector $|\psi(t)\rangle$ should be solved with the initial condition that at time $t = -\infty$, where $V(\boldsymbol{r}(t)) = 0$, the nucleus is in its ground state $|0\rangle$, i.e.

$$|\psi(-\infty)\rangle = |0\rangle. \tag{7}$$

Solving eq. (6) one finds at time $t = +\infty$ the state vector $|\psi(+\infty)\rangle$. The expansion coefficients a_n of this state on the eigenstates of the free nucleus $|n\rangle$ are the excitation amplitudes

$$a_n = \langle n|\psi(+\infty)\rangle, \tag{8}$$

where the state $|n\rangle$ is defined by

$$H_o\,|n\rangle = E_n\,|n\rangle. \tag{9}$$

A general discussion of the methods by which one may solve eq. (6) is given in chapter II. These methods include, besides the perturbation expansion, a number of other methods which will be briefly mentioned below. Once the amplitudes a_n have been determined the evaluation of cross sections, angular distributions and polarizations is a straightforward geometrical problem of angular momentum algebra. One finds e.g. that the excitation probability is given by

$$P_{0\to n} = |a_n|^2. \tag{10}$$

It should be noted that n must specify the level and the magnetic quantum number and that for the evaluation of the cross section (5) appropriate sums and averages over the magnetic quantum numbers should be performed. For the calculation of angular distributions of γ-rays following Coulomb excitation it is convenient to introduce the density matrix formalism as discussed in detail in chapter III.

Above we have neglected the fact that the projectile loses the energy $\Delta E_n = E_n - E_0$ to the target nucleus during the collision. This energy loss cannot be included in any accurate way since we do not know at which point of the orbit the energy is transferred. For the applicability of the classical picture it is therefore essential that this energy loss does not modify the orbit significantly, i.e. we must assume

$$\Delta E_n/E \ll 1, \tag{11}$$

where E is the centre-of-mass energy $\frac{1}{2}\, m_o v^2$.

High-lying states which may violate the condition (11) are excited only weakly due to the fact that the excitation process for such states becomes adiabatic. Thus it is well known, that a pulsed field like $V(\boldsymbol{r}(t))$ in (6) can only excite a state $|n\rangle$ if the time duration of the pulse (the collision time) is short or of the same order of magnitude as the nuclear period characteristic for the transition from $0 \to n$ i.e. $\hbar/\Delta E_n$. The product of the nuclear frequency $\Delta E_n/\hbar$

and the collision time τ is denoted by ξ i.e. $\xi_{0\to n} = \Delta E_n \tau/\hbar$. The quantum transition $|0\rangle \to |n\rangle$ can only be induced if this quantity is of the order of, or smaller than, unity. The collision time can be estimated as the time it takes for the projectile to travel a distance of the order of

$$b(\vartheta) = a(1 + 1/\sin \tfrac{1}{2}\vartheta), \tag{12}$$

which is the distance of closest approach when the scattering angle is ϑ. We thus define

$$\tau(\vartheta) = \frac{b(\vartheta)}{2v} = \frac{a}{2v}\left(1 + \frac{1}{\sin \frac{1}{2}\vartheta}\right) \tag{13}$$

and

$$\xi_{0\to n}(\vartheta) = \xi_{0\to n}\frac{1}{2}\left(1 + \frac{1}{\sin \frac{1}{2}\vartheta}\right), \tag{14}$$

where

$$\xi_{0\to n} = \xi_{0\to n}(\pi) = \frac{\Delta E_n}{\hbar}\frac{a}{v}. \tag{15}$$

The ratio $\Delta E_n/E$ can now be written

$$\Delta E_n/E = 2\xi_{0\to n}/\eta, \tag{16}$$

where η is given by (1). Since the condition for the applicability of the classical treatment is $\eta \gg 1$ and since $\xi \lesssim 1$ for one step processes the condition (11) is mostly fulfilled for the cases of interest. Typical values of η for heavy ions are of the order of magnitude 10 to 500 and it is thus noted that according to (16), ξ may become quite large without violating condition (11). However, for large values of ξ the excitation probabilities vanish exponentially with ξ.

The magnitude of the excitation probability depends of course also on the strength of the interaction $V(\boldsymbol{r}(t))$. A measure of this strength is given by the matrix elements of the action integral,

$$\langle n| \int_{-\infty}^{+\infty} V(\boldsymbol{r}(t))\, \mathrm{d}t\, |0\rangle \approx \langle n|\, V(b(\vartheta))\, |0\rangle\, \tau, \tag{17}$$

which has been estimated by the value of V at closest approach and the collision time. If one measures this quantity in units of $\hbar$ one obtains the dimensionless parameter

$$\chi_{0\to n}(\vartheta) = \frac{\langle n|\, V(b(\vartheta))\, |0\rangle\, b(\vartheta)}{2\hbar v}, \tag{18}$$

which measures the number of quanta which are exchanged during the collision in the excitation of the state $|n\rangle$. If χ is a small number there is only a small chance that a quantum is exchanged, and one may treat the excitation of the state by perturbation theory. For large χ the state is strongly excited if at the same time ξ is not too large.

As we shall see in chapter II one usually expands the interaction $V(\boldsymbol{r})$ in multipole components. The parameter χ is correspondingly decomposed in a sum of partial χ's

$$\chi = \sum_{\lambda} \chi^{(\lambda)}, \tag{19}$$

each term belonging to the part of $V(\boldsymbol{r})$ which has multipole order λ. It is noted that one may, to a good approximation, neglect the transverse field. This is due to the fact that this field is of the order v/c times the longitudinal field and that the condition of non-penetration implies that $v/c \lessapprox 0.1$. For the quadrupole part of the Coulomb interaction between the charge of the projectile Z_1e and a nucleus of quadrupole moment Q one estimates for $\vartheta = 180°$

$$V(b) \approx Z_1 e Q/b^3 \tag{20}$$

and

$$\chi^{(2)} \approx Z_1 Q e/2\hbar b^2 v. \tag{21}$$

For actual nuclei, and for energies below the barrier one sees that $\chi^{(2)}$ is at most of the order of 10. For the other multipole orders one finds similarly

$$\begin{aligned} \chi^{(1)} &\lessapprox 10 \\ \chi^{(2)} &\lessapprox 10 \\ \chi^{(3)} &\lessapprox 0.5 \\ \chi^{(4)} &\lessapprox 0.1 \end{aligned} \tag{22}$$

and it is seen that high multipole excitation becomes less important. The estimates (22) are upper limits and more typical values are discussed in chapter X.

It is noted finally that the monopole part of the interaction $Z_1Z_2e^2/r$ (which has only diagonal matrix elements) leads to a parameter

$$\chi^{(0)} = Z_1 Z_2 e^2/\hbar v, \tag{23}$$

which is identical to η. The parameter η thus measures the strength of the monopole interaction in exactly the same way as $\chi^{(1)}, \chi^{(2)} \ldots$ measure the strengths of the higher multipole interactions.

Since the parameter $\chi^{(\lambda)}$ measures the number of quanta of angular momentum $\lambda\hbar$ which are exchanged during the collision, the total transfer of angular momentum Δl to the nucleus can be estimated by $\Delta l \leqslant \lambda\chi^{(\lambda)}\hbar$. In the discussion of the applicability of the classical picture, we neglected the fact that the angular momentum transfer changes the orbit. In order that the simple classical picture applies one must assume that $\Delta l/l \ll 1$, where l is the orbital angular momentum. This leads to the condition $\chi^{(\lambda)}\lambda/\eta \ll 1$, which is somewhat more restrictive than the simple condition (2). It is noted that the ratio $\chi^{(\lambda)}/\eta = \chi^{(\lambda)}/\chi^{(0)}$ is independent of $\hbar$. This indicates that at least part of the effects of the angular momentum transfer on the trajectory are due to the classical distortion that the higher multipole moments have on the classical orbit.

The solution of the Schrödinger equation (6) is especially simple when the parameters χ are small compared to unity, so that one may use first order perturbation theory. This treatment is discussed in detail in chapter IV. The excitation amplitude (8) is then readily found to be

$$a_n = \frac{1}{\mathrm{i}\hbar}\int_{-\infty}^{+\infty} \langle n|\ V(\boldsymbol{r}(t))\ |0\rangle \exp(\mathrm{i}\Delta Et/\hbar)\,\mathrm{d}t. \tag{24}$$

When one expands $V(\boldsymbol{r}(t))$ into multipole components the amplitude (24) is factorized into a part which depends on the nucleus only through the matrix element of the multipole moments, and an integral which only depends on ΔE and the kinematics of the classical orbit. These integrals can be expressed by elementary functions if the excitation energy ΔE of the state $|n\rangle$ can be neglected. The integrals are then functions of the deflection angle ϑ, behaving as ϑ^λ for small values of ϑ. The limit $\Delta E = 0$ corresponds to the situation in which the nuclear period is long compared to the collision time, i.e. where the process can be considered as a sudden impact. For non-vanishing energy loss the main dependence of the integrals is given by the parameter $\xi(\vartheta) = \Delta E\tau/\hbar$ (see eq. (14)). This parameter measures the extent to which the process is adiabatic. Thus for $\xi(\vartheta) \ll 1$ the sudden approximation applies, while $\xi(\vartheta) \gg 1$ represents the adiabatic limit. In this limit the integrals (24) vanish mainly as $\exp[-\xi(\vartheta)]$. For finite energy losses the process thus always becomes adiabatic for small deflection angles and the Coulomb excitation cross section vanishes exponentially in the forward direction.

It is noted that the amplitude (24) for $\Delta E = 0$ is almost identical to the parameter $\chi_{0\to n}(\vartheta)$ of eq. (18). To remove the dependence on the magnetic quantum numbers it is convenient to define the χ parameter between two energy levels n and m by

$$\chi_{n\to m}(\vartheta) = \pm\sqrt{P_{n\to m}(\vartheta, \xi = 0)}. \tag{25}$$

The quantity $P_{n\to m}$ is the transition probability in first order perturbation theory between the levels n and m for vanishing energy difference ΔE. As is seen from eq. (18) the χ-parameters are largest for $\vartheta = \pi$ and one may therefore write

$$\chi_{n\to m}(\vartheta) = \chi_{n\to m}R(\vartheta), \tag{26}$$

where

$$\chi_{n\to m} = \pm\sqrt{P_{n\to m}(\pi, 0)}. \tag{27}$$

The factor $R(\vartheta)$ which is smaller than unity can be expressed in terms of the above mentioned orbital integrals. The sign in (25) and (27) is chosen to be the same as the sign of the reduced matrix element of the multipole operator.

A necessary condition for the validity of the perturbation treatment is that the computed excitation probabilities be small numbers for all of the excited states. In actual cases this is well fulfilled for projectiles of low charge such as protons, but may be violated especially for collective nuclear states excited by particles of higher charge.

Under circumstances where the condition for the applicability of first-order perturbation theory is not too heavily violated it may be sufficient to take into account the deviations from the first-order theory by carrying the perturbation treatment to second order. This gives rise to two types of modifications. Firstly, one obtains a correction to the excitation amplitudes of the states which are already strongly excited in first order. Some of these corrections depend on the static multipole moments of these states. Secondly, by double excitation, one obtains the possibility of exciting states which were otherwise inaccessible or only accessible by excitations of high multipolarity. The second-order semiclassical treatment of Coulomb excitation is discussed in detail in chapter V.

In cases where the condition for the first-order theory is strongly violated the perturbation expansion becomes completely impractical and multiple excitation occurs. A given nuclear state can then be populated in a number of different ways by virtual excitations through many intermediate states. Two experiments in which the inelastic cross section is of the same order of magnitude as the elastic one are illustrated in fig. 2. A general discussion of the methods which one may use is given in chapter VI.

One can obtain a qualitative picture of the excitation process by arranging

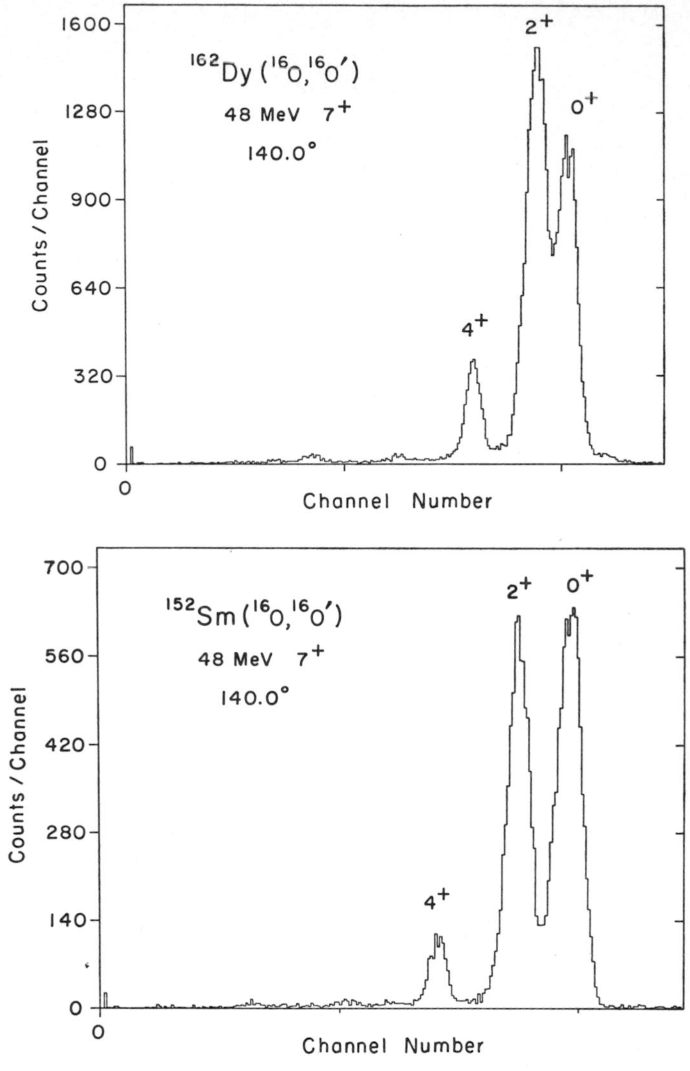

Fig. 2. Spectrum of ^{16}O ions, elastically and inelastically scattered through $\vartheta = 140°$ from ^{162}Dy and ^{152}Sm. (From I. Yang Lee and J. X. Saladin, 1974, Phys. Rev. **9c,** 2406).

the parameters $\chi_{n \to m}$ (see eq. (27)) connecting any pair of nuclear states in a matrix. This χ-matrix is then resolved into submatrices in such a way that states within a submatrix are connected by large χ's ($\chi \gtrsim 1$) while all the matrix elements between the submatrices are small. The states within a submatrix, which are connected by large χ's, will mix during the collision and multiple Coulomb excitation will occur. The transitions between the groups is weak and may be treated by perturbation theory. Only if all χ's connecting to the ground state as well as all χ's connecting to the final state are small the simple perturbation theory applies. In actual cases one finds for heavy-ion Coulomb excitation that the conditions for the perturbation theory will practically never be fulfilled. The computation of excitation probabilities for a group of states with $\chi \geqslant 1$ is in general quite complicated. However, large values of χ occur mostly for collective nuclear states with low excitation energies, thus leading to small values of the parameters ξ. This suggests that as a first approximation one might neglect the energy differences between all the states involved in the multiple excitation. In this sudden approximation it is possible to give a simple explicit expression for the excitation amplitude

$$a_n = \langle n| \exp\left\{ \int_{-\infty}^{+\infty} -\frac{i}{\hbar} V(r(t))\, dt \right\} |0\rangle . \tag{28}$$

It is noted that the series development of the exponential function leads to the perturbation expansion for $\xi = 0$ of which the first term is given by eq. (24).

In eq. (28) the excitation amplitude is expressed as a matrix element of a complicated, but well known operator, which depends on the scattering parameters such as projectile energy, deflection angle etc. If the nuclear wave functions of the initial and final states are known the evaluation of a_n is simply a matter of evaluating definite integrals. For the excitation of surface vibrational states (as discussed in chapter VII) and for the excitation of rotational states in deformed nuclei (see chapter VIII) one finds by inserting the interaction energy in terms of the collective nuclear degrees of freedom, that (28) can be evaluated in terms of rather elementary functions, which depend on the parameters of the nuclear model. An example of the multiple excitation of a rotational band is shown in fig. 3.

As was stressed earlier the only nuclear properties which enter into the calculation of Coulomb excitation amplitudes are the electric multipole matrix elements. Since these may be known from gamma spectroscopy, it may be convenient to express the excitation amplitudes directly in terms of these parameters. This can actually be done by evaluating (28) for a limited number of nuclear states by means of a diagonalization procedure as discussed in chapter VI.

The analysis of actual experiments should, however, be based on model-independent calculations for finite values of ξ. Such calculations have been performed in two different ways. The first method relies on the diagonaliza-

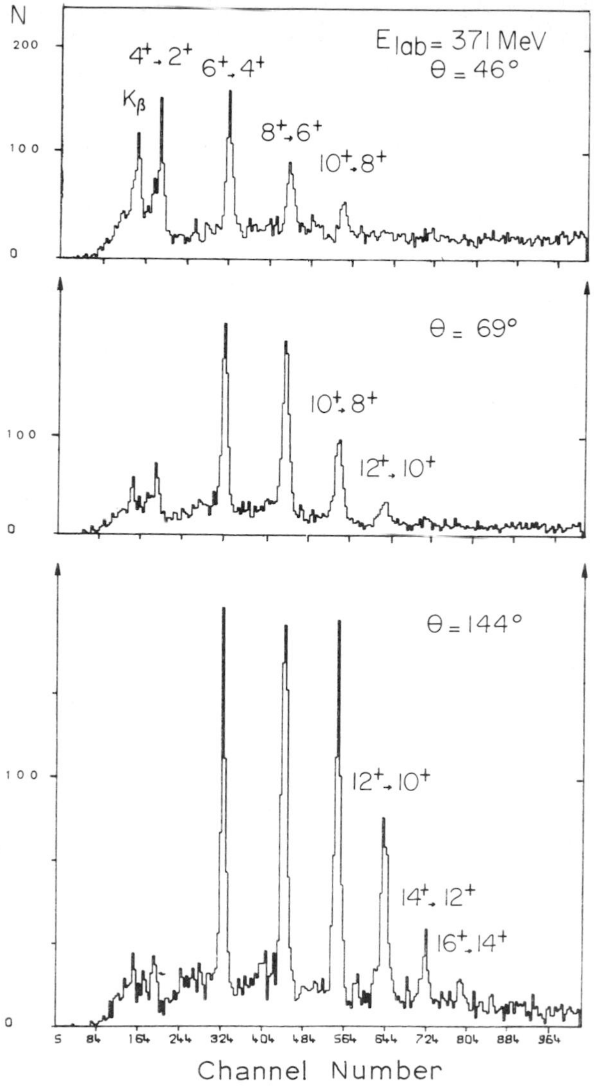

Fig. 3. Relative yields (corrected for conversion and detection efficiency) of the de-excitation gamma rays in the ground state rotational band of ^{232}Th excited by 317 MeV ^{84}Kr ions scattered through various angles. (From Colombani et al., 1973, *Proc. Intern. Conf. on Nuclear Physics*, Munich (North-Holland Publ. Co., Amsterdam) p. 376, and private communication.)

tion procedure in conjunction with a generalization of eq. (28) for finite values of ξ. Another approach is the direct numerical integration of the coupled differential equations which describe the amplitudes of the nuclear levels during the collision.

For the actual execution of these general methods one must rely on large high speed computers. A computer program by which one may calculate cross sections and angular distributions for multiple Coulomb excitation in arbitrary nuclei is discussed in [WIN 65].

The above discussion has been based on the semiclassical treatment which is expected to be quite accurate since the parameter η is usually much larger than unity. In the pure semiclassical treatment one does not distinguish between the initial velocity v_i and the final velocity v_f of the projectile since one assumes that the relative energy loss of the projectile is small enough for the classical orbit to be well defined. If the difference between v_i and v_f is so large that significantly different results are obtained by assuming $v = v_i$ or $v = v_f$, it is suggestive that improved results are obtained by inserting an average of the initial and final velocity for v.

To obtain a more systematic improvement of the semiclassical theory one has to study quantum mechanical effects, taking into account the finite value of the parameter η. In the quantum theory of Coulomb excitation one describes the projectile by a wave function which is the solution of the Schrödinger equation for a particle in a Coulomb field, and which behaves asymptotically as a plane wave (Coulomb wave). If the interaction can be considered as a perturbation one can easily write down an expression for the excitation cross section (distorted wave Born approximation). This calculation has been carried out and has shown that, by introducing a proper symmetrization, the quantum mechanical effects are usually of minor importance. The complete quantum mechanical treatment of multiple excitation is discussed in chapter IX.

In chapter X we consider a number of subjects which are of practical interest for the analysis of experiments on Coulomb excitation.

CHAPTER II

Semiclassical Equations of Motion

The semiclassical approximation forms the fundamental basis for the description of Coulomb excitation processes in the present review. A quantum mechanical description is presented in chapter IX mainly in order to investigate the accuracy of the semiclassical treatment. Even with the simplification of the classical description of the relative motion the solution of the Coulomb excitation problem is difficult. In this chapter we shall mainly be concerned with a general discussion of the approximation methods for solving the Schrödinger equation describing semiclassical Coulomb excitation. This discussion will be the basis for the following chapters in which we carry out in detail the calculation of cross sections for various cases of interest.

§ 1. *The electromagnetic interaction*

In the following we consider pure Coulomb excitation processes of two in general extended nuclei. We thus only study collisions in which the nuclear forces between the two nuclei can be neglected (see fig. I.1) and where the interaction is purely electromagnetic. It also follows from the condition of non-penetration that the relative velocity can not be too large. One finds in practice that the maximum velocity fulfills the relation $(v/c)^2 \lesssim 0.01$. Within this accuracy (cf. § X.7) we may thus neglect retardation and other relativistic effects and we may separate the centre-of-mass motion. In the collision between heavy ions the relativistic corrections for the centre-of-mass motion can be of the same order of magnitude as the retardation effects.

The collision process can then be described by the following Hamiltonian

$$H = H_0'(1) + H_0'(2) + W(1, 2). \tag{1}$$

Here $H_0'(1)$ and $H_0'(2)$ are the Hamiltonians of the free projectile and target nuclei while $W(1, 2)$ is the mutual electromagnetic interaction Hamiltonian to first order in v/c. Neglecting retardation one finds

$$W(1, 2) = \iint \frac{\rho(\boldsymbol{r}_1)\rho(\boldsymbol{r}_2) - \boldsymbol{j}(\boldsymbol{r}_1)\boldsymbol{j}(\boldsymbol{r}_2)/c^2}{|\boldsymbol{r}_1 - \boldsymbol{r}_2|} \, \mathrm{d}\tau_1 \, \mathrm{d}\tau_2, \tag{2}$$

where $\rho(\boldsymbol{r}_1)$, $\boldsymbol{j}(\boldsymbol{r}_1)$ are the charge and current densities at the point $\boldsymbol{r}_1$ for the projectile, while $\rho(\boldsymbol{r}_2)$, $\boldsymbol{j}(\boldsymbol{r}_2)$ are the corresponding quantities for the target nucleus (see appendix A).

We want to express $W(1, 2)$ in terms of the electric and magnetic multipole moments of the two nuclei in their respective rest systems. The electromagnetic multipole moments are defined by (see e.g. [BOH 69])

$$\mathcal{M}(\mathrm{E}\lambda, \mu) = \int \rho(\boldsymbol{r}) r^{\lambda} Y_{\lambda\mu}(\hat{\boldsymbol{r}})\, \mathrm{d}\tau \tag{3}$$

and

$$\mathcal{M}(\mathrm{M}\lambda, \mu) = \frac{-\mathrm{i}}{c(\lambda + 1)} \int \boldsymbol{j}(\boldsymbol{r}) r^{\lambda} \boldsymbol{L} Y_{\lambda\mu}(\hat{\boldsymbol{r}})\, \mathrm{d}\tau, \tag{4}$$

where

$$\boldsymbol{L} = -\mathrm{i}\boldsymbol{r} \times \boldsymbol{\nabla}. \tag{5}$$

The integration is performed over the coordinates $\boldsymbol{r}$ of the charge-current distribution $\rho(\boldsymbol{r})$, $\boldsymbol{j}(\boldsymbol{r})$ usually measured with respect to the centre of mass of the nucleus. Since retardation is neglected, and since the charge-current densities are assumed to be completely separated in space, one finds the following expression for $W(1, 2)$:

$$W(1, 2) = W_{\mathrm{E}}(1, 2) + W_{\mathrm{M}}(1, 2) + W_{\mathrm{EM}}(1, 2). \tag{6}$$

Here $W_{\mathrm{E}}(1, 2)$ is the mutual electric multipole-multipole interaction and $W_{\mathrm{M}}(1, 2)$ the mutual magnetic multipole-multipole interaction, while $W_{\mathrm{EM}}(1, 2)$ is the interaction between the electric and the magnetic multipole moments caused by the relative motion of the two systems. The explicit form of these three terms is derived in appendix A.

One finds that the electric multipole-multipole interaction may be written

$$W_{\mathrm{E}}(1, 2) = \sum_{\substack{\lambda_1\lambda_2 \\ \mu_1\mu_2}} c(\lambda_1, \lambda_2) \begin{pmatrix} \lambda_1 & \lambda_2 & \lambda_1 + \lambda_2 \\ \mu_1 & \mu_2 & \mu \end{pmatrix} \times \mathcal{M}_1(\mathrm{E}\lambda_1, \mu_1) \mathcal{M}_2(\mathrm{E}\lambda_2, \mu_2) \frac{1}{r^{\lambda_1 + \lambda_2 + 1}} Y_{\lambda_1 + \lambda_2, \mu}(\theta, \phi), \tag{7}$$

where r, θ, ϕ are the polar coordinates of the centre of 1 with respect to the centre of 2. The numerical coefficient $c(\lambda_1, \lambda_2)$ is given by

$$c(\lambda_1, \lambda_2) = (4\pi)^{3/2}(-1)^{\lambda_2} \left[\frac{(2\lambda_1 + 2\lambda_2)!}{(2\lambda_1 + 1)!\,(2\lambda_2 + 1)!} \right]^{1/2} \tag{8}$$

For the 3-j symbol in eq. (7) we use the notation of Edmonds [EDM 57].

The magnetic multipole-multipole interaction $W_{\mathrm{M}}(1, 2)$ is obtained from (7) by substituting the electric multipole moments $\mathcal{M}_1(\mathrm{E}\lambda_1, \mu_1)$ and $\mathcal{M}_2(\mathrm{E}\lambda_2, \mu_2)$

of the two nuclei by the corresponding magnetic multipole moments $\mathscr{M}(\mathrm{M}\lambda, \mu)$.

The interaction $W_{\mathrm{EM}}(1, 2)$ between the electric and magnetic moments is proportional to the relative velocity $\dot{\boldsymbol{r}}$ of system 1 with respect to system 2. The explicit expression may be written

$$W_{\mathrm{EM}}(1, 2) = \sum_{\substack{\lambda_1\lambda_2\\ \mu_1\mu_2}} ic(\lambda_1, \lambda_2)\{\mathscr{M}_1(\mathrm{E}\lambda_1\mu_1)\mathscr{M}_2(\mathrm{M}\lambda_2\mu_2) - \mathscr{M}_1(\mathrm{M}\lambda_1\mu_1)\mathscr{M}_2(\mathrm{E}\lambda_2\mu_2)\}$$
$$\times \frac{\dot{\boldsymbol{r}}}{c}\frac{1}{r^{\lambda_1+\lambda_2+1}}\left\{\begin{pmatrix}\lambda_1 & \lambda_2 & \lambda_1+\lambda_2\\ \mu_1 & \mu_2 & \mu\end{pmatrix}\frac{1}{\lambda_1+\lambda_2}\boldsymbol{L}Y_{\lambda_1+\lambda_2,\mu}(\theta, \phi)\right.$$
$$\left. - \begin{pmatrix}\lambda_1 & \lambda_2 & \lambda_1+\lambda_2-1\\ \mu_1 & \mu_2 & \mu\end{pmatrix}\sqrt{\frac{\lambda_1}{\lambda_2(\lambda_1+\lambda_2)}}\,\boldsymbol{\Phi}_{\lambda_1+\lambda_2,\lambda_1+\lambda_2-1,\mu}(\theta, \phi)\right\}, \tag{9}$$

where $\boldsymbol{\Phi}_{lJ\mu}$ is the vector spherical harmonic defined in [EDM 57].

The main term in eq. (6) is the electric monopole-monopole interaction in $W_{\mathrm{E}}(1, 2)$ which is the point Coulomb energy $Z_1Z_2e^2/r$. This interaction does not give rise to any excitation, since it does not depend on the intrinsic degrees of freedom of the two nuclei. It essentially determines the relative motion of the two nuclear centres.

Next in importance are the terms in eq. (7) that describe the interaction of the monopole moment of the projectile with the electric multipole moments of the target and which give rise to Coulomb excitation of the target. Similarly there will be terms that describe the interaction of the monopole moment of the target with the electric multipole moments of the projectile, and which give rise to projectile excitation.

Finally eq. (7) contains interactions between higher electric multipole moments which give rise to dipole-multipole, quadrupole-multipole etc. excitations. These excitations are denoted mutual excitations and will be treated in § IV.6. They are quite small and can usually be neglected.

Of the terms in eq. (6) which contain the magnetic multipole moments, the largest are usually the monopole-multipole terms of $W_{\mathrm{EM}}(1, 2)$ in eq. (9). They are proportional to the electric monopole moment of projectile or target and give rise to magnetic excitation of target or projectile, respectively. Since, however, these terms are proportional to v/c, magnetic excitation of similar order of magnitude may for light ions arise from the main term of $W_{\mathrm{M}}(1, 2)$, i.e. the magnetic dipole-multipole interaction.

In the following we shall write the interaction between the monopole moments and the electric and magnetic multipole moments in the form

$$W(1, 2) = V(1, \boldsymbol{r}) + V(2, \boldsymbol{r}) + Z_1Z_2e^2/r, \tag{10}$$

where e.g.

$$V(2, \boldsymbol{r}) = V_{\mathrm{E}}(2, \boldsymbol{r}) + V_{\mathrm{M}}(2, \boldsymbol{r}) \tag{11}$$

gives rise to target excitation. Explicitly one finds

$$V_{\mathrm{E}}(2, \boldsymbol{r}) = \sum_{\substack{\lambda \geqslant 1 \\ \mu}} \frac{4\pi Z_1 e}{(2\lambda + 1)} \mathcal{M}_2(\mathrm{E}\lambda, -\mu)(-1)^{\mu} r^{-\lambda-1} Y_{\lambda\mu}(\theta, \phi) \tag{12}$$

and

$$V_{\mathrm{M}}(2, \boldsymbol{r}) = \sum_{\substack{\lambda \geqslant 1 \\ \mu}} \frac{4\pi Z_1 e}{(2\lambda + 1)} \frac{\mathrm{i}}{\lambda} \mathcal{M}_2(\mathrm{M}\lambda, -\mu)(-1)^{\mu} \frac{\dot{\boldsymbol{r}}}{c} r^{-\lambda-1} \boldsymbol{L} Y_{\lambda\mu}(\theta, \phi). \tag{13}$$

The term $V(1, \boldsymbol{r})$ which is obtained from $V(2, \boldsymbol{r})$ by interchanging the indices 1 and 2, gives rise to projectile excitation. Magnetic excitations caused by the term (13) are always quite weak and will be treated by perturbation theory in § IV.5.

§ 2. *Semiclassical picture*

In Coulomb excitation below the Coulomb barrier the monopole-monopole interaction $Z_1 Z_2 e^2/r$ is by far the most important interaction. It does not give rise to any excitations but essentially determines the relative motion of the two nuclei. The magnitude of this point charge Coulomb interaction is also responsible for a great simplification in the theory since it ensures that in most cases of practical interest one is justified in using a classical description for the relative motion. This can be seen qualitatively from the fact that the parameter η defined by

$$\eta = Z_1 Z_2 e^2/\hbar v \tag{1}$$

in all heavy ion collisions with bombarding energy below the Coulomb barrier satisfies the condition

$$\eta \gg 1. \tag{2}$$

In (1) v denotes the relative velocity of the two ions at large distances. The validity of (2) for different combinations of heavy ions can be seen from fig. X.4 in chapter X. The quantity η is defined as the ratio of half the distance of closest approach in a head-on collision

$$a = Z_1 Z_2 e^2/m_0 v^2, \tag{3}$$

where m_0 is the reduced mass of the two ions, and the wavelength ($\lambda\!\!\!^{-} = \hbar/m_0 v$) in the relative motion.

The condition (2) ensures that one may form a wavepacket containing several waves and still having a size which is small compared to the dimensions of the classical trajectory. Such a wavepacket will move according to the classical equations of motion. A more accurate discussion of the validity of this argument is given in appendix B. It should be noted that condition (2) not only ensures the classical description of the relative motion but also that no penetration will take place through the Coulomb barrier.

In a strict semiclassical description the relative motion of the two heavy ions in the Coulomb field will follow a well defined hyperbola and the multipole interaction (1.6) will, through its dependence on the relative position vector $\boldsymbol{r}$, become a function of time which may cause excitations in the two ions. The probability that the ions are scattered by an angle ϑ into the solid angle $\mathrm{d}\Omega$ (both measured in the centre-of-mass system) is expressed by the Rutherford cross section,

$$(\mathrm{d}\sigma/\mathrm{d}\Omega)_{\mathrm{R}} = \tfrac{1}{4}a^2 \sin^{-4}(\tfrac{1}{2}\vartheta). \tag{4}$$

If in the collision the probability of excitation from the ground state $|0\rangle$ to the state $|n\rangle$ is P_n the inelastic scattering cross section to this state is given by

$$(\mathrm{d}\sigma/\mathrm{d}\Omega)_n = (\mathrm{d}\sigma/\mathrm{d}\Omega)_{\mathrm{R}} P_n. \tag{5}$$

This conclusion can only be correct in so far as one may neglect the reaction on the orbit which is caused by the excitation process. There are two types of such reactions, namely those that are associated with energy transfer from relative motion to intrinsic motion, and those that are related to angular momentum transfer.

In order that the energy transfer has negligible influence on the motion, the excitation energy ΔE_n must be small compared to the total energy E in the relative motion, i.e.

$$\Delta E_n/E \ll 1. \tag{6}$$

Although this condition is obviously not satisfied for all states $|n\rangle$ it is in practice fulfilled for all those states that have an appreciable excitation probability. This is due to the general feature which holds both classically and quantum mechanically for weak inelastic processes that one can only excite a degree of freedom if the corresponding frequency is smaller than or of the order of magnitude of $1/\tau$, where τ is the collision time. In Coulomb excitation we estimate τ by the quantity a/v. Since quantum mechanically the nuclear frequency corresponding to the transition $0 \to n$ is $\Delta E_n/\hbar$, the state $|n\rangle$ can only be excited if the relation $\xi_{0\to n} \lesssim 1$ is fulfilled where $\xi_{0\to n}$ is defined by (I.15). For processes that cannot be considered weak, i.e. where many quanta are exchanged, the excitation energy is rather limited by the relation $\xi_{0\to n} \lesssim \chi$ where χ defined by (I.18) is a typical strength parameter in the reaction.

Since we may write the ratio (6) in the form

$$\Delta E_n/E = 2\xi_{0\to n}/\eta, \tag{7}$$

it is seen that for weak excitations the relation (6) is fulfilled automatically according to (2) for all states that have an appreciable excitation probability. For cases of strong excitation ($\chi > 1$) the condition (6) is only fulfilled automatically if

$$\eta \gg \chi. \tag{8}$$

In order that the angular momentum transfer Δl has a negligible influence on the motion of the ions, the following relation should be fulfilled

$$\Delta l/l \ll 1. \tag{9}$$

In (9) l is the total orbital angular momentum, which can be estimated by

$$l \approx m_0 va = \hbar\eta. \tag{10}$$

For weak transitions where perturbation theory applies Δl can be estimated by $\Delta l \leqslant \lambda\hbar$, where λ is the multipolarity of the interaction potential, and it is seen that (9) is fulfilled automatically if (2) is satisfied. For strong excitations an angular momentum

$$\Delta l \sim \lambda\chi^{(\lambda)}\hbar \tag{11}$$

can be transferred since the strength parameter $\chi^{(\lambda)}$, defined in (I.18), measures the number of λ-pole quanta that are transferred. In this case the relation (9) is only fulfilled automatically if

$$\eta \gg \chi^{(\lambda)}\lambda. \tag{12}$$

In the semiclassical description discussed above the classical motion of the heavy ions was assumed to take place on hyperbolic orbits characteristic for elastic scattering in the monopole field. This is a reasonable approach in Coulomb excitation where the strength parameters for the higher multipole interactions rarely become much larger than unity (see eq. (I.22)). Actually, the conditions (8) and (12) express that the multipole field should be weak compared to the monopole field whose strength is described by η (c.f. eq. (I.23)).

If the parameters $\chi^{(\lambda)}$ (for $\lambda \geqslant 1$) should ever become large compared to unity the excitation mechanism itself becomes partly classical, and it is possible to formulate a semiclassical theory where the energy loss and angular momentum transfer along the trajectory is incorporated in the classical motion. In a situation of relatively weak excitations where $\chi \lesssim 1$ it is in principle not possible to tell at which point of the orbit the energy (or angular momentum) transfer takes place. The best one can do, in this more common situation, is to use for each transition an orbit corresponding to an average between initial and final state.

This will be shown in chapter IX, where the semiclassical limit of the quantal description will be given. An alternative approach in terms of a wavepacket description is given in appendix C.

Since this symmetrization of the orbit can be introduced at a later stage (c.f. § IV.7 and § VI.8) we shall in this chapter neglect the reaction of the energy and angular momentum transfer on the orbit. In this case the motion of the two centres of mass of the colliding ions is completely separated from the intrinsic motion, and the intrinsic wavefunction $|\psi_{\text{int}}\rangle$ for the two nuclei is governed by the equation

$$i\hbar \frac{\partial}{\partial t} |\psi_{\text{int}}(t)\rangle = \left(H_0(1) + H_0(2) + W(1, 2, \boldsymbol{r}(t)) - \frac{Z_1 Z_2 e^2}{r(t)} \right) |\psi_{\text{int}}(t)\rangle . \quad (13)$$

The Hamiltonians $H_0(1)$ and $H_0(2)$ are the intrinsic free Hamiltonians while $\boldsymbol{r}(t)$ is the relative centre-of-mass position vector. Satisfying the classical equations of motion, $\boldsymbol{r}(t)$ is a given function of time. Since the kinetic energies of the centres of mass of the two nuclei as well as the monopole-monopole interaction was included in the determination of this classical motion the corresponding terms were left out in the Hamiltonian (13) as compared to the total Hamiltonian H in (1.1). Since $\boldsymbol{r}(t)$ is the relative centre-of-mass coordinate one should in the multipole-multipole expansion of the interaction $W(1, 2, \boldsymbol{r})$ (given in § 1) measure all multipole moments from the respective centres of mass of the two nuclei.

The solution of (13) is uniquely determined from the initial condition at $t = -\infty$ where the two nuclei are far apart, and where $|\psi_{\text{int}}\rangle$ can be written as the product of the intrinsic eigenstates $|\psi_n^{(1)}\rangle$ and $|\psi_m^{(2)}\rangle$ of target and projectile, respectively, i.e.

$$|\psi_{\text{int}}(-\infty)\rangle = |\psi_0^{(1)}\rangle |\psi_0^{(2)}\rangle . \quad (14)$$

By the label 0 we have indicated that under usual circumstances the two nuclei will be in their ground states before the collision.

The main terms of $W(1, 2)$ in (13) are the interactions between the monopole moment of one of the nuclei with the multipole moments of the second nucleus. Including only these terms, the Schrödinger equation (13) separates into two equations each of which depends on the intrinsic degrees of freedom of one of the nuclei only. These equations are of the form

$$i\hbar\frac{\partial}{\partial t}|\psi\rangle = (H_0 + V(t))|\psi\rangle, \tag{15}$$

where $V(t)$ is given by eqs. (1.12) and (1.13). We shall call particle 1 the projectile and particle 2 the target nucleus. It is obvious that the role of target and projectile can be interchanged.

The explicit expression for the multipole expansion of $V(t)$ can be written in the form

$$V(t) = \sum_{\lambda=1,\mu}^{\infty} \frac{4\pi Z_1 e}{(2\lambda+1)}(-1)^{\mu}\bar{S}_{\lambda\mu}(t)\mathscr{M}(\lambda, -\mu). \tag{16}$$

We shall mostly be concerned with electric excitations where

$$\bar{S}_{\lambda\mu}(t) = \bar{S}_{E\lambda\mu}(t) = [r(t)]^{-\lambda-1}Y_{\lambda\mu}(\theta(t), \phi(t)). \tag{17}$$

For magnetic excitations we introduce correspondingly

$$\bar{S}_{\lambda\mu}(t) = \bar{S}_{M\lambda\mu}(t) = \frac{i}{\lambda}\frac{1}{r^{\lambda+1}(t)}\frac{\dot{\boldsymbol{r}}(t)\boldsymbol{L}}{c}\,Y_{\lambda\mu}(\theta(t), \phi(t)). \tag{18}$$

We denote by $r(t)$, $\theta(t)$ and $\phi(t)$ the time dependent spherical coordinates of the position vector $\boldsymbol{r}(t)$ of the projectile. The spherical coordinates are referred to a coordinate system with its origin in the target centre of mass and with a fixed direction of the polar axis.

In the following we shall mostly neglect the multipole-multipole interaction.

§ 3. *S-matrix formalism*

In the semiclassical approximation the Coulomb excitation process is completely determined by the time dependent Schrödinger equation (2.13). For convenience we consider in the following only the monopole-multipole interaction (1.11) in which case the Schrödinger equation is of the form (2.15) with $V(t)$ given by (2.16) to (2.18). In these equations $V(t)$ is a well known function of the time t and of the nuclear degrees of freedom. If the Schrödinger equation is solved with the initial condition of the nucleus being in its ground state before the collision, the wave function at the time $t = +\infty$ is a superposition of all the nuclear states. It may happen that some component of the wave function may describe particle unstable states decaying during the collision. In the present section we shall disregard this complication which will be discussed in § 7 below.

Despite these simplifying assumptions a complete solution for real nuclei is not possible. We shall therefore discuss in this section some formal solutions of the Schrödinger equation which can be used as basis for approximation methods.

For the following discussion* it will be convenient to use the interaction representation, where the nuclear state vector $|\Phi(t)\rangle$ is defined in terms of the state vector $|\psi\rangle$ appearing in (2.15) by the relation

$$|\psi\rangle = \exp(-iH_0 t/\hbar)\,|\Phi(t)\rangle. \tag{1}$$

Before and after the collision this state vector is time independent and it satisfies the equation

$$i\hbar \frac{\partial}{\partial t}|\Phi(t)\rangle = \tilde{V}(t)\,|\Phi(t)\rangle, \tag{2}$$

where

$$\tilde{V}(t) = \exp(iH_0 t/\hbar) V(t) \exp(-iH_0 t/\hbar). \tag{3}$$

With this definition of $|\Phi(t)\rangle$ the amplitudes $a_n(t)$ on the nuclear eigenstates:

$$a_n(t) = \langle n|\Phi(t)\rangle \tag{4}$$

become time independent before and after the collision. Here $|n\rangle$ is the time independent eigenstate belonging to the eigenvalue E_n of H_0, i.e.

$$H_0\,|n\rangle = E_n\,|n\rangle. \tag{5}$$

We introduce the time-development operator of the state vector $|\Phi(t)\rangle$ by the equation

$$|\Phi(t)\rangle = U(t, t_1)\,|\Phi(t_1)\rangle. \tag{6}$$

Inserting this definition into (2) it is seen that the unitary operator $U(t, t_1)$ satisfies the integral equation

$$U(t, t_1) = 1 + \frac{1}{i\hbar}\int_{t_1}^{t} \tilde{V}(t')U(t', t_1)\,dt'. \tag{7}$$

* For more details on several aspects of this discussion see e.g. [ROM 65].

If the interaction energy $V(t)$ is weak this equation can be solved by an iteration which leads to the usual perturbation expansion

$$|\Phi(t)\rangle = U(t, -\infty)\,|\Phi(-\infty)\rangle$$
$$= \left(1 - \frac{\mathrm{i}}{\hbar}\int_{-\infty}^{t} \tilde{V}(t')\,\mathrm{d}t' + \left(\frac{-\mathrm{i}}{\hbar}\right)^2 \int_{-\infty}^{t} \mathrm{d}t'\,\tilde{V}(t') \int_{-\infty}^{t'} \mathrm{d}t''\,\tilde{V}(t'') + \cdots\right) |\Phi(-\infty)\rangle. \tag{8}$$

This series can not be summed because the operators $\tilde{V}(t')$ and $\tilde{V}(t'')$ in general do not commute if the two times are different. If they do commute one can utilize the following formula

$$\int_{-\infty}^{t} \mathrm{d}t_1 \cdots \int_{-\infty}^{t} \mathrm{d}t_n\, f(t_1 \cdots t_n) = n! \int_{-\infty}^{t} \mathrm{d}t_1 \int_{-\infty}^{t_1} \mathrm{d}t_2 \cdots \int_{-\infty}^{t_{n-1}} \mathrm{d}t_n\, f(t_1 \cdots t_n), \tag{9}$$

which holds for any function $f(t_1 \cdots t_n)$ which is symmetric in all variables, and the expression (8) can then be written

$$|\Phi(t)\rangle = \exp\left\{-\frac{\mathrm{i}}{\hbar}\int_{-\infty}^{t} \tilde{V}(t')\,\mathrm{d}t'\right\} |\Phi(-\infty)\rangle. \tag{10}$$

The simplification (10) can be used if all nuclear states $|n\rangle$ can be considered to have the same energy or, equivalently, if the collision time is short compared to the nuclear periods. In this sudden approximation the integrands in eq. (8) are nonvanishing only for a short time interval around $t = 0$ where the exponential functions in (3) may be neglected, i.e. $\tilde{V}(t) \approx V(t)$.

In the general case it is sometimes convenient to perform the summation leading to eq. (10) in a formal way by introducing the time ordering operator $\mathscr{T}$. This operator is defined by

$$\mathscr{T} A(t_1) \cdots A(t_n) = A(t_{p_1}) A(t_{p_2}) \cdots, \tag{11}$$

where

$$t_{p_1} \geqslant t_{p_2} \geqslant \cdots \geqslant t_{p_n}.$$

Using the operator $\mathscr{T}$ one finds that eq. (8) can be written

$$|\Phi(t)\rangle = \mathscr{T} \exp\left\{-\frac{\mathrm{i}}{\hbar}\int_{-\infty}^{t} \tilde{V}(t')\,\mathrm{d}t'\right\} |\Phi(-\infty)\rangle. \tag{12}$$

If the nucleus before the collision is in the ground state $|0\rangle$ of eq. (5) the solution (12) leads to the following expression for the amplitudes on the various excited states after the collision,

$$a_n(\infty) = \langle n|\, \mathscr{T} \exp\left\{-\frac{\mathrm{i}}{\hbar}\int_{-\infty}^{+\infty} \tilde{V}(t)\,\mathrm{d}t\right\} |0\rangle. \tag{13}$$

The expression (13) offers a formal solution of the Schrödinger equation which in most cases can only be evaluated through the series expansion (8) in powers of the interaction energy.

The matrix elements of the terms in the perturbation expansion are in general complex numbers. In the following we shall choose the phases of the nuclear wave functions (see appendix E) and the coordinate systems (see § 8 below) such that the first-order matrix elements of $V(t)$ are real, i.e.

$$\langle n| \int_{-\infty}^{+\infty} \tilde{V}(t)\, \mathrm{d}t\, |m\rangle^* = \langle n| \int_{-\infty}^{+\infty} \tilde{V}(t)\, \mathrm{d}t\, |m\rangle . \tag{14}$$

In this case the first-order term of (8) is purely imaginary.

The second-order term is only real in the sudden approximation, where, according to eq. (9), it may be written as a product of two matrix elements of the form (14). In general one may separate the real from the imaginary part by introducing the step function

$$\theta(t-t') = \tfrac{1}{2}[1+\varepsilon(t-t')] = \begin{cases} 1, & t > t' \\ 0, & t < t' \end{cases} \tag{15}$$

in the double integral. Thus one finds by inserting (15) into the second-order term of (8)

$$\int_{-\infty}^{+\infty} \mathrm{d}t'\, \tilde{V}(t') \int_{-\infty}^{t'} \mathrm{d}t''\, \tilde{V}(t'')$$
$$= \tfrac{1}{2}\left\{\int_{-\infty}^{+\infty} \mathrm{d}t'\, \tilde{V}(t')\right\}^2 + \tfrac{1}{2}\int_{-\infty}^{+\infty} \mathrm{d}t' \int_{-\infty}^{+\infty} \mathrm{d}t''\, \varepsilon(t'-t'')\, \tilde{V}(t')\, \tilde{V}(t''). \tag{16}$$

While the matrix elements of the first term according to (14) are real, the matrix elements of the second term are purely imaginary, since $\varepsilon(x)$ is an odd function of x.

It is often convenient to express the S-matrix

$$S = U(\infty, -\infty) = \mathscr{T} \exp\left\{-\frac{\mathrm{i}}{\hbar}\int_{-\infty}^{+\infty} \tilde{V}(t)\, \mathrm{d}t\right\} \tag{17}$$

in terms of Hermitian operators which have real matrix elements with the conventions leading to (14). Two such expressions for the S-matrix are:

$$S = (1+\mathrm{i}K)/(1-\mathrm{i}K), \tag{18}$$

in terms of the Hermitian K-operator, and

$$S = \mathrm{e}^{2\mathrm{i}Q}, \tag{19}$$

in terms of the Hermitian phase shift operator Q. As is seen from the definition the two operators K and Q are connected by the formula

$$Q = \arctan K = K - \tfrac{1}{3}K^3 + \tfrac{1}{5}K^5 \cdots . \tag{20}$$

A series expansion for the K-operator in powers of the interaction $\tilde{V}(t)$ can be obtained in the following way.* We first consider the integral equation (7), which by using the function (15) can be written in the form

$$U(t, t_1) = \tfrac{1}{2} + \tfrac{1}{2}U(t_2, t_1) + \frac{1}{2i\hbar}\int_{t_1}^{t_2} \varepsilon(t - t')\, \tilde{V}(t')\, U(t', t_1)\, dt', \tag{21}$$

where $t_2 \geqslant t$.

Introducing the operator $g_{t_1}^{t_2}(t)$ by the equation

$$g_{t_1}^{t_2}(t) = \frac{1}{\hbar}\int_{t_1}^{t_2} \varepsilon(t - t')\, \tilde{V}(t')\, dt' + \frac{1}{2i\hbar}\int_{t_1}^{t_2} \varepsilon(t - t')\, \tilde{V}(t')\, g_{t_1}^{t_2}(t')\, dt' \tag{22}$$

one finds that the solution of (21) can be written in terms of this operator by

$$U(t, t_1) = \tfrac{1}{2} + \tfrac{1}{2}U(t_2, t_1) + \frac{1}{2i}\, g_{t_1}^{t_2}(t)[\tfrac{1}{2} + \tfrac{1}{2}U(t_2, t_1)]. \tag{23}$$

Solving this equation for g with $t = t_2$ one may verify that

$$K = -\tfrac{1}{2}g_{-\infty}^{+\infty}(+\infty). \tag{24}$$

The series expansion for K is now found directly by iterating (22). The result is

$$K = -\frac{1}{2\hbar}\int_{-\infty}^{\infty} dt'\, \tilde{V}(t') - \frac{1}{i}\left(\frac{1}{2\hbar}\right)^2 \int_{-\infty}^{\infty} dt' \int_{-\infty}^{\infty} dt''\, \varepsilon(t' - t'')\, \tilde{V}(t')\, \tilde{V}(t'')$$
$$+ \cdots + \frac{1}{i}\frac{1}{(2i\hbar)^n}\int_{-\infty}^{\infty} dt' \cdots \int_{-\infty}^{\infty} dt^{(n)}$$
$$\times\, \varepsilon(t' - t'') \cdots \varepsilon(t^{(n-1)} - t^{(n)})\, \tilde{V}(t')\, \tilde{V}(t'') \cdots \tilde{V}(t^{(n)}). \tag{25}$$

It is seen that with the conventions leading to (14) the matrix elements of all terms in this expansion are real numbers. It is noted furthermore that the perturbation expansion up to second order of (18) receives partly a contribution from the second-order term in K which is identical to the second term in the matrix elements of (16), and partly a contribution from the square of the first-order term which is the first term in (16). It should finally be pointed out

* This proof was given to us by P. Kristensen, University of Aarhus, Denmark.

that the expression (18) is unitary, even if one inserts for K only a finite perturbation expansion, which means that in contrast to the perturbation expansion (8) the total probability is exactly conserved.

For the applications we shall introduce the following notation

$$2K = R + G + H + \cdots, \tag{26}$$

where

$$R = -\frac{1}{\hbar}\int_{-\infty}^{\infty} \tilde{V}(t)\,\mathrm{d}t, \tag{27}$$

$$G = \frac{\mathrm{i}}{2\hbar^2}\int_{-\infty}^{\infty} \mathrm{d}t' \int_{-\infty}^{\infty} \mathrm{d}t''\, \varepsilon(t' - t'')\, \tilde{V}(t')\, \tilde{V}(t''), \tag{28}$$

and

$$H = \frac{1}{4\hbar^3}\int_{-\infty}^{\infty} \mathrm{d}t' \int_{-\infty}^{\infty} \mathrm{d}t'' \int_{-\infty}^{\infty} \mathrm{d}t'''\, \varepsilon(t' - t'')\, \varepsilon(t'' - t''')\, \tilde{V}(t')\, \tilde{V}(t'')\, \tilde{V}(t'''). \tag{29}$$

From the series expansions (26) and (20) one obtains the following expression for the phase shift operator Q

$$2\mathrm{i}Q = \mathrm{i}[R + G + (H - \tfrac{1}{12}R^3) + \cdots]. \tag{30}$$

An explicit formula for the nth term in the expansion (30) has been given by Bialynicki-Birula et al. [BIA 69]. Their result may be written in the form

$$2\mathrm{i}Q = \sum_{n=1}^{\infty} \left(\frac{1}{\mathrm{i}\hbar}\right)^n \int_{-\infty}^{\infty} \mathrm{d}t_1 \int_{-\infty}^{\infty} \mathrm{d}t_2 \cdots \int_{-\infty}^{\infty} \mathrm{d}t_n$$

$$\times\, \tilde{V}(t_1)\, \tilde{V}(t_2) \cdots \tilde{V}(t_n)\, \frac{(-1)^{n-1-\theta_n}}{n} \binom{n-1}{\theta_n}^{-1}, \tag{31}$$

where $\theta_1 = 0$ and

$$\theta_n = \theta(t_n - t_{n-1}) + \theta(t_{n-1} - t_{n-2}) + \cdots + \theta(t_2 - t_1). \tag{32}$$

Again with the above mentioned conventions all matrix elements of the operator Q are real numbers. The S-matrix (19) is always unitary even if the series (30) is cut off after a finite number of terms.

An important property of the expansion (30) of the Q-operator is that its first term leads to an S-matrix which contains multiple excitations and which becomes exact for short collision times. This can be seen explicitly from (30), since the terms can be written as multiple commutators, i.e.

$$G = \frac{\mathrm{i}}{2\hbar^2}\int_{-\infty}^{\infty} \mathrm{d}t' \int_{-\infty}^{t'} \mathrm{d}t''[\tilde{V}(t'), \tilde{V}(t'')], \tag{33}$$

and

$$H - \tfrac{1}{12}R^3 = \frac{1}{6\hbar^3}\int_{-\infty}^{\infty} dt' \int_{-\infty}^{t'} dt'' \int_{-\infty}^{t''} dt'''\{[\tilde{V}(t'), [\tilde{V}(t''), \tilde{V}(t''')]] + [\tilde{V}(t'''), [\tilde{V}(t''), \tilde{V}(t')]]\}. \tag{34}$$

A general formula for the phaseshift operator in terms of multiple commutators has been given by Bialynicki-Birula et al. [BIA 69].

Since the sudden approximation implies $\tilde{V}(t) \approx V(t)$, all multiple commutators vanish and we find for the S-matrix

$$S = \exp\left\{-\frac{i}{\hbar}\int_{-\infty}^{\infty} V(t)\, dt\right\}, \tag{35}$$

in accordance with (10). Equations (19) and (30) thus offer an expansion in the parameters ξ, as well as in the strength of the coupling. In special cases it may happen that the series (30) automatically terminates and (19) offers in such cases a closed expression for the excitation amplitudes (see chapter VII).

From the expansion (30) we may reconstruct the perturbation expansion to third order in $V(t)$. One finds

$$S = e^{i2Q} \approx 1 + iR + (-\tfrac{1}{2}R^2 + iG) + [i(H - \tfrac{1}{4}R^3) - \tfrac{1}{2}(RG + GR)] + \cdots. \tag{36}$$

Since the operators R, G and H are Hermitian the form (36) offers a decomposition of the S-matrix elements in real and imaginary parts.

If one includes only a finite number of terms in (30) one obtains an approximation to the S-matrix, which encompasses the sudden approximation as well as perturbation theory with conserved total probability. The result (27)–(29) for R, G and H can also be obtained directly from the definition (19) and (30) by expanding S in powers of V and comparing term by term with the power series expansion (8).

While (36) is an expansion in powers of V it is for many applications important to keep the major part of V in the exponent and expanding only in powers of the smaller contributions. Thus, if we write

$$2Q = A + B, \tag{37}$$

where B is small compared to A, one may expand the S-matrix in powers of B. The first terms in this expansion are given by

$$S = e^{2iQ} = e^{i(A+B)} = e^{iA} + i\int_0^1 dx\, e^{ixA} B\, e^{i(1-x)A} + \cdots. \tag{38}$$

This formula can be proved easily by expanding both sides in powers of A and B and using the expression

$$\int_0^1 x^n(1-x)^m \,\mathrm{d}x = \frac{n!\,m!}{(n+m+1)!}. \tag{39}$$

A similar expansion can be obtained from eq. (17) where the S-matrix is expressed in terms of the time ordering operator

$$S = \mathscr{T} \exp\left\{\mathrm{i}\int_{-\infty}^{+\infty} [A(t) + B(t)]\,\mathrm{d}t\right\}, \tag{40}$$

where again the time-dependent operator $B(t)$ is assumed to be small compared to $A(t)$. The series expansion of (40) in powers of B yields

$$S = \mathscr{T} \exp\left\{\mathrm{i}\int_{-\infty}^{+\infty} A(t)\,\mathrm{d}t\right\} + \mathrm{i}\int_{-\infty}^{+\infty} \mathrm{d}t \left[\mathscr{T} \exp\left\{\mathrm{i}\int_t^{\infty} A(t)\,\mathrm{d}t\right\}\right.$$
$$\left. \times\, B(t)\left(\mathscr{T} \exp\left\{\mathrm{i}\int_{-\infty}^{t} A(t)\,\mathrm{d}t\right\}\right)\right] + \cdots. \tag{41}$$

The expansions (38) and (41) are mainly used for the evaluation of the excitation amplitudes in chapters VI and VIII.

§ 4. *Coupled differential equations*

The expansions of the S-matrix which we have discussed in the previous section are especially well suited for the discussion of the Coulomb excitation in specific nuclear models. For the evaluation of the Coulomb excitation process in a more general, model-independent way, it may be advantageous to solve the Schrödinger equation (2.13) or (2.15) directly by means of computers.

For this purpose one may formulate the Schrödinger equation

$$\mathrm{i}\hbar \frac{\partial}{\partial t} |\psi(t)\rangle = H(t)|\psi(t)\rangle, \tag{1}$$

where

$$H(t) = H_0 + V(t), \tag{2}$$

as a system of coupled differential equations by expanding the nuclear state vector $|\psi(t)\rangle$ in terms of a convenient complete set of orthogonal states, as e.g. the eigenstates of the free nuclear Hamiltonian H_0. We define the eigenstates $|n\rangle$ of H_0 by

$$H_0\,|n\rangle = E_n\,|n\rangle, \tag{3}$$

and introduce the time dependent excitation amplitudes

$$a_n(t) = \langle n|\psi\rangle \exp(\mathrm{i}E_n t/\hbar). \tag{4}$$

These excitation amplitudes become time independent for $t \to \pm\infty$, where the interaction $V(t)$ vanishes. From these definitions one finds that the Schrödinger equation (1) is equivalent to the following set of linear differential equations

$$\mathrm{i}\hbar\dot{a}_n = \sum_m \langle n|V(t)|m\rangle \exp[\mathrm{i}\,(E_n - E_m)t/\hbar]a_m(t). \tag{5}$$

These equations should be solved with the initial condition that at time $t = -\infty$, where the projectile is far away, the nucleus is in its ground state, i.e.

$$a_n(-\infty) = \delta_{0n}. \tag{6}$$

The values of a_n at $t = +\infty$ are then the excitation amplitudes after the collision, and the excitation probabilities are thus given by

$$P_n = |a_n|^2. \tag{7}$$

The coupled equations (5) have as many independent solutions as the number of states included in the summation. We may ennumerate the solution by the initial conditions. Thus we may define a solution $a_{n,k}(t)$ by the property

$$a_{n,k}(-\infty) = \delta_{nk}, \tag{8}$$

corresponding to an initial population in state k only. If one considers $a_{n,k}$ as a matrix it follows from the Hermitian character of $V(t)$ that this matrix is unitary at all times, i.e.

$$\sum_k a^*_{n,k}(t)\, a_{m,k}(t) = \delta_{nm}, \tag{9}$$

and

$$\sum_n a^*_{n,k}(t)\, a_{n,k'}(t) = \delta_{kk'}. \tag{10}$$

These equations hold also if only a finite number of states N is included in (5), provided the summations in (9) and (10) are performed over all N states. The last equation shows that the excitation probabilities (7) as functions of time satisfy the conservation of total probability, i.e.

$$\sum_n P_n(t) = \sum_n |a_n|^2 = 1. \tag{11}$$

The summation over m in eq. (5) should in principle be performed over all nuclear states $|m\rangle$. For a practical solution one can, of course, only take a finite number of states into account. Since we know already from the general argument of adiabaticity that states with high excitation energy can not be populated appreciably it seems justified to leave out these high energy states, i.e. to include in the sum of eq. (5) only those states $|n\rangle$ for which the parameters ξ satisfy the inequality (cf. § 2)

$$\xi_{0\to n} = \frac{\Delta E_n}{\hbar}\frac{a}{v} \lessapprox 1. \tag{12}$$

In spite of the fact that the high-lying states are not populated, they may nevertheless have an influence on the excitation process in changing the excitation amplitudes of the low-lying states.

The excitation of high-lying states which violate (12) may conveniently be described in the adiabatic representation of the Schrödinger equation. The state vector $|\psi(t)\rangle$ is here expanded in terms of the complete set of eigenstates for the total Hamiltonian $H(t)$, where t is considered as a fixed parameter. We thus introduce the state vector $|n(t)\rangle$ by

$$H(t)\,|n(t)\rangle = E_n(t)\,|n(t)\rangle, \tag{13}$$

where $E_n(t)$ is the energy of the nucleus in the presence of the projectile at the position $\boldsymbol{r}(t)$. The complete set of states $|n(t)\rangle$ and the energies $E_n(t)$ thus change in time as the projectile moves along its orbit. At the time $t = \pm\infty$ they coincide with the energy eigenstates of the free nucleus. In analogy to eq. (4) we introduce the excitation amplitude $c_n(t)$ by

$$c_n(t) = \langle n(t)|\psi\rangle \exp\left\{\frac{i}{\hbar}\int_0^t E_n(t')\,dt'\right\}. \tag{14}$$

Differentiating eq. (14) with respect to t and utilizing eqs. (1) and (13) one finds

$$\dot{c}_n = \sum_m \frac{\partial\langle n(t)|}{\partial t}\,|m(t)\rangle\, c_m(t) \exp\left\{\frac{i}{\hbar}\int_0^t [E_n(t') - E_m(t')]\,dt'\right\}. \tag{15}$$

One may reformulate this equation using the orthogonality of $|n(t)\rangle$. Since $\langle n(t)|m(t)\rangle = \delta_{nm}$ it follows that

$$\frac{\partial\langle n(t)|}{\partial t}\,|m(t)\rangle + \langle n(t)|\,\frac{\partial}{\partial t}\,|m(t)\rangle = 0, \tag{16}$$

and one may write eq. (15) in the form

$$\dot{c}_n = -\sum_m \langle n(t)|\,\frac{\partial}{\partial t}\,|m(t)\rangle \exp\left\{\frac{i}{\hbar}\int_0^t [E_n(t') - E_m(t')]\,dt'\right\} c_m(t). \tag{17}$$

In the summation over the states $|m(t)\rangle$ one may leave out the state $|n(t)\rangle$ if one chooses the phases of the states such that

$$\langle n(t)| \frac{\partial}{\partial t} |n(t)\rangle = 0. \tag{18}$$

This can be done, since according to eq. (16) it follows that the matrix element

$$\langle n(t)| \frac{\partial}{\partial t} |n(t)\rangle = \mathrm{i}\alpha(t) \tag{19}$$

is purely imaginary. A new state vector $|n'(t)\rangle$ defined by

$$|n'(t)\rangle = |n(t)\rangle \exp\{-\mathrm{i}\int_0^t \alpha(t')\,\mathrm{d}t'\} \tag{20}$$

will satisfy eq. (18).

Another convenient form of the coupled adiabatic equations is obtained by taking the derivative of eq. (13) with respect to t and multiplying the resulting expression by $\langle m(t)|$. Inserting the result in (17) one finds

$$\dot{c}_n = \sum_{m \neq n} \frac{\langle n(t)|\,(\partial H/\partial t)\,|m(t)\rangle}{E_n(t) - E_m(t)} \exp\left\{\frac{\mathrm{i}}{\hbar}\int_0^t [E_n(t') - E_m(t')]\,\mathrm{d}t'\right\} c_m(t), \tag{21}$$

where the sum should be carried out over all states $m \neq n$. The equations (17) or (21) should be solved with the initial condition

$$c_n(-\infty) = \delta_{n0}, \tag{22}$$

which means, that one defines

$$|n(-\infty)\rangle = |n\rangle. \tag{23}$$

The excitation probabilities for $t = +\infty$ are given directly by

$$P_n = |c_n(+\infty)|^2. \tag{24}$$

It is seen from eq. (21) that for very slow collisions, where $\partial H/\partial t \to 0$ the nucleus stays in the adiabatic ground state $|0(t)\rangle$ and that all excitation probabilities vanish. The adiabatic ground state energy $E_0(t)$ in general differs from E_0, and the difference $E_0(t) - E_0$ may be included as a polarization potential for the elastic scattering. A detailed discussion of the polarization effect will be given in § 6 below.

We shall conclude this section by a discussion of expansion methods which apply to the situation where the interaction $V(t)$ besides a major component

$V_0(t)$ contains a small contribution $V_1(t)$, i.e. the coupled equations have the form

$$\mathrm{i}\hbar\dot{a}_n = \sum_m \langle n|\, V_0(t) + V_1(t)\, |m\rangle \exp\left\{\frac{\mathrm{i}}{\hbar}(E_n - E_m)t\right\} a_m, \tag{25}$$

where

$$\langle n|\, V_1(t)\, |m\rangle \ll \langle n|\, V_0(t)\, |m\rangle . \tag{26}$$

Let us write the amplitudes $a_n(t)$ in a series expansion in powers of V_1 as

$$a_n(t) = a_n^{(0)}(t) + a_n^{(1)}(t) + \cdots, \tag{27}$$

where $a_n^{(0)}(t)$ satisfies the zero-order equations

$$\mathrm{i}\hbar\dot{a}_n^{(0)} = \sum_m \langle n|\, V_0(t)\, |m\rangle \exp\left\{\frac{\mathrm{i}}{\hbar}(E_n - E_m)t\right\} a_m^{(0)} . \tag{28}$$

Neglecting higher order terms one finds that the amplitudes $a_n^{(1)}(t)$ obey the following inhomogeneous equations

$$\mathrm{i}\hbar\dot{a}_n^{(1)} = \sum_m \langle n|\, V_0(t)\, |m\rangle \exp\left\{\frac{\mathrm{i}}{\hbar}(E_n - E_m)t\right\} a_m^{(1)}(t)$$

$$+ \sum_m \langle n|\, V_1(t)\, |m\rangle \exp\left\{\frac{\mathrm{i}}{\hbar}(E_n - E_m)t\right\} a_m^{(0)}(t) . \tag{29}$$

If a complete solution for the corresponding homogeneous equation, which is identical to (28), is known, an explicit solution of (29) can be found by the method of the variation of the constants. Thus we write the general solution of (28) in the form

$$a_n(t) = \sum_k \lambda_k a_{n,k}^{(0)}(t), \tag{30}$$

where $a_{n,k}^{(0)}$ are the solutions defined by (8) and λ_k are arbitrary constants. Considering now $\lambda_k(t)$ to be functions of time, we use (30) as an "ansatz" for $a_n^{(1)}$. Inserting into eq. (29) we get with the initial condition (6)

$$\mathrm{i}\hbar \sum_k \dot{\lambda}_k a_{n,k}^{(0)}(t) = \sum_m \langle n|\, V_1(t)\, |m\rangle \exp\left\{\frac{\mathrm{i}}{\hbar}(E_n - E_m)t\right\} a_{m,0}^{(0)}(t) . \tag{31}$$

Utilizing the orthogonality property (10) one may solve the equation with the following result

$$\lambda_k(t) = \frac{1}{\mathrm{i}\hbar} \sum_{lm} \int_{-\infty}^{t} a_{l,k}^{(0)*}(t')\, \langle l|\, V_1(t')\, |m\rangle \exp\left\{\frac{\mathrm{i}}{\hbar}(E_l - E_m)t'\right\} a_{m,0}^{(0)}(t')\, \mathrm{d}t' . \tag{32}$$

Inserting (32) in (30) we obtain at $t = +\infty$

$$a_n^{(1)}(\infty) = \frac{1}{i\hbar} \sum_{lkm} a_{n,k}^{(0)}(\infty) \int_{-\infty}^{+\infty} a_{l,k}^{(0)*}(t') \langle l | V_1(t') | m \rangle \times \exp\left\{\frac{i}{\hbar}(E_l - E_m)t'\right\} a_{m,0}^{(0)}(t')\, dt'. \tag{33}$$

This expression offers an explicit formula for the first-order correction that is linear in $V_1(t)$. To evaluate the integral one must know the complete set of solutions of the unperturbed equation (28), corresponding to all initial situations where the nucleus is in the kth excited state before the collision. It is possible, however, to perform the summation over the index k. Thus, the function

$$\bar{a}_{l,n}(t) = \sum_k a_{n,k}^{(0)*}(\infty)\, a_{l,k}^{(0)}(t) \tag{34}$$

is also a solution of the differential equation (28) and according to eq. (9) it has the property that for $t = +\infty$

$$\bar{a}_{l,n}(\infty) = \delta_{nl}. \tag{35}$$

The solution $\bar{a}_{l,n}(t)$ can thus be obtained from the differential equation (28) by integrating backwards in time from $t = +\infty$ with the initial condition (35). In terms of this backward solution one finds

$$a_n^{(1)}(\infty) = \frac{1}{i\hbar} \sum_{lm} \int_{-\infty}^{\infty} \bar{a}_{l,n}^{(0)*}(t') \langle l | V_1(t') | m \rangle\, a_{m,0}^{(0)}(t') \exp\left\{\frac{i}{\hbar}(E_l - E_m)t'\right\} dt'. \tag{36}$$

This equation is identical to the result (3.41).

§ 5. *Coupled equations in Fourier components*

In the semiclassical picture of Coulomb excitation the excitation amplitudes in first-order perturbation theory are proportional to the Fourier components of the time-dependent interaction $V(t)$ with frequency $\omega = \Delta E/\hbar$, where ΔE is the excitation energy (see eq. (I.24)). For those problems in Coulomb excitation, where it is important to keep track of the energy which is transferred to the nucleus, as e.g. in the excitation of decaying states, it is convenient to formulate the equations of motion in terms of these Fourier components.

The Fourier transformed nuclear state $|\hat{\psi}(\omega)\rangle$ is connected to the time-dependent state $|\psi(t)\rangle$ of eq. (4.1) by the relation

$$|\psi(t)\rangle = \int_{-\infty}^{\infty} |\hat{\psi}(\omega)\rangle \exp(-i\omega t)\, d\omega. \tag{1}$$

Expanding the state vector $|\hat{\psi}(\omega)\rangle$ in terms of the eigenstates $|n\rangle$ defined in (4.3) we get

$$|\hat{\psi}(\omega)\rangle = \sum_n \beta_n(\omega)\,|n\rangle. \tag{2}$$

The connection between these expansion coefficients and the coefficients $a_n(t)$ of eq. (4.4) is

$$a_n(t) = \int_{-\infty}^{\infty} \beta_n(\omega) \exp\left\{\frac{\mathrm{i}}{\hbar}(E_n - \hbar\omega)t\right\} \mathrm{d}\omega. \tag{3}$$

Introducing (1) and (2) into the Schrödinger equation (4.1) we obtain the following coupled integral equation for $\beta_n(\omega)$:

$$(\hbar\omega - E_n)\beta_n(\omega) = \sum_m \int_{-\infty}^{\infty} \mathrm{d}\omega'\, \hat{V}_{n,m}(\omega - \omega')\beta_m(\omega'), \tag{4}$$

where

$$\hat{V}_{n,m}(x) = \frac{1}{2\pi}\int_{-\infty}^{\infty} \langle n|\, V(t)\, |m\rangle\, \mathrm{e}^{\mathrm{i}xt}\, \mathrm{d}t \tag{5}$$

are the Fourier components of the matrix elements of $V(t)$.

The integral equations (4) do not have a unique solution because of the singularities for $\hbar\omega = E_n$. A unique solution is only obtained if one specifies the path of integration around the poles which result from these singularities. This ambiguity corresponds to the freedom in the choice of initial conditions in the differential equations (4.5).

In order to satisfy the initial condition (4.6), i.e.

$$a_n(-\infty) = \delta_{n0} \tag{6}$$

it is convenient to introduce instead of $\beta_n(\omega)$ the new variables $\alpha_n(\omega)$ defined by

$$\alpha_n(\omega) \doteq \beta_n(\omega) - \delta(\omega - E_0/\hbar)\delta_{n0}. \tag{7}$$

The δ-function corresponds to the solution of eq. (4) for the case where there is no interaction ($V(t) = 0$). The amplitudes $\alpha_n(\omega)$ satisfy the inhomogeneous integral equations

$$(\hbar\omega^+ - E_n)\alpha_n(\omega) = \hat{V}_{n,0}\left(\omega - \frac{E_0}{\hbar}\right) + \int_{-\infty}^{\infty} \sum_m \hat{V}_{n,m}(\omega - \omega')\alpha_m(\omega')\,\mathrm{d}\omega'. \tag{8}$$

In eq. (8) all singularities lie below the real axis, which has been indicated by ascribing to ω on the left-hand side a small positive imaginary part. This prescription ensures retarded solutions.

The equations (8) are the semiclassical equations of motion in Fourier components. From the amplitudes $\alpha_m(\omega)$, which have singularities at $\omega = \omega_n = E_n/\hbar$, one may evaluate the excitation amplitudes $a_n(t = \infty)$ by means of the relation

$$a_n(+\infty) = \delta_{no} + \frac{2\pi}{i\hbar} \lim_{\omega \to \omega_n} (\hbar\omega^+ - E_n)\alpha_n(\omega). \tag{9}$$

The equations (8) may also be formulated in terms of the quantities

$$\tau_n(\omega) = \frac{2\pi}{i\hbar} (\hbar\omega^+ - E_n)\alpha_n(\omega), \tag{10}$$

which have no singularities. The result is

$$\tau_n(\omega) = \frac{2\pi}{i\hbar} \hat{V}_{n,0}(\omega - \omega_0) + \sum_m \int_{-\infty}^{+\infty} d\omega' \frac{\hat{V}_{n,m}(\omega - \omega')}{\hbar\omega'^+ - E_m} \tau_m(\omega'). \tag{11}$$

This is the equation for the classical T-matrix. It can be solved by iteration, which leads to the usual perturbation expansion for the excitation amplitude. One finds

$$\begin{aligned} a_n(\infty) &= \delta_{no} + \tau_n(\omega_n) \\ &= \delta_{no} + \frac{2\pi}{i\hbar} \hat{V}_{n,0}(\omega_n - \omega_0) \\ &\quad + \frac{2\pi}{i\hbar} \sum_m \int_{-\infty}^{+\infty} d\omega' \frac{\hat{V}_{n,m}(\omega_n - \omega')\hat{V}_{m,0}(\omega' - \omega_0)}{\hbar\omega'^+ - E_m} + \cdots. \end{aligned} \tag{12}$$

The connection with the perturbation expansion (3.8) is established by noting that the small positive imaginary part of ω' in the denominator of (12) leads to a time ordering of the double integral over time, when the quantities $\hat{V}$ are expressed in terms of $V(t)$. If on the other hand the integral over ω' in (12) is decomposed into a δ-function plus a principal part integral, we obtain the result of eq. (3.16).

We may note that according to (12) the S-matrix is related to the T-matrix τ_n on the energy shell by the familiar looking formula

$$\langle n| S |0\rangle = \delta_{no} + \tau_n(\omega_n). \tag{13}$$

§ 6. *Effect of high-lying states*

For the practical solution of the coupled equations (4.5) it is essential that one may confine the calculation to a finite number of states. In this truncation one should of course include all states which are strongly excited in the Coulomb excitation process. The excitation of states which are coupled weakly to the strongly excited states can be evaluated separately in a perturbation treatment (see e.g. § 4). Also states of high excitation energy may be left out if the parameters ξ are so large, that the excitation probabilities become vanishingly small. If the high-lying states are coupled strongly to the states that are excited, this may, however, be rather dangerous. Although the high states are not excited, they may nevertheless influence the excitation of the lower states by the fact that virtual transitions through the high states may be important. This effect is known as a polarization effect. In this section we shall see how one may modify the coupled equations (4.5) so that they incorporate the polarization effect.

We denote by $|z\rangle$ or $|z'\rangle$ a group of high-lying states that are strongly coupled to the low-lying excited states $|n\rangle$ and $|m\rangle$ and that are mutually strongly coupled. The differential equations (4.5) can thus be written in the form

$$\begin{aligned} \mathrm{i}\hbar\dot{a}_n &= \sum_m \langle n|\, V(t)\, |m\rangle \exp\left\{\frac{\mathrm{i}}{\hbar}(E_n - E_m)t\right\} a_m \\ &\quad + \sum_z \langle n|\, V(t)\, |z\rangle \exp\left\{\frac{\mathrm{i}}{\hbar}(E_n - E_z)t\right\} a_z, \end{aligned} \tag{1}$$

and

$$\begin{aligned} \mathrm{i}\hbar\dot{a}_z &= \sum_{z'} \langle z|\, V(t)\, |z'\rangle \exp\left\{\frac{\mathrm{i}}{\hbar}(E_z - E_{z'})t\right\} a_{z'} \\ &\quad + \sum_m \langle z|\, V(t)\, |m\rangle \exp\left\{\frac{\mathrm{i}}{\hbar}(E_z - E_m)t\right\} a_m. \end{aligned} \tag{2}$$

We shall assume that $(E_z - E_n)$ is much larger than the energy differences $|E_z - E_{z'}|$ and $|E_n - E_m|$.

With the aim in mind to eliminate the excitation amplitudes $a_z(t)$ in (1) we solve formally the inhomogeneous equations (2) by the method described in (4.25)–(4.33). We thus introduce the solutions $a^{(0)}_{z,k}(t)$ of the homogeneous equation (2), i.e.

$$\mathrm{i}\hbar\dot{a}^{(0)}_{z,k} = \sum_{z'} \langle z|\, V(t)\, |z'\rangle \exp\left\{\frac{\mathrm{i}}{\hbar}(E_z - E_{z'})t\right\} a^{(0)}_{z',k}. \tag{3}$$

The index k enumerates a complete set of solutions. In terms of these quantities the solution of (2) can be written

$$\begin{aligned} a_z(t) &= \frac{1}{\mathrm{i}\hbar} \sum_{\substack{kz' \\ m}} a^{(0)}_{z,k}(t) \int_{-\infty}^{t} a^{(0)*}_{z',k}(t')\, \langle z'\, |V(t')|\, m\rangle \\ &\quad \times \exp\left\{\frac{\mathrm{i}}{\hbar}(E_{z'} - E_m)t'\right\} a_m(t')\, \mathrm{d}t'. \end{aligned} \tag{4}$$

Inserting this result in (1) we obtain an integro-differential equation for the amplitudes a_n:

$$
\begin{aligned}
\mathrm{i}\hbar \dot{a}_n(t) = {} & \sum_m \langle n| V(t) |m\rangle \exp\left\{\frac{\mathrm{i}}{\hbar}(E_n - E_m)t\right\} a_m(t) \\
& + \frac{1}{\mathrm{i}\hbar} \sum_{\substack{zz' \\ m}} \int_{-\infty}^{t} \mathrm{d}t' \langle n| V(t) |z\rangle \langle z'| V(t') |m\rangle \\
& \times \sum_k a_{z,k}^{(0)}(t) a_{z',k}^{(0)*}(t') \exp\left\{\frac{\mathrm{i}}{\hbar}(E_n - E_m)t\right\} \exp\left\{\frac{\mathrm{i}}{\hbar}(E_{z'} - E_z)t'\right\} \\
& \times \exp\left\{\frac{\mathrm{i}}{\hbar}(E_m - E_z)(t - t')\right\} a_m(t'). \qquad (5)
\end{aligned}
$$

We now assume that $(E_z - E_m)$ is so large that the variation of the integrand in (5) can be neglected over a period of the function $\exp[\mathrm{i}(E_m - E_z)(t - t')/\hbar]$. The integral then receives a contribution only close to the upper limit $t' = t$. In fact one may expand the integrals in inverse powers of $(E_z - E_m)/\hbar$ according to the formula

$$
\int_{-\infty}^{t} \mathrm{e}^{\mathrm{i}x(t-t')} f(t')\, \mathrm{d}t' = -\frac{1}{\mathrm{i}x} f(t) - \frac{1}{\mathrm{i}x^2} f'(t) \cdots, \qquad (6)
$$

assuming $f(-\infty) = 0$. This formula can be obtained by repeated partial integration. Utilizing eq. (4.9) one obtains

$$
\mathrm{i}\hbar \dot{a}_n = \sum_m \langle n| V(t) + V_{\mathrm{pol}}(t) |m\rangle \exp\left\{\frac{\mathrm{i}}{\hbar}[E_n - E_m]t\right\} a_m(t), \qquad (7)
$$

where to first order

$$
\langle n| V_{\mathrm{pol}}(t) |m\rangle = -\sum_z \frac{\langle n| V(t) |z\rangle \langle z| V(t) |m\rangle}{E_z - E_m}. \qquad (8)
$$

It should be noted that the polarization potential $V_{\mathrm{pol}}(t)$ has diagonal as well as non-diagonal matrix elements. As was mentioned earlier the diagonal matrix element in the ground state gives rise to a slight change of the elastic scattering which could be included as a modification to the Coulomb potential.

The higher-order terms for $V_{\mathrm{pol}}(t)$ could readily be evaluated by means of (6) using again eq. (4.9) to eliminate the quantities $a_{z,k}^{(0)}$. It is remarkable that the strong transitions $z \to z'$ (with a strength parameter $\chi_{z\to z'} \gtrsim 1$) only enter in these higher-order terms. The expansion, of which (8) is the first term, is thus an expansion in powers of the quantities $\chi_{z\to z'}/\xi_{z\to m}$ and $1/\xi_{z\to m}$.

The polarization potential (8) can also be obtained from eq. (5.11) for the

T-matrix $\tau_m(\omega)$ by eliminating the high-lying states. Finally it can be obtained by truncating the adiabatic equations (4.21), evaluating the solution of (4.13) for the high-lying states by second-order stationary perturbation theory.

§ 7. *Effects of decaying states*

We have hitherto assumed that all nuclear states $|n\rangle$ were true stationary states, which could only decay by a small decay rate after the Coulomb excitation process. If the lifetime of the state is comparable to, or smaller than the collision time, an essential modification of the above description must be introduced. This may well happen for high excited states of actual nuclei which decay by particle emission [BRE 60, GLU 60, NAK 66, SPE 70, WEI 71].

The difficulties in describing the excitation of decaying states in the time dependent semiclassical picture can be appreciated by considering the adiabatic limit, where in principle also states in the continuum must be included in describing the adiabatic ground state. This admixture of states in the continuum only leads to decay if the energy of the adiabatic ground state is above the threshold of particle emission. More generally, a state in the continuum can only decay if the excitation amplitude has a frequency which corresponds to an energy above threshold.

From this discussion it seems natural that for the present purpose one uses the description of Coulomb excitation in terms of the Fourier components as introduced in § 5.

We shall be mainly interested in the excitation of resonance states, which can be described as bound states weakly coupled to continuum states. For the description of these states we use the formalism of [MAH 69]. We therefore write the free Hamiltonian of the target nucleus in the form

$$H_0 = \sum_m |\phi_m\rangle\, E'_m\, \langle\phi_m| + \sum_{c=1}^{\Lambda} \int_{E_c}^{\infty} \mathrm{d}E'\, |\chi^c_{E'}\rangle\, E'\, \langle\chi^c_{E'}| + W. \tag{1}$$

Here $|\phi_m\rangle$ denotes bound states and states in the continuum which in the absence of the coupling W would be perfectly stable. The channel wave functions $|\chi^c_{E'}\rangle$ describe scattering states of total energy E' which contain standing waves in channel c only. The threshold energy in channel c is denoted by E_c. The states $|\phi_m\rangle$ and $|\chi^c_E\rangle$ are assumed to form a complete and orthonormal set i.e.

$$\langle\phi_m|\,\phi_n\rangle = \delta_{mn}, \qquad \langle\phi_m|\,\chi^c_E\rangle = 0,$$

and

$$\langle\chi^c_E|\chi^{c'}_{E'}\rangle = \delta_{cc'}\delta(E - E'). \tag{2}$$

We write the interaction W in the form

$$W = \sum_{jc} \int_{E_c}^{\infty} \mathrm{d}E' \, |\chi_{E'}^{c}\rangle \, W_j^c(E') \, \langle \phi_j|$$

$$+ \sum_{c'c''} \int_{E_{c'}}^{\infty} \mathrm{d}E' \int_{E_{c''}}^{\infty} \mathrm{d}E'' \, |\chi_{E'}^{c'}\rangle \, W^{c'c''}(E', E'') \, \langle \chi_{E''}^{c''}| + \text{h.c.} \tag{3}$$

We have here assumed that the states $|\phi_m\rangle$ do not couple with each other, i.e. the Hamiltonian H_0 has been diagonalized in the space spanned by the functions $\{\phi_m\}$. We shall assume that the quantities $W_j^c(E)$ are smooth functions of the energy, i.e. that no narrow single particle resonances occur.

In analogy to (5.2) we expand the nuclear state $|\hat{\psi}(\omega)\rangle$ in the following way

$$|\hat{\psi}(\omega)\rangle = \sum_m \beta_m(\omega) \, |\phi_m\rangle + \sum_{c=1}^{\Lambda} \int \mathrm{d}E' \beta_{E'}^c(\omega) \, |\chi_{E'}^c\rangle. \tag{4}$$

Since the perturbing potential in this case has a constant part W in addition to the time-dependent interaction $V(t)$ the equations of motion for $\beta(\omega)$ have the following form

$$(\hbar\omega - E_n')\beta_n(\omega) = \sum_{c=1}^{\Lambda} \int_{E_c}^{\infty} \mathrm{d}E' \, W_n^c(E') \, \beta_{E'}^c(\omega)$$

$$+ \sum_m \int_{-\infty}^{\infty} \mathrm{d}\omega' \hat{V}_{n,m}(\omega - \omega')\beta_m(\omega') \tag{5}$$

and

$$(\hbar\omega - E')\beta_{E'}^c(\omega) = \sum_m W_m^c(E') \, \beta_m(\omega), \tag{6}$$

where we have neglected the direct excitation of the continuum states as well as the continuum-continuum coupling. From these equations we may now eliminate the amplitudes $\beta_{E'}^c(\omega)$ on the continuum states. Since $|\psi(t)\rangle$ should contain only outgoing waves in all channels, we introduce a small positive imaginary part in $\hbar\omega$ and find

$$\beta_{E'}^c(\omega) = \sum_m \frac{W_m^c(E')\beta_m(\omega)}{\hbar\omega^+ - E'}, \tag{7}$$

which inserted into (5) leads to

$$(\hbar\omega - E_n')\beta_n(\omega) = \sum_{mc} F_{nm}^c(\omega)\beta_m(\omega) + \sum_m \int_{-\infty}^{\infty} \hat{V}_{n,m}(\omega - \omega')\beta_m(\omega') \, \mathrm{d}\omega'. \tag{8}$$

The quantities $F_{nm}^c(\omega)$ defined by

$$F_{nm}^c(\omega) = \int_{E_c}^{\infty} \mathrm{d}E' \, \frac{W_n^c(E') \, W_m^c(E')}{\hbar\omega^+ - E'} \tag{9}$$

describe the coupling of the bound states via the continuum states $|\chi_E^c\rangle$. The real part of F is related to the level shift arising from the coupling to the channel c, while the imaginary part of F gives rise to a partial width for the decay of the states $|\phi_n\rangle$ and $|\phi_m\rangle$ into channel c. It is noted that the imaginary part of $F_{nm}^c(\omega)$ vanishes for $\hbar\omega < E_c$. For $\hbar\omega > E_c$ one finds

$$\operatorname{Im} F_{nm}^c(\omega) = -\pi W_n^c(\hbar\omega) W_m^c(\hbar\omega). \tag{10}$$

In the absence of the external field $V(t)$, eq. (8) determines the energies of the true bound states as the roots ω_n of the equation

$$\det\left\{(E_n' - \hbar\omega)\delta_{nm} + \sum_c F_{nm}^c(\omega)\right\} = 0. \tag{11}$$

In the following we shall neglect the non-diagonal elements of $F_{nm}^c(\omega)$ which corresponds to the assumption of non-overlapping resonances. In this case eq. (11) simplifies to

$$\hbar\omega = E_n' + \sum_c F_{nn}^c(\omega) \equiv E_n' + F_n(\omega). \tag{12}$$

For ω values below the lowest threshold, $F_n(\omega)$ is real and signifies the level shift, i.e. the true energies of the nuclear states are given by

$$E_n = E_n' + F_n(E_n/\hbar). \tag{13}$$

For energies above this threshold, $F_n(\omega)$ receives an imaginary part which we denote by $-\frac{1}{2}\Gamma_n(\omega)$. The real level shift at the energy E_n we incorporate in the left-hand side of eq. (8) and we may therefore write

$$(\hbar\omega - E_n)\beta_n(\omega) = -\tfrac{1}{2}\mathrm{i}\,\Gamma_n(\omega)\beta_n(\omega) + \sum_m \int_{-\infty}^{\infty} \hat{V}_{n,m}(\omega - \omega')\,\beta_m(\omega')\,\mathrm{d}\omega'. \tag{14}$$

We have here also neglected the real part of $F_n(\omega) - F_n(E_n/\hbar)$ which is known [MAH 69] to be quite small.

In analogy to § 5 we may introduce the initial condition (5.6) and write the equation for the amplitudes $\alpha_n(\omega)$ defined in (5.7) in the following way

$$(\hbar\omega - E_n + \tfrac{1}{2}\mathrm{i}\,\Gamma_n(\omega))\alpha_n(\omega) = \hat{V}_{n,0}\left(\omega - \frac{E_0}{\hbar}\right) + \sum_m \int_{-\infty}^{\infty} \hat{V}_{n,m}(\omega - \omega')\,\alpha_m(\omega')\,\mathrm{d}\omega'. \tag{15}$$

The excitation amplitudes for bound states are given by (5.9). The amplitudes $\alpha_n(\omega)$ for resonance states give information about the particle emission. One may evaluate the particle current in channel c and one finds for the probability of particle emission at the energy E in the interval $\mathrm{d}E$ the following result (see [WEI 71])

$$P^c(E)\,\mathrm{d}E = \frac{2\pi}{\hbar^2}\sum_n \Gamma_n^c(E/\hbar)|\alpha_n(E/\hbar)|^2\,\mathrm{d}E, \tag{16}$$

where the partial width Γ_n^c is defined by $\Gamma_n^c = -2\,\mathrm{Im}\,F_{nn}^c$. The total probability for particle emission in channel c is then

$$P^c = \frac{2\pi}{\hbar^2}\int_{E_c}^{\infty}\sum_n \Gamma_n^c(E/\hbar)|\alpha_n(E/\hbar)|^2\,\mathrm{d}E. \tag{17}$$

The equations (15)–(17) constitute a set of equations for the excitation of decaying states. In the following we shall discuss some applications of this formalism.

First we consider the case of narrow resonances, i.e.

$$\Gamma_n(E/\hbar) \ll \hbar/\tau, \tag{18}$$

where τ is the collision time (I.13). In this case eq. (15) becomes almost identical to the equation (5.8) for bound states, as was to be expected. The decay probability can be evaluated from (17), i.e.

$$P^c = \frac{2\pi}{\hbar^2}\sum_n\int_{E_c}^{\infty}\mathrm{d}E\,\Gamma_n^c(E/\hbar)\frac{1}{(E-E_n)^2+\frac{1}{4}|\Gamma_n|^2}$$
$$\times\left|\hat{V}_{n,0}\left(\frac{E-E_0}{\hbar}\right)+\sum_m\int_{-\infty}^{\infty}\hat{V}_{n,m}\left(\frac{E}{\hbar}-\omega'\right)\alpha_m(\omega')\,\mathrm{d}\omega'\right|^2. \tag{19}$$

Considering (18) we may assume that the last factor is constant over the small region of the resonance and can be taken outside the integral over E. This leads to

$$P^c = \sum_n |\tau_n(E_n/\hbar)|^2\Gamma_n^c/\Gamma_n, \tag{20}$$

where the summation should be extended over all decaying states. The quantities τ_n which are defined by (5.10) are at the energy E_n exactly the excitation amplitudes. Eq. (20) illustrates the obvious result, that all the virtual states at some time will decay and that in the limit (18) the excitation and the de-excitation processes are independent. This result is applicable for all γ-decays.

In the case, where the relation (18) is not fulfilled, it is necessary to consider in more detail the energy dependence of the width. It is seen from eq. (19)

that since the excitation amplitudes decrease very strongly for increasing energies E, the decay probability will depend most sensitively on the values of Γ for energies in the neighbourhood of the threshold. Close to the threshold Γ_n^c is known to be proportional to the penetration factor Q for the free particles in channel c. Thus for neutrons

$$Q(E) \sim (E - E_c)^{l+1/2}, \tag{21}$$

where l is the orbital angular momentum of the neutron. For charged particles one finds similarly

$$Q(E) \sim \frac{kR}{|F_l(kR)|^2 + |G_l(kR)|^2}, \tag{22}$$

where $F_l(kR)$ and $G_l(kR)$ are the regular and irregular Coulomb wave functions at the nuclear surface R (see e.g. § IX.2) and k is the wave number

$$k^2 = 2m_c(E - E_c)/\hbar^2, \tag{23}$$

m_c being the reduced mass in channel c.

For higher values of the energy the function $\Gamma(E/\hbar)$ goes through a maximum and starts oscillating with decreasing amplitude. Characteristically the first zero of $\Gamma(E/\hbar)$ may be at 100 MeV. Since the high energy dependence is irrelevant in Coulomb excitation we may represent the energy dependence of Γ by the formula

$$\Gamma(E/\hbar) \sim Q(E) \exp(-\lambda E), \tag{24}$$

where λ is of the order of magnitude of 30 MeV^{-1}.

Finally, it may be interesting to transform the equation of motion (15) back into the time-dependent picture. One obtains

$$\begin{aligned} i\hbar\dot{a}_n(t) = \sum_m \langle n|\, V(t)\, |m\rangle \exp\left\{\frac{i}{\hbar}(E_n - E_m)t\right\} a_m(t) \\ + \int_{-\infty}^{\infty} dt'\, K_n(t' - t)\, a_n(t'), \end{aligned} \tag{25}$$

where the kernel K is the Fourier transform of $\Gamma_n(\omega + E_n/\hbar)$, i.e.

$$K_n(t' - t) = -\frac{i}{4\pi}\int_{-\infty}^{\infty} d\omega \exp\{i\omega(t' - t)\}\Gamma_n(\omega + E_n/\hbar). \tag{26}$$

The function K is complex and has the general property that it vanishes for $t' > t$.

With the "ansatz" (24) for $\Gamma(E)$ one finds for neutrons the following simple expression for K

$$K(t' - t) = \begin{cases} -\dfrac{\mathrm{i}K_o}{(1 + \mathrm{i}(t' - t)/T)^{l+3/2}}, & \text{for } t > t' \\ 0, & \text{for } t < t', \end{cases} \tag{27}$$

where K_o is a constant and the characteristic time T is defined by $T = \lambda\hbar$.

In the extreme adiabatic limit, where the nucleus stays in the adiabatic ground state $|0(t)\rangle$, the time-dependent equations can be solved with the result (see [WEI 71])

$$c_n(t) = \delta_{n0} \exp\left\{-\frac{1}{2\hbar}\int_{-\infty}^{t} \Gamma_0(E_0(t')/\hbar)\,\mathrm{d}t'\right\}. \tag{28}$$

Since the width $\Gamma_0(\omega)$ vanishes for $\hbar\omega < E_c$, the depopulation of the adiabatic ground state will only happen within the time interval where the adiabatic ground state energy $E_0(t)$ is above the lowest threshold.

§ 8. *Symmetries, invariances and coordinate systems*

For the evaluation of the Coulomb excitation amplitudes it is useful to realize the symmetries in the problem. These symmetries lead to selection rules or relations between different amplitudes which can be used to simplify the calculation. In some cases even, the result of an experiment will be uniquely determined by the invariance properties.

We first consider the invariances of the nuclear Hamiltonian H_o and the transformation properties of the nuclear wave functions which can be derived from these symmetries.

Denoting the operator for the rotation through the angle ω around an arbitrary axis $\hat{\boldsymbol{\omega}}$ by $R_{\hat{\omega}}(\omega)$ the rotational invariance of H_o is expressed by

$$R_{\hat{\omega}}(\omega)\, H_o\, R_{\hat{\omega}}^{-1}(\omega) = H_o. \tag{1}$$

The consequences of this invariance for the transformation of the simultaneous eigenstates $|I, M\rangle$ of H_o, and the total angular momentum operators $\boldsymbol{I}^2$ and I_z are discussed in appendix D. In this appendix we have for reference purposes collected also a number of properties of the rotation matrices $D^I_{MM'}(\alpha, \beta, \gamma)$.

The invariance of H_o under parity reflection in the origin P is expressed by the relation

$$PH_oP = H_o. \tag{2}$$

Correspondingly the eigenstates $|IM\rangle$ satisfy the relation

$$P\,|IM\rangle = (-1)^{\pi}\,|IM\rangle, \tag{3}$$

where π is the (even or odd) parity of the state $|IM\rangle$. The nuclear Hamiltonian is also assumed to be invariant under time reversal T, i.e.

$$TH_0T^{-1} = H_0. \tag{4}$$

This relation can be used to define the phases of the nuclear state vectors in a convenient way. We shall use the phase convention that the nuclear state $|IM\rangle$ is invariant under the combined operation RPT where R is the rotation $R_y(\pi)$ of 180 degrees in a positive sense around the y-axis, i.e.

$$RPT\,|IM\rangle = |IM\rangle. \tag{5}$$

Since the multipole operators (1.3) and (1.4) are invariant under RPT, this convention ensures that all matrix elements of these operators are real numbers. A more detailed discussion of this phase convention is given in appendix E.

Next we consider the interaction Hamiltonian $W(1, 2)$. Being an electromagnetic interaction $W(1, 2)$ it is symmetric under rotation, parity reflections and time reversal. In the semiclassical description two degrees of freedom in W, namely the relative centre-of-mass distance $\boldsymbol{r}$ and corresponding velocity $\dot{\boldsymbol{r}}$ are c-numbers which are not influenced by the operators R, P and T. The symmetry of $W(1, 2)$ under e.g. rotation is therefore only fulfilled if one performs together with the transformation under the operator R the corresponding rotation of the classical quantities $\boldsymbol{r}$ and $\dot{\boldsymbol{r}}$. For an arbitrary classical planar trajectory the only classical symmetry is the reflection symmetry in the plane of the orbit. This operation which we denote by $P_{\perp}$, can be written as the product of the parity operation in the nuclear centre and a rotation $R_{\perp}(\pi)$ around an axis perpendicular to the plane through this centre, i.e.

$$P_{\perp}W(1, 2; \boldsymbol{r}(t), \dot{\boldsymbol{r}}(t))P_{\perp}^{-1} = W(1, 2; \boldsymbol{r}(t), \dot{\boldsymbol{r}}(t)), \tag{6}$$

where we have specified explicitly the dependence on the classical variables $\boldsymbol{r}(t)$ and $\dot{\boldsymbol{r}}(t)$. In the special case of backwards scattering one has the exact symmetry of rotational invariance around the projectile orbit.

For a symmetrical orbit, as e.g. for the hyperbolic Rutherford orbit, where the energy loss of the projectile during the collision is neglected, one has the additional symmetry that a time reflection T followed by a rotation of 180 degrees $R_{\|}(\pi)$ around the symmetry axis reconstitutes the physical situation. The interaction Hamiltonian thus shows the property

$$\begin{aligned}(TR_{\|}(\pi))W(1, 2; \boldsymbol{r}(t), \dot{\boldsymbol{r}}(t))(TR_{\|}(\pi))^{-1} &= R_{\|}(\pi)W(1, 2; \boldsymbol{r}(t), -\dot{\boldsymbol{r}}(t))R_{\|}^{-1}(\pi) \\ &= W(1, 2; \boldsymbol{r}(-t), \dot{\boldsymbol{r}}(-t)).\end{aligned} \tag{7}$$

In the following we consider only the monopole-multipole part $V(t)$ of the total interaction $W(1, 2)$.

Through a combination of the invariance properties of H_0 and $V(t)$ we may conclude that $\tilde{V}(t)$ has the same invariance properties as $V(t)$. For the time integral of $\tilde{V}(t)$ one finds the following symmetries:

$$(R_{\perp}(\pi)P)\int_{-\infty}^{\infty} \tilde{V}(t')\,\mathrm{d}t'\,(R_{\perp}(\pi)P)^{-1} = \int_{-\infty}^{\infty} \tilde{V}(t')\,\mathrm{d}t', \tag{8}$$

and

$$(R_{\|}(\pi)T)\int_{-\infty}^{\infty} \tilde{V}(t')\,\mathrm{d}t'\,(R_{\|}(\pi)T)^{-1} = \int_{-\infty}^{\infty} \tilde{V}(t')\,\mathrm{d}t'. \tag{9}$$

It is noted that in the sudden approximation the action integral for electric interaction possesses the additional symmetries of time reversal invariance and invariance under the rotation $R_{\|}(\pi)$ separately, i.e.

$$T\int_{-\infty}^{\infty} V_{\mathrm{E}}(t')\,\mathrm{d}t'\,T^{-1} = \int_{-\infty}^{\infty} V_{\mathrm{E}}(t')\,\mathrm{d}t' \tag{10}$$

and

$$R_{\|}(\pi)\int_{-\infty}^{\infty} V_{\mathrm{E}}(t')\,\mathrm{d}t'\,(R_{\|}(\pi))^{-1} = \int_{-\infty}^{\infty} V_{\mathrm{E}}(t')\,\mathrm{d}t'. \tag{11}$$

Thus if the collision can be considered as a sudden impact the excitation process is independent of the direction in which the projectile moves along its orbit.

To derive the consequences of the above symmetries for the Coulomb excitation amplitudes it is most convenient to consider the formal solution of the Schrödinger equation as it is given by the S-matrix formulation of § 3. It is thus seen directly from the expression (3.20) and (3.25) that K and Q possess the symmetries

$$\begin{aligned}(R_{\perp}(\pi)\,P)Q(R_{\perp}(\pi)\,P)^{-1} &= Q, \\ (R_{\|}(\pi)\,T)Q(R_{\|}(\pi)\,T)^{-1} &= Q\end{aligned} \tag{12}$$

and in addition in the sudden approximation

$$R_{\|}(\pi)\,Q\,R_{\|}(\pi)^{-1} = Q \tag{13}$$

and similar for K.

For the S-matrix (3.17)–(3.19) one finds finally

$$\begin{aligned}(R_{\perp}(\pi)P)\, S\, (R_{\perp}(\pi)P)^{-1} &= S, \\ (R_{\parallel}(\pi)T)\, S\, (R_{\parallel}(\pi)T)^{-1} &= S^{\dagger} = S^{-1}\end{aligned} \tag{14}$$

and in addition for electric interaction in the sudden approximation

$$\begin{aligned}R_{\parallel}(\pi)\, S\, (R_{\parallel}(\pi))^{-1} &= S, \\ TST^{-1} &= S^{\dagger} = S^{-1}.\end{aligned} \tag{15}$$

The above mentioned symmetries can be expressed in a very convenient way if one chooses the quantization axis for the nuclear states in one of the two following ways:

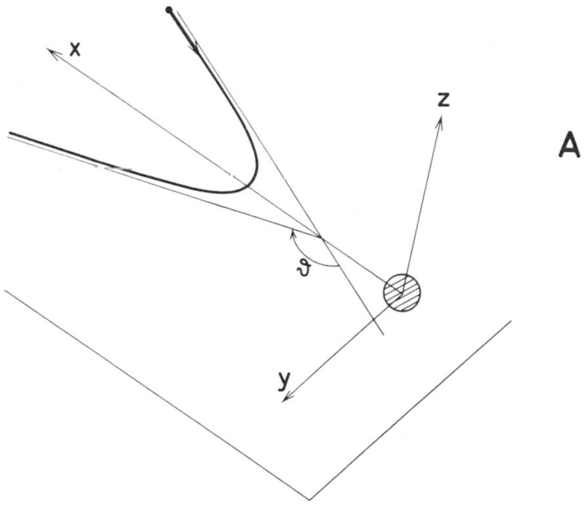

Fig. 1. The relative coordinate system A which is often used for the description of Coulomb excitation. The origin is chosen in the centre of mass of the target nucleus. The z-axis is perpendicular to the plane of the orbit, the x-axis is along the apex-line towards the projectile, while the y-axis is chosen such, that the y-component of the projectile velocity is positive. The scattering angle ϑ of the projectile is indicated.

In the coordinate *system A* (see fig. 1) one chooses the z-axis perpendicular to the plane of the orbit and the x-axis along the apex line towards the projectile. The direction of the y-axis is chosen such that the y-component of the projectile velocity is positive. The origin is chosen in the centre of mass of the target nucleus.

Another convenient coordinate system is the *system B* (see fig. 2) in which one chooses the z-axis along the apex-line towards the projectile and the

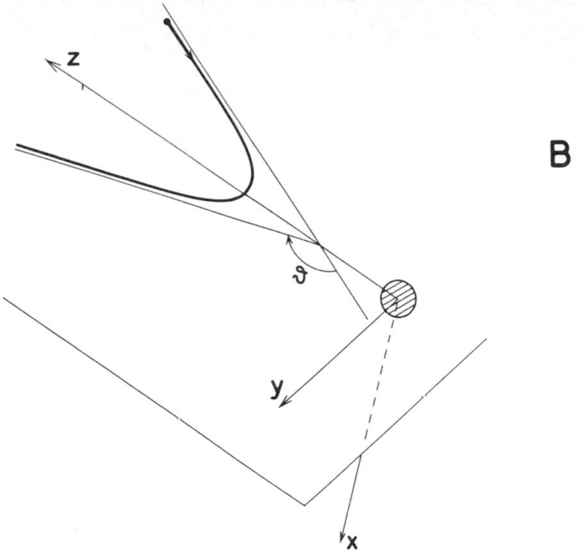

Fig. 2. The relative coordinate system B which is often used in Coulomb excitation. The origin is chosen in the centre of mass of the target nucleus. The x-axis is perpendicular to the plane of the orbit, the z-axis is along the apex-line towards the projectile, while the y-axis is chosen such, that the y-component of the projectile velocity is positive. The scattering angle ϑ of the projectile is indicated.

x-axis perpendicular to the orbit. The y-axis is the same as above, and the origin is again in the centre of mass of the target nucleus.

We denote by the symbol $a_{I_f M_f, I_i M_i}$ the excitation amplitude of the state $|I_f M_f\rangle$ when the nucleus in its initial state has the quantum numbers $I_i M_i$, i.e.

$$a_{I_f M_f, I_i M_i} = \langle I_f M_f | \, S \, | I_i M_i \rangle . \tag{16}$$

Utilizing the above mentioned transformation properties one finds directly the following rules.

In *coordinate system A*, reflexion in the plane of the orbit leads to

$$a_{I_f M_f, I_i M_i} = (-1)^{M_f - M_i + \pi_f - \pi_i} a_{I_f M_f, I_i M_i}, \tag{17}$$

where $(-1)^{\pi_i}$ and $(-1)^{\pi_f}$ are the parity quantum numbers of initial and final states, respectively. From this relation one may conclude that for excitations with no parity change the magnetic quantum number can only change by an even number, while for states with a parity different from that of the ground state the magnetic quantum number must change by an odd integer.

For electric excitations where the parameter ξ vanishes one finds

$$a_{I_f M_f, I_i M_i} = (-1)^{I_f - I_i} a_{I_f - M_f, I_i - M_i}. \tag{18}$$

In *coordinate system B* one finds from the reflection symmetry in the plane of the orbit

$$a_{I_f M_f, I_i M_i} = (-1)^{I_f - I_i + \pi_f - \pi_i} a_{I_f - M_f, I_i - M_i}. \tag{19}$$

This relation shows that in the system B it is sufficient to compute the excitation amplitudes for situations in which the magnetic quantum number in the initial state is positive or zero.

In the limit of a sudden impact one finds for electric excitations

$$a_{I_f M_f, I_i M_i} = (-1)^{M_f - M_i} a_{I_f M_f, I_i M_i}, \tag{20}$$

which means that in this limit the quantum number M can only change by an even integer.

The time reflection properties are most easily obtained by combining time reversal with a reflection in the orbit. In both *coordinate systems A and B* one finds that

$$(R_{\parallel} T)(R_{\perp} P) = R_y TP, \tag{21}$$

which leads to the symmetry

$$a_{I_f M_f, I_i M_i} = a_{I_i M_i, I_f M_f}. \tag{22}$$

In terms of the coupled differential equations (4.5) this relation has the significance that with the initial condition that the nucleus is in the excited state $|I_f M_f\rangle$ the amplitude for ending in the state $|I_i M_i\rangle$ is the same as the amplitude for ending in the state $|I_f M_f\rangle$ if the nucleus was initially in the state $|I_i M_i\rangle$. It is noted that from the relation (22) one finds in the perturbation theory the relation (3.14) utilizing that $V(t)$ is a Hermitian operator, i.e.

$$\langle I_f M_f| \int_{-\infty}^{+\infty} \tilde{V}(t)\, \mathrm{d}t\, |I_i M_i\rangle^* = \langle I_f M_f| \int_{-\infty}^{+\infty} \tilde{V}(t)\, \mathrm{d}t\, |I_i M_i\rangle. \tag{23}$$

In the *coordinate system B* the special symmetry for backwards scattering leads to the simple condition that the magnetic quantum number is conserved, i.e.

$$M_i = M_f. \tag{24}$$

Although this relation only holds strictly for backwards scattering one finds in practice that the magnetic quantum number in the coordinate system B is approximately conserved also for scattering angles different from 180 degrees. As we shall see later in chapter VI this approximation is most accurate for small values of ξ. The approximation method based on this observation is called the $\chi(\vartheta)$-approximation. It is only in the coordinate system B that this method takes a simple form and we shall therefore in the following use this coordinate system quite frequently. Corresponding formulae in coordinate system A can always be obtained by rotating the reference system by -90 degrees around the y-axis. Note that all the equations in the present section can be used also for magnetic interaction, except eqs. (10), (11) and the results which follow from these in the sudden approximation, i.e. eqs. (18) and (20). With the magnetic interaction of the form (1.13) it follows from (7) that eq. (10) takes the form

$$T\left(\int_{-\infty}^{+\infty} V_{\mathrm{M}}(t')\,\mathrm{d}t'\right)T^{-1} = -\int_{-\infty}^{+\infty} V_{\mathrm{M}}(t')\,\mathrm{d}t'. \tag{25}$$

For excitation which proceeds solely through magnetic interactions the amplitudes in the sudden approximation thus satisfy the relation

$$a_{I_\mathrm{f}M_\mathrm{f},I_\mathrm{i}M_\mathrm{i}} = (-1)^{I_\mathrm{f}-I_\mathrm{i}+1}a_{I_\mathrm{f}-M_\mathrm{f},I_\mathrm{i}-M_\mathrm{i}} \tag{26}$$

in coordinate system A instead of (18), and

$$a_{I_\mathrm{f}M_\mathrm{f},I_\mathrm{i}M_\mathrm{i}} = (-1)^{M_\mathrm{f}-M_\mathrm{i}+1}a_{I_\mathrm{f}M_\mathrm{f},I_\mathrm{i}M_\mathrm{i}} \tag{27}$$

in coordinate system B instead of (20).

§ 9. *Parametrization of orbital motion*

In § 2 it was shown how one may separate the motion of the two centres of mass of the colliding nuclei from the intrinsic excitation in the semiclassical description. The motion of the nuclear centres is determined as the solutions of the classical equations of motion for the two nuclear centres under the influence of the mutual Coulomb interaction $Z_1Z_2e^2/r$ while the intrinsic motion is determined by the time-dependent Schrödinger equation (2.13).

In principle one might improve the classical equations of motion by including other parts of the interaction (1.6). As was discussed in § 2, it is not possible to take these corrections accurately into account, since the interaction depends on the relative orientation of the nuclei, which we have to treat quantum mechanically. One may, at most, include the effect of the average interaction and of the average energy transfer to intrinsic motion. We shall later include the main part of this important effect by a symmetrization procedure, where one chooses an orbit corresponding to elastic scattering, but with an energy intermediate between initial and final energy. High-lying states, which are not excited, but which are strongly coupled to the ground state will give rise to a modification of the ground state energy which we may include in the scattering potential (see § 6). As we shall see in § V.7 below, this polarization potential is, however, quite weak compared to the point Coulomb field.

In this section we shall consider in some detail the solution of the classical equation for the relative motion in a repulsive Coulomb field. The effects associated with the transformation from the relative motion to the laboratory system are discussed in § X.1 below.

We define the hyperbolic orbit in the classical Kepler problem by the scattering angle ϑ and energy E, both measured in the centre-of-mass system. The energy E is connected with the relative velocity v at large distances by

$$E = \tfrac{1}{2} m_0 v^2. \tag{1}$$

Besides the total energy E

$$E = \tfrac{1}{2} m_0 \dot{\boldsymbol{r}}^2 + Z_1 Z_2 e^2 / r \tag{2}$$

there are two other constants of motion. One is the relative angular momentum vector

$$\boldsymbol{l} = \boldsymbol{r} \times \boldsymbol{p}, \tag{3}$$

where $\boldsymbol{p}$ is the relative momentum

$$\boldsymbol{p} = m_0 \dot{\boldsymbol{r}}. \tag{4}$$

The second constant of motion, which is characteristic of the Coulomb field, is the Lenz vector $\boldsymbol{A}$ defined by (c.f. e.g. [BIE 65])

$$\boldsymbol{A} = \eta_{\text{cl}} \boldsymbol{r}/r - \boldsymbol{l} \times \boldsymbol{p}/p, \tag{5}$$

where the parameter η_{cl} is the action of the Coulomb field (see eq. (I.23)), i.e.

$$\eta_{\text{cl}} = \eta \hbar = Z_1 Z_2 e^2 / v. \tag{6}$$

It is noted that while $\boldsymbol{l}$ is perpendicular to the plane of the orbit the vector $\boldsymbol{A}$ points along the main axis of the hyperbola towards the projectile (see fig. 3). The magnitude of the orbital angular momentum is related to the impact parameter ρ and the scattering angle ϑ by

$$|\boldsymbol{l}| = m_0 v \rho = m_0 v a \cot(\tfrac{1}{2}\vartheta) = \eta_{\text{cl}} \cot(\tfrac{1}{2}\vartheta) \tag{7}$$

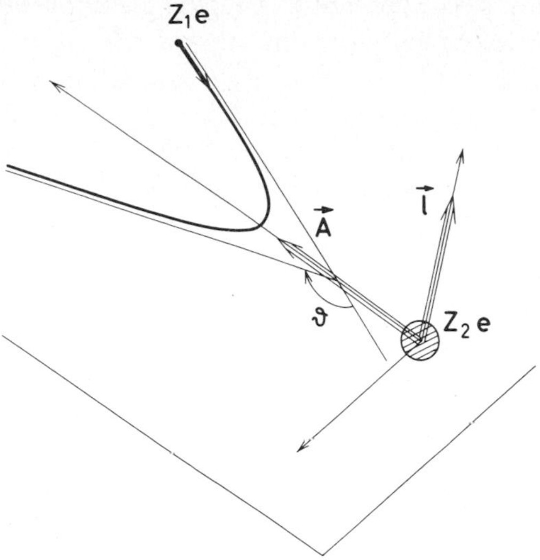

Fig. 3. The classical hyperbolic orbit in the relative coordinate system with the origin in the target nucleus. The relative angular momentum vector $\boldsymbol{l}$ and the Lenz vector $\boldsymbol{A}$ are indicated.

where a is given by (2.3). The magnitude of $\boldsymbol{A}$ is given by

$$|\boldsymbol{A}| = \sqrt{\eta_{\text{cl}}^2 + \boldsymbol{l}^2} = \eta_{\text{cl}}/\sin(\tfrac{1}{2}\vartheta) = \eta_{\text{cl}}\varepsilon, \tag{8}$$

where instead of the scattering angle ϑ we have used the excentricity ε defined by

$$\varepsilon = 1/\sin(\tfrac{1}{2}\vartheta) = \sqrt{1 + l^2/\eta_{\text{cl}}^2}. \tag{9}$$

It is convenient to introduce a parametric representation of the hyperbola which simultaneously determines the position of the projectile and the time in terms of a dimensionless parameter. We thus introduce the parameter w by the relations

$$r = a[\varepsilon \cosh w + 1], \qquad t = \frac{a}{v}[\varepsilon \sinh w + w]. \tag{10}$$

In the *coordinate system A* where the z-axis is chosen along $\boldsymbol{l}$ and the x-axis along $\boldsymbol{A}$ the projectile coordinates are given by

$$\begin{aligned} x_{\text{A}} &= a[\cosh w + \varepsilon], \\ y_{\text{A}} &= a\sqrt{\varepsilon^2 - 1}\,\sinh w, \\ z_{\text{A}} &= 0. \end{aligned} \tag{11}$$

In the *coordinate system B* the z-axis is chosen along the vector $\boldsymbol{A}$ and the x-axis along $-\boldsymbol{l}$ and one finds here similarly

$$\begin{aligned} x_B &= 0, \\ y_B &= a\sqrt{\varepsilon^2 - 1}\,\sinh w, \\ z_B &= a[\cosh w + \varepsilon]. \end{aligned} \tag{12}$$

In both coordinate systems the magnitude of the velocity vector $\dot{\boldsymbol{r}}(t)$ is given by

$$|\dot{\boldsymbol{r}}| = v\sqrt{(\varepsilon\cosh w - 1)/(\varepsilon\cosh w + 1)}. \tag{13}$$

When the parameter w varies from $-\infty$ to $+\infty$ the particle moves along the hyperbola in such a way that the point of closest approach is reached for $w = t = 0$.

The collision functions (2.17) and (2.18) can be written in terms of the parameter w by means of the expressions (10)–(12). In the *coordinate system A* we find

$$\begin{aligned} \theta(t) &= \tfrac{1}{2}\pi, \\ \sin\phi(t) &= \sqrt{\varepsilon^2 - 1}\,\sinh w/(\varepsilon\cosh w + 1) \end{aligned} \tag{14}$$

and the electric collision function may thus be written

$$\bar{S}^{A}_{E\lambda\mu}(t) = Y_{\lambda\mu}(\tfrac{1}{2}\pi, 0)\,\frac{[\cosh w + \varepsilon + i\sqrt{\varepsilon^2 - 1}\,\sinh w]^{\mu}}{a^{\lambda+1}[\varepsilon\cosh w + 1]^{\lambda+\mu+1}}. \tag{15}$$

It is noted that $\bar{S}^{A}_{E\lambda\mu}$ vanishes for $\lambda + \mu$ odd since

$$Y_{\lambda\mu}(\tfrac{1}{2}\pi, 0) = \begin{cases} \sqrt{\dfrac{2\lambda + 1}{4\pi}}\,(-1)^{(\lambda+\mu)/2}\,\dfrac{[(\lambda-\mu)!\,(\lambda+\mu)!]^{1/2}}{(\lambda+\mu)!!\,(\lambda-\mu)!!}, & \text{for } \lambda + \mu \text{ even} \\ 0, & \text{for } \lambda + \mu \text{ odd.} \end{cases} \tag{16}$$

In the *coordinate system B* the polar angles θ and ϕ are given by

$$\sin\theta(t) = \sqrt{\varepsilon^2 - 1}\,\sinh w/(\varepsilon\cosh w + 1) \quad \text{and} \quad \phi = \tfrac{1}{2}\pi. \tag{17}$$

Instead of the collision function $\bar{S}_{E\lambda\mu}(t)$ defined in eq. (2.17) it will often be convenient to introducc a normalized and dimensionless collision function $Q_{\lambda\mu}(\varepsilon, w)$. It is defined by the relation

$$Q_{\lambda\mu}(\varepsilon, w) = a^{\lambda}\,\frac{(2\lambda - 1)!!}{(\lambda - 1)!}\sqrt{\frac{\pi}{2\lambda + 1}}\,r(w)\,\bar{S}_{E\lambda\mu}(t(w)). \tag{18}$$

This function is normalized in such a way that for backwards scattering

($\varepsilon = 1$) (see § IV.2 below)

$$\sum_{\mu} \left| \int_{-\infty}^{+\infty} Q_{\lambda\mu}(\varepsilon = 1, w)\, dw \right|^2 = 1. \tag{19}$$

The explicit form of the dimensionless collision function $Q_{\lambda\mu}$ in *coordinate system A* is according to (15) and (18) given by

$$Q^{A}_{\lambda\mu}(w) = \frac{(2\lambda - 1)!!}{(\lambda - 1)!} \sqrt{\frac{\pi}{2\lambda + 1}}\, Y_{\lambda\mu}(\tfrac{1}{2}\pi, 0)\, \frac{[\cosh w + \varepsilon + i\sqrt{\varepsilon^2 - 1} \sinh w]^{\mu}}{[\varepsilon \cosh w + 1]^{\lambda+\mu}}. \tag{20}$$

It is seen from (16) that this collision function satisfies the following symmetry relations

$$Q^{A}_{\lambda\mu}(\varepsilon, -w) = Q^{A}_{\lambda\mu}(\varepsilon, w)^* = (-1)^{\mu} Q^{A}_{\lambda-\mu}(\varepsilon, w). \tag{21}$$

For backwards scattering ($\varepsilon = 1$) one finds from (20) that all $Q^{A}_{\lambda\mu}(w)$ have the same w-dependence independent of μ.

In the *coordinate system B* the collision function looks more complicated and explicit expressions for $\lambda = 1, 2, 3$ and 4 are given in table 1. It is noted that $Q^{B}_{\lambda\mu}(\varepsilon = 1, w)$ vanishes for $\mu \neq 0$, which is a consequence of the rotational symmetry around the z-axis. From the symmetry in the plane of the orbit and from time reversal symmetry one may conclude that the collision functions $Q^{B}_{\lambda\mu}(\varepsilon, w)$, in general, satisfy the following relations in the coordinate system B:

$$\begin{aligned} Q^{B}_{\lambda\mu}(\varepsilon, w) &= Q^{B}_{\lambda-\mu}(\varepsilon, w), \\ Q^{B}_{\lambda\mu}(\varepsilon, -w) &= (-1)^{\mu} Q^{B}_{\lambda\mu}(\varepsilon, w) = Q^{B}_{\lambda\mu}(\varepsilon, w)^*. \end{aligned} \tag{22}$$

For magnetic excitations it is convenient to rewrite the collision function (2.18) by transforming the operator in the following way

$$\bar{S}_{M\lambda\mu}(t) = -\frac{1}{c\lambda m_0 r^{\lambda+1}(t)}\, \boldsymbol{l}\nabla(Y_{\lambda\mu}(\theta(t), \phi(t))), \tag{23}$$

where the constant relative orbital angular momentum $\boldsymbol{l}$ of the projectile with respect to the target is defined in eq. (3). Inserting (7) in (23) one finds

$$\bar{S}_{M\lambda\mu}(t) = -\frac{av}{c\lambda r^{\lambda+1}(t)} \cot(\tfrac{1}{2}\vartheta)\, \nabla_{l}(Y_{\lambda\mu}(\theta(t), \phi(t))), \tag{24}$$

where ∇_l is the gradient in the direction of $\boldsymbol{l}$.

TABLE 1

The dimensionless collision function $Q^{B}_{\lambda\mu}(\varepsilon, w)$ in the coordinate system B for $\lambda = 1, 2, 3$ and 4. (As is indicated the function is symmetric under a change of sign of the index μ.)

λ	μ	$Q^{B}_{\lambda\mu}(\varepsilon, w)$
1	0	$\frac{1}{2}\frac{\cosh w + \varepsilon}{(\varepsilon \cosh w + 1)^2}$
1	±1	$-i\frac{1}{2}\sqrt{\frac{1}{2}}\frac{\sqrt{\varepsilon^2 - 1}\sinh w}{(\varepsilon \cosh w + 1)^2}$
2	0	$\frac{3}{4}\frac{2(\cosh w + \varepsilon)^2 - (\varepsilon^2 - 1)\sinh^2 w}{(\varepsilon \cosh w + 1)^4}$
2	±1	$-i\frac{3}{2}\sqrt{\frac{3}{2}}\frac{(\cosh w + \varepsilon)\sqrt{\varepsilon^2 - 1}\sinh w}{(\varepsilon \cosh w + 1)^4}$
2	±2	$-\frac{3}{4}\sqrt{\frac{3}{2}}\frac{(\varepsilon^2 - 1)\sinh^2 w}{(\varepsilon \cosh w + 1)^4}$
3	0	$\frac{15}{8}\frac{(\cosh w + \varepsilon)[2(\cosh w + \varepsilon)^2 - 3(\varepsilon^2 - 1)\sinh^2 w]}{(\varepsilon \cosh w + 1)^6}$
3	±1	$-i\frac{15}{16}\sqrt{3}\frac{[4(\cosh w + \varepsilon)^2 - (\varepsilon^2 - 1)\sinh^2 w]\sqrt{\varepsilon^2 - 1}\sinh w}{(\varepsilon \cosh w + 1)^6}$
3	±2	$-\frac{15}{8}\sqrt{\frac{15}{2}}\frac{(\cosh w + \varepsilon)(\varepsilon^2 - 1)\sinh^2 w}{(\varepsilon \cosh w + 1)^6}$
3	±3	$i\frac{15}{16}\sqrt{5}\frac{(\varepsilon^2 - 1)^{3/2}\sinh^3 w}{(\varepsilon \cosh w + 1)^6}$
4	0	$\frac{35}{32}\frac{8(\cosh w + \varepsilon)^4 - 24(\cosh w + \varepsilon)^2(\varepsilon^2 - 1)\sinh^2 w + 3(\varepsilon^2 - 1)^2\sinh^4 w}{(\varepsilon \cosh w + 1)^8}$
4	±1	$-i\frac{35}{16}\sqrt{5}\frac{(\cosh w + \varepsilon)[4(\cosh w + \varepsilon)^2 - 3(\varepsilon^2 - 1)\sinh^2 w]\sqrt{\varepsilon^2 - 1}\sinh w}{(\varepsilon \cosh w + 1)^8}$
4	±2	$-\frac{35}{16}\sqrt{\frac{5}{2}}\frac{[6(\cosh w + \varepsilon)^2 - (\varepsilon^2 - 1)\sinh^2 w](\varepsilon^2 - 1)\sinh^2 w}{(\varepsilon \cosh w + 1)^8}$
4	±3	$i\frac{35}{16}\sqrt{35}\frac{(\cosh w + \varepsilon)(\varepsilon^2 - 1)^{3/2}\sinh^3 w}{(\varepsilon \cosh w + 1)^8}$
4	±4	$\frac{35}{32}\sqrt{\frac{35}{2}}\frac{(\varepsilon^2 - 1)^2\sinh^4 w}{(\varepsilon \cosh w + 1)^8}$

In the coordinate system A, where $\boldsymbol{l}$ is in the direction of the z-axis one finds

$$\nabla_z Y_{\lambda\mu}(\tfrac{1}{2}\pi, \phi) = -\frac{1}{r}\left(\frac{2\lambda + 1}{2\lambda + 3}[(\lambda + 1)^2 - \mu^2]\right)^{1/2} Y_{\lambda+1,\mu}(\tfrac{1}{2}\pi, \phi). \quad (25)$$

Inserting this into (24) the collision function takes the form

$$\bar{S}^{A}_{M\lambda\mu}(t) = \frac{av}{\lambda c} \cot(\tfrac{1}{2}\vartheta) \left(\frac{2\lambda + 1}{2\lambda + 3} [(\lambda + 1)^2 - \mu^2]\right)^{1/2} \bar{S}^{A}_{E\lambda+1,\mu}(t), \tag{26}$$

where the last factor is given by (15). In analogy to (18) one defines the dimensionless collision function $Q_{M\lambda\mu}(\varepsilon, w)$ by the relation

$$\bar{S}_{M\lambda\mu}(t) = \frac{v}{ca^{\lambda}} \sqrt{\frac{2\lambda + 1}{\pi}} \frac{(\lambda - 1)!}{(2\lambda - 1)!!} \frac{1}{r(w)} Q_{M\lambda\mu}(\varepsilon, w). \tag{27}$$

Combining (27) and (26) with definition (18) one obtains the result

$$Q^{A}_{M\lambda\mu}(\varepsilon, w) = \frac{[(\lambda + 1)^2 - \mu^2]^{1/2}}{2\lambda + 1} \cot(\tfrac{1}{2}\vartheta)\, Q^{A}_{\lambda+1,\mu}(\varepsilon, w). \tag{28}$$

In the *coordinate system B*, the collision functions $\bar{S}^{B}_{M\lambda\mu}$ and $Q^{B}_{M\lambda\mu}$ can be obtained by rotating the coordinate system by an angle $-\frac{1}{2}\pi$ around the y-axis utilizing that the functions transform like spherical tensors of order λ, μ. The dimensionless collision functions $Q^{B}_{M\lambda\mu}(\varepsilon, w)$ are given explicitly for $\lambda = 1$ and $\lambda = 2$ in table 2.

TABLE 2

The dimensionless collision functions $Q^{B}_{M\lambda\mu}(\varepsilon, w)$ in coordinate system B for $\lambda = 1$ and 2.

λ	μ	$Q^{B}_{M\lambda\mu}(\varepsilon, w)$
1	0	0
1	± 1	$\mp \frac{1}{2\sqrt{2}} \frac{\sqrt{\varepsilon^2 - 1}}{(\varepsilon \cosh w + 1)^2}$
2	0	0
2	± 1	$\mp \frac{3\sqrt{6}}{8} \frac{\sqrt{\varepsilon^2 - 1}(\cosh w + \varepsilon)}{(\varepsilon \cosh w + 1)^4}$
2	± 2	$\pm \frac{3\sqrt{6}}{8} i \frac{(\varepsilon^2 - 1) \sinh w}{(\varepsilon \cosh w + 1)^4}$

The functions are antisymmetric under a change of sign of the index μ.

CHAPTER III

Cross Sections and Angular Distributions

In the previous chapter we have discussed methods for the evaluation of Coulomb excitation amplitudes. From these amplitudes one may calculate all measurable quantities like cross sections and angular distributions of γ-quanta or conversion electrons which are emitted after the excitation process. The states which are populated in the Coulomb excitation process usually have a lifetime which is long enough so that the decay takes place after the collision. The more complicated case where the decay happens during the collision was discussed in § II.7. In the first-mentioned simple case one may separate the decay process completely from the excitation and one may evaluate the measurable quantities by quadratic expressions in the excitation amplitudes. The explicit calculation of the angular distributions is thus mainly a problem of angular momentum algebra. The technique for solving this type of problems is discussed in detail in the literature [FRA 65, ROS 67, DEV 57]. If the decay process takes place through a number of stages, e.g. by emission of a cascade of γ-quanta it is convenient to use the concept of density matrices and statistical tensors whereby the process can be separated into independent steps.

In this chapter we shall give a survey over the relevant formulae for the evaluation of the measurable quantities in Coulomb excitation. They can be used in all the examples given later, independent of the way in which the amplitudes were evaluated. Since the results are even essentially independent of the semiclassical description we shall write the formulae so that they can directly be used also in a quantum mechanical description.

§ 1. *Differential and total cross sections*

We have already mentioned in § II.2 how one may evaluate the differential cross section from the Coulomb excitation probability $P_{\rm f}$ (see (II.2.5))

$$\left(\frac{\mathrm{d}\sigma}{\mathrm{d}\Omega}\right)_{\rm f} = \tfrac{1}{4}a^2 \frac{1}{\sin^4(\frac{1}{2}\vartheta)} P_{\rm f}. \tag{1}$$

In order to evaluate the excitation probability P_{f} of an energy level f from the excitation amplitudes one must perform a summation over the magnetic quantum numbers. If the ground state of spin I_0 is unpolarized, an average over the initial polarization should also be performed. In terms of the excitation amplitudes $a_{I_{\mathrm{f}}M_{\mathrm{f}},I_0M_0}$ (see (II.8.16)) from the state of spin I_0 to the states of spin I_{f} the excitation probability is thus given by

$$P_{\mathrm{f}} = \frac{1}{2I_0+1} \sum_{M_0 M_{\mathrm{f}}} |a_{I_{\mathrm{f}}M_{\mathrm{f}},I_0M_0}|^2 . \tag{2}$$

The total cross section for exciting the energy level f is obtained by integrating (1) over all scattering angles, i.e.

$$\begin{aligned} \sigma_{\mathrm{f}} &= \tfrac{1}{4}a^2 \int P_{\mathrm{f}}(\vartheta) \frac{\mathrm{d}\Omega}{\sin^4(\frac{1}{2}\vartheta)} \\ &= a^2\pi \int_0^{\pi} P_{\mathrm{f}}(\vartheta) \frac{\cos(\frac{1}{2}\vartheta)\,\mathrm{d}\vartheta}{\sin^3(\frac{1}{2}\vartheta)} . \end{aligned} \tag{3}$$

These expressions apply directly if the excitation cross section is measured by observing the inelastically scattered particle of a definite energy. If, however, one observes the excitation through the yield of subsequent γ-rays or conversion electrons, one must take into consideration that these decay products show, in general, a non-isotropic distribution. Also one must realise that the state in question may be populated through cascades of unobserved γ-rays or conversion electrons. In the following sections we shall study these effects in more detail.

It may be interesting to compare the expression (1) with the corresponding expression which would follow from a quantum mechanical description of the Coulomb excitation process. It is well known that for inelastic scattering the cross section for exciting the state f is given by [MES 64]

$$\left(\frac{\mathrm{d}\sigma}{\mathrm{d}\Omega}\right)_{\mathrm{f}} = \frac{m_0^{\,2}}{4\pi^2\hbar^4} \frac{v_{\mathrm{f}}}{v_0} \frac{1}{(2I_0+1)} \sum_{M_0 M_{\mathrm{f}}} |\langle I_{\mathrm{f}} M_{\mathrm{f}} \boldsymbol{p}_{\mathrm{f}}| \, T \, |I_0 M_0 \boldsymbol{p}_0\rangle|^2, \tag{4}$$

where m_0 is the reduced mass, while v_{f} and v_0 are the relative velocities of the projectile before and after the collision. The T-matrix describes the transition amplitude between the incoming projectile of momentum $\boldsymbol{p}_0$ on the ground state and the outgoing projectile with momentum $\boldsymbol{p}_{\mathrm{f}}$ leaving the nucleus in the excited state.

It is noted that in a first-order perturbation treatment the T-matrix is equal to the matrix element of the interaction energy. The T-matrix plays a similar role in the quantum mechanical description as the excitation amplitudes in the semiclassical theory. The exact correspondence will be investigated in § IX.3, but we may note here already that in the semiclassical limit T may be written

$$\langle I_f M_f \boldsymbol{p}_f | T | I_0 M_0 \boldsymbol{p}_0 \rangle = \frac{\pi \hbar^2 a}{m_o} \frac{1}{\sin^2(\frac{1}{2}\vartheta)} \exp\{-i\eta \log \sin^2(\frac{1}{2}\vartheta) + 2i\sigma_0\} a_{I_f M_f, I_0 M_0}, \tag{5}$$

where σ_o is the Coulomb phase shift

$$\sigma_o = \arg \Gamma(1 + i\eta). \tag{6}$$

The relation (5) only holds if one can neglect the difference between the initial and final velocities and it is then also seen that the expression (4) becomes identical to (1) and (2). An improved expression (5) for a finite difference between v_0 and v_f can be obtained by symmetrization between initial and final states as discussed in chapters IV and VI.

§ 2. *Polarization in Coulomb excitation*

In the Coulomb excitation process, as well as during the decay processes that may happen after the excitation has taken place, we deal with an ensemble of target nuclei which should be described by a density matrix or the equivalent statistical tensors. In appendix F we have summarized the theory of density matrices as it applies to the theory of angular distributions and correlations. In the semiclassical theory of Coulomb excitation we consider the situation in which the target nucleus is bombarded with a heavy projectile whose deflection angle ϑ is detected. The interaction with the target is then a completely well defined function of time and the initial density matrix of the target nucleus changes according to the equation of motion (F.10). Introducing the time-dependent excitation amplitudes according to (II.3.4) and (II.3.6)

$$a_{n,0}(t) = \langle n | U(t, -\infty) | 0 \rangle, \tag{1}$$

one finds the density matrix at time t

$$\langle n | \rho(t) | n' \rangle = \sum_{00'} a_{n,0}(t) \langle 0 | \rho_i | 0' \rangle a^*_{n',0'}(t). \tag{2}$$

The summation should be performed over the magnetic substates of the ground state $|0\rangle$, the matrix $\langle 0| \rho_i |0'\rangle$ indicating the initial state of polarization. It is noted that the trace of (2) is unity as can be seen from the orthogonality relation (II.4.10) utilizing that the trace of ρ_i is also unity.

The time development of the density matrix (2) can thus be illustrated in the following way. At time $t = -\infty$ only the sub-matrix $\langle 0| \rho_i |0'\rangle$ is non-vanishing with trace unity. Gradually other matrix elements emerge and after the collision nonvanishing matrix elements appear between all states, which can be reached in the multiple excitation process, in such a way that the trace is conserved.

It is a characteristic feature of the semiclassical theory that the density matrix contains elements which connect different energy levels. This is due to the fact that in the theory the time is completely well defined in the sense that we know that at time $t = 0$ the projectile is exactly at the distance of closest approach. Accordingly, the energy of the target nucleus after the collision is undefined. In practice one will not be able to perform experiments in which interference terms between different energy levels are of importance, except those which may arise by hyperfine splitting of the magnetic substates of one nuclear energy level. We may therefore neglect the non-diagonal matrix elements and write the density matrix after the Coulomb excitation process ($t = +\infty$) as

$$\langle IM| \rho |IM'\rangle = \sum_{M_0 M_0'} a_{IM,I_0M_0} \langle I_0M_0| \rho_i |I_0M_0'\rangle a^*_{IM',I_0M_0'}, \tag{3}$$

where the excitation amplitudes are defined by (II.8.16). If the target nucleus is initially unpolarized the density matrix ρ_i is given by

$$\langle I_0M_0| \rho_i |I_0M_0'\rangle = \frac{1}{2I_0+1} \delta_{M_0M_0'} \tag{4}$$

and the density matrix then takes the form

$$\langle IM| \rho |IM'\rangle = \frac{1}{2I_0+1} \sum_{M_0} a^*_{IM',I_0M_0} a_{IM,I_0M_0}. \tag{5}$$

It is noted that the quantity

$$P_I = \sum_M \langle IM| \rho |IM\rangle = \frac{1}{2I_0+1} \sum_{M_0 M} |a_{IM,I_0M_0}|^2 \tag{6}$$

denotes the excitation probability of the state of spin I.

If one does not observe the scattering angle of the projectile, one must extend the ensemble to include the projectiles which hit the target in random points. The bombarding condition is mostly such that the ensemble of projectiles corresponds to a uniform distribution over the target, i.e. the projectile flux ϕ_0 is a constant over the target. The density matrix (3) depends on the impact parameter ρ which is connected with the deflection angle ϑ in the Rutherford scattering by the relation $\tan(\frac{1}{2}\vartheta) = a/\rho$, where a is half the distance of closest approach in a head-on collision. For the density matrix of the target nuclei one therefore finds

$$\begin{aligned}\langle IM|\,\rho_{\mathrm{tot}}\,|IM'\rangle &= \phi_0 \int \langle IM|\,\rho(\vartheta,\varphi)\,|IM'\rangle\,\rho\,\mathrm{d}\rho\,\mathrm{d}\varphi \\ &= \phi_0 \frac{a^2}{2} \int \langle IM|\,\rho(\vartheta,\varphi)\,|IM'\rangle \frac{\cos(\frac{1}{2}\vartheta)}{\sin^3(\frac{1}{2}\vartheta)}\,\mathrm{d}\vartheta\,\mathrm{d}\varphi, \qquad (7)\end{aligned}$$

where we have indicated explicitly that the density matrix (3) depends on ϑ as well as the azimuthal scattering angle φ. The diagonal matrix elements of ρ_{tot} are the cross sections for exciting the various nuclear state times ϕ_0.

It is thus often convenient to relax the exact normalization condition of trace unity and instead consider the matrix

$$\langle IM|\,\rho_\sigma\,|IM'\rangle = \langle IM|\,\rho\,|IM'\rangle\,\tfrac{1}{4}a^2/\sin^4(\tfrac{1}{2}\vartheta) \qquad (8)$$

whose expectation values directly give the cross sections for Coulomb excitation.

In the quantum mechanical description of Coulomb excitation one similarly introduces a density matrix (see (F.27))

$$\begin{aligned}\langle IM|\,\rho_\sigma\,|IM'\rangle = \frac{m_0^2}{4\pi^2\hbar^4}\frac{v_{\mathrm{f}}}{v_0} \sum_{M_0M_0'} &\langle IM\boldsymbol{p}_{\mathrm{f}}|\,T\,|I_0M_0\boldsymbol{p}_0\rangle \\ &\times \langle I_0M_0|\,\rho_{\mathrm{i}}\,|I_0M_0'\rangle\,\langle I_0M_0'\boldsymbol{p}_0|\,T^\dagger\,|IM'\boldsymbol{p}_{\mathrm{f}}\rangle, \qquad (9)\end{aligned}$$

where the expectation values are equal to the cross sections according to (1.4).

The state of polarization after a Coulomb excitation to a level of spin I is conveniently described by means of the statistical tensor. According to (F.37) one finds

$$\rho_{k\kappa}(I,I) = \sqrt{2I+1}\sum_{MM'}(-1)^{I-M'}\begin{pmatrix} I & k & I \\ -M' & \kappa & M\end{pmatrix}\langle IM|\,\rho\,|IM'\rangle. \qquad (10)$$

For unpolarized target nuclei one finds according to (5)

$$\rho_{k\kappa}(I,I) = \frac{\sqrt{2I+1}}{2I_0+1}\sum_{M_0MM'}(-1)^{I-M'}\begin{pmatrix} I & k & I \\ -M' & \kappa & M\end{pmatrix} a^*_{IM',I_0M_0}a_{IM,I_0M_0}. \qquad (11)$$

It is noted that

$$\rho_{00}(I;I) = P_I \qquad (12)$$

are the excitation probabilities for scattering angle ϑ. The polarization of the state of spin I is in this case described by the tensor (F.41)

$$P_{k\kappa}(I) = \rho_{k\kappa}(I, I)/P_I. \tag{13}$$

Besides the general symmetry relation

$$P^*_{k\kappa}(I) = (-1)^\kappa P_{k,-\kappa}(I) \tag{14}$$

the spherical tensors (11) or (13) satisfy a number of relations which follow from the symmetries discussed in § II.8. These relations which depend on the coordinate system used, are summarized in table 1. It is noted furthermore that in first-order perturbation theory according to (II.3.14) the excitation amplitudes in the coordinate systems A and B are purely imaginary. The statistical tensors $\rho_{k\kappa}$ and $P_{k\kappa}$ are therefore real numbers.

From the quantities $P_{k\kappa}$ one may evaluate the vector and tensor polarization of the excited nucleus. The vector polarization $\boldsymbol{P}$ is defined by the expectation value of the spin vector $\boldsymbol{I}$ as

$$\boldsymbol{P} = \langle \boldsymbol{I} \rangle / I = \mathrm{Tr}(\boldsymbol{I}\rho)/I. \tag{15}$$

TABLE 1

Symmetry properties of the statistical tensors in Coulomb excitation.

Coordinate system	General symmetry in orbit	$\xi = 0$	$\vartheta = 180°$
A	$\rho_{k\kappa} = (-1)^\kappa \rho_{k\kappa}$ or κ even	$\rho_{k\kappa} = (-1)^k \rho_{k-\kappa}$ or real for k even imag. for k odd	—
B	$\rho_{k\kappa} = (-1)^k \rho_{k-\kappa}$ or real for $k+\kappa$ even imag. for $k+\kappa$ odd	$\rho_{k\kappa} = (-1)^\kappa \rho_{k\kappa}$ or κ even and real for k even imag. for k odd	$\rho_{k\kappa} = \delta_{\kappa 0}\rho_{k0}$ and real and k even
C	$\rho_{k\kappa} = (-1)^{k+\kappa} \rho_{k-\kappa}$ or real for k even imag. for k odd	—	$\rho_{k\kappa} = \delta_{\kappa 0}\rho_{k0}$ and real and k even

In each of the three coordinate systems A, B, and C the relations are listed which follow from the general reflection symmetry in the plane of the orbit, as well as from the special symmetries which are satisfied for $\xi = 0$ and $\vartheta = 180°$. The reality properties follow from the general relation $\rho^*_{k\kappa} = (-1)^\kappa \rho_{k-\kappa}$. The coordinate system C is defined in fig. 2 below.

Evaluating the matrix elements of $\boldsymbol{I}$ one finds

$$\langle I_{1\mu} \rangle = \sqrt{I(I+1)}P_{1\mu}, \tag{16}$$

where $I_{1\mu}$ are the tensor components of $\boldsymbol{I}$. For the components of the vector polarization one thus obtains

$$\begin{aligned} P_z &= \sqrt{\frac{I+1}{I}}\, P_{10}, \\ P_x &= -\sqrt{2}\sqrt{\frac{I+1}{I}}\, \mathrm{Re}\, P_{11}, \\ P_y &= -\sqrt{2}\sqrt{\frac{I+1}{I}}\, \mathrm{Im}\, P_{11}. \end{aligned} \tag{17}$$

From the symmetry relations of table 1 follows that the vector polarization is perpendicular to the plane of the orbit. In the coordinate system A one finds for the vector polarization the following explicit expression

$$P_z = \frac{1}{I} \sum_{MM_0} M\, |a^{\mathrm{A}}_{IM,I_0M_0}|^2 \Big/ \sum_{MM_0} |a^{\mathrm{A}}_{IM,I_0M_0}|^2. \tag{18}$$

The tensor polarization may be defined through the expectation values of higher powers of the components of angular momentum. Thus, for second order one defines the real polarization tensor

$$P_{ij} = \frac{1}{2I^2} \langle I_i I_j + I_j I_i \rangle. \tag{19}$$

The diagonal elements are all positive and less or equal to unity. The sum of these elements is constant and equal to $(I+1)/I$. The non-diagonal elements are numerically less than unity. The quantities (19) can be evaluated in terms of the polarization tensors $P_{2\mu}$, since

$$\frac{\langle I_{2\mu} \rangle}{I(I+1)} = \sqrt{\frac{(2I-1)(2I+3)}{2I(2I+2)}}\, P_{2\mu}(I), \tag{20}$$

where $I_{2\mu}$ are the spherical tensor operators which can be formed from the symmetrized products $I_i I_j$ and where

$$I_{20} = \tfrac{1}{2}(3I_z^2 - I^2). \tag{21}$$

The explicit expressions for the quantities (19) in terms of the polarization tensor $P_{2\mu}$ is then given by

$$\begin{aligned} 2P_{zz} - P_{xx} - P_{yy} &= 2f(I)P_{20}(I), \\ P_{xx} - P_{yy} &= \sqrt{\tfrac{8}{3}}f(I)\,\mathrm{Re}\,P_{22}(I), \\ P_{xz} &= -\sqrt{\tfrac{2}{3}}f(I)\,\mathrm{Re}\,P_{21}(I), \\ P_{yz} &= -\sqrt{\tfrac{2}{3}}f(I)\,\mathrm{Im}\,P_{21}(I), \\ P_{xy} &= \sqrt{\tfrac{2}{3}}f(I)\,\mathrm{Im}\,P_{22}(I) \end{aligned} \tag{22}$$

and

$$P_{xx} + P_{yy} + P_{zz} = (I+1)/I, \tag{23}$$

where

$$f(I) = \sqrt{(I+1)(I-\tfrac{1}{2})(I+\tfrac{3}{2})/I^3}\,. \tag{24}$$

In order to illustrate the tensor polarization one may introduce a polarization ellipsoid defined by the equation

$$\sum_{ij} P_{ij}x_i x_j = 1\,. \tag{25}$$

The distance ξ from the origin to a point on the ellipsoid in the direction $\boldsymbol{\xi}$ is given by

$$\xi = 1/\sqrt{P_{\xi\xi}}, \tag{26}$$

where $P_{\xi\xi}$ is the expectation value of I_ξ^2 divided by I^2. In the case of Coulomb excitation it is seen from table 1 that in coordinate system B

$$P_{xz} = P_{xy} = 0\,. \tag{27}$$

In the sudden approximation or in first-order perturbation theory one finds furthermore $P_{yz} = 0$. This means that one of the main axes of the polarization ellipsoid is along the x-direction perpendicular to the plane of the orbit. For $\xi = 0$ or in first-order perturbation theory the other two main axes are along the y- and z-axis. In general there will be a finite angle between the main axes and the coordinate axes in the plane of the orbit. For backward scattering it is seen from table 1 that $P_{xx} = P_{yy}$ which means that the ellipsoid has rotational symmetry around the beam axis. For even-even nuclei, where only the states with $M = 0$ are excited, the ellipsoid degenerates to a rotational cylinder.

§ 3. *Angular distribution of gamma rays*

In Appendix G we have summarized a general formalism which allows us to compute the angular distribution and correlation of γ-quanta or conversion electrons emitted after a Coulomb excitation process. In this section we shall first apply the formalism to the simplest case where we observe only one quantum, emitted after a Coulomb excitation in which the energy of the scattered particle has been observed in a definite scattering angle (see fig. 1).

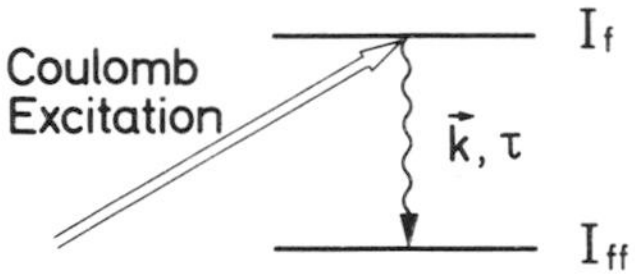

Fig. 1. The figure illustrates the Coulomb excitation of the state of spin I_f and the subsequent detection of a γ-transition to the state of spin I_{ff}. The vector $\boldsymbol{k}$ is the direction of the γ-quantum and τ indicates the circular polarization quantum number.

If initially the target nucleus is unpolarized the density matrix is given by (2.4) while after the Coulomb excitation it is changed into (2.5). The probability P_f of observing the projectile with an energy loss corresponding to a definite state $|f\rangle$ is given by (2.6) and the density matrix is then, according to (F.23) given by

$$\langle I_f M_f | \rho | I_f M_f' \rangle = \frac{1}{P_f} \frac{1}{2I_0 + 1} \sum_{M_0} a^*_{I_f M_f', I_0 M_0} a_{I_f M_f, I_0 M_0} . \tag{1}$$

This density matrix now develops in time according to the equations (F.28) and (F.29).

The probability P_γ of observing the γ-quantum before the time t in a counter which is placed in a definite direction $\boldsymbol{k}$, is obtained from this density matrix by computing the trace of $\rho\varepsilon$ (see (F.24)) over the magnetic quantum numbers of the final state M_{ff} and over the polarization quantum number τ. According to (2.13) and (G.17) the probability is given by

$$\mathrm{d}P_\gamma(\boldsymbol{k}) = \frac{1 - \exp(-\gamma_f t)}{4\pi\gamma_f} \sum_{k\ \mathrm{even}} \delta_L \delta^*_{L'} F_k(LL'I_{ff}I_f)\sqrt{2k+1}$$
$$\times D^{k*}_{\kappa 0}(z \to \boldsymbol{k}) P_{k\kappa}(I_f)\, \mathrm{d}\Omega_\gamma , \tag{2}$$

where $P_{k\kappa}(I_f)$ is the polarization tensor corresponding to (1) which was discussed in the previous section. The total decay constant of the state f is denoted by γ_f (see (F.31)). Note that the partial decay constant for γ-emission to the state ff is given by $\sum_L |\delta_L|^2$ which is in general smaller than γ_f.

The probability of observing both the scattered particle and the γ-quantum is the product of P_f and P_γ. Instead of this probability one would often be interested in the differential cross section $d^2\sigma/d\Omega_p\,d\Omega_\gamma$ for this event. This is obtained from the above probability by multiplying with the Rutherford cross section, i.e.

$$\frac{d^2\sigma}{d\Omega_p\,d\Omega_\gamma} = \tfrac{1}{4}a^2\frac{1}{\sin^4(\frac{1}{2}\vartheta)}\frac{1}{\gamma_f}\frac{1}{\sqrt{4\pi}}\sum_{k\,\text{even}}\rho^*_{k\kappa}(I_f)\sum_{LL'}\delta_L\delta^*_{L'}$$

$$\times F_k(LL'I_{ff}I_f)\,Y_{k\kappa}(\vartheta_\gamma,\varphi_\gamma), \tag{3}$$

where $\rho_{k\kappa}$ is given by (2.11).

In this expression we have left out the exponential time factor and we have indicated the direction of the γ-quantum by the polar angles ϑ_γ and φ_γ. The expressions (2) and (3) are valid in any coordinate system with origin in the recoiling target nucleus.

In a laboratory system the angular distribution will be modified due to Doppler shift and aberration. For heavy ions this can be a very important correction which will be discussed in § X.6. Since the recoiling nucleus is approximately moving in a direction parallel to the z-axis in coordinate system B, cf. eq. (X.3.5), it is convenient to use this coordinate system for the evaluation of the angular distribution of gamma quanta. Also in system B (or A), the statistical tensor $\rho_{k\kappa}$ has simple symmetry properties as was discussed above. The disadvantage in using system B (or A) is that it is not fixed in space when the scattering angle is changed, and the integration over a finite particle detector becomes impractical. It is therefore convenient to transform the expression (3) for the angular distribution to a space-fixed coordinate system still evaluating $\rho_{k\kappa}$ in system B. In this form it is also easy to integrate over all scattering angles to obtain the γ-angular distribution when the scattered particle is not observed.

Through the beam direction of the incoming particles one has a convenient definition of the polar axis for the space fixed system. In coincidence experiments the particle detector together with this axis defines a plane from which to measure the azimuthal angle. We thus define a coordinate system C (see fig. 2) in which the z-axis is along the direction of the incoming projectile, the x-axis is chosen in the plane of orbit in such a way that the x-component of the impact parameter is positive.

The coordinate transformation from coordinate system B to system C is

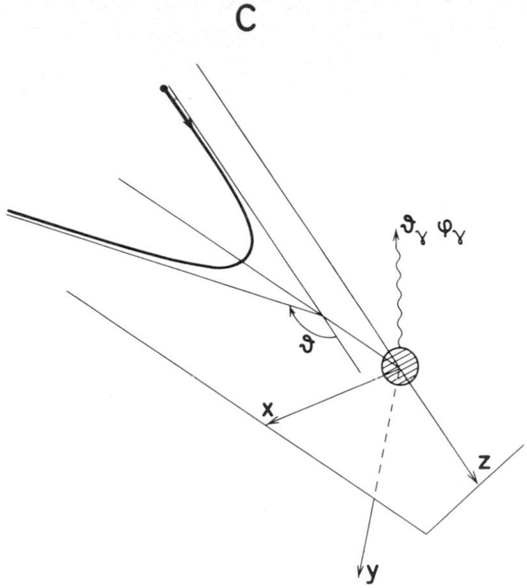

Fig. 2. The coordinate system C which is often used for the description of γ angular distribution after Coulomb excitation. The z-axis is chosen along the incoming beam direction, the x-axis is in the plane of the orbit such that the x-component of the velocity is positive. A γ-decay in the direction ϑ_γ, φ_γ is indicated.

generated by a rotation with the following Eulerian angles

$$(\alpha, \beta, \gamma) = (\tfrac{1}{2}\pi, \tfrac{1}{2}(\pi + \vartheta), \pi). \tag{4}$$

To obtain the angular distribution (3) in the new system C, we use the transformation properties of the spherical harmonics $Y_{k\kappa}$. One finds

$$\frac{d^2\sigma}{d\Omega_p\, d\Omega_\gamma} = \frac{a^2}{4} \frac{1}{\sin^4(\frac{1}{2}\vartheta)} \frac{1}{\gamma_f} \frac{1}{\sqrt{4\pi}} \sum_{\substack{k\ \text{even} \\ \kappa\kappa'}} (\rho^{B}_{k\kappa})^* D^{k}_{\kappa\kappa'}(\tfrac{1}{2}\pi, \tfrac{1}{2}(\pi + \vartheta), \pi)$$
$$\times \sum_{LL'} \delta_L \delta^*_{L'} F_k(LL'I_{ff}I_f)\, Y_{k\kappa'}(\vartheta^C_\gamma, \varphi^C_\gamma), \tag{5}$$

where the statistical tensor is evaluated in coordinate system B. The D-functions are defined in appendix D. The polar angles ϑ^C_γ, φ^C_γ are measured in the coordinate system C with origin in the target nucleus.

The angular distribution in the coordinate system C can, of course, also be expressed in terms of the statistical tensor in this system. For the normalized distribution (2) one finds

$$W^{\mathrm{C}} = \frac{\sqrt{4\pi}}{\gamma_{\mathrm{f}\to\mathrm{ff}}} \sum_{\substack{k\ \mathrm{even}\\ \kappa}} P^{\mathrm{C}}_{k\kappa} \sum_{LL'} \delta_L \delta^*_{L'} F_k(LL'I_{\mathrm{ff}}I_{\mathrm{f}})\, Y_{k\kappa}(\vartheta^{\mathrm{C}}_\gamma, \varphi^{\mathrm{C}}_\gamma), \tag{6}$$

where the normalized particle parameters $P^{\mathrm{C}}_{k\kappa}$ are defined by

$$P^{\mathrm{C}}_{k\kappa} = \sum_{\kappa'} P^{\mathrm{B}}_{k\kappa'} \mathrm{i}^{-\kappa'} D^k_{\kappa'\kappa}(0, \tfrac{1}{2}(\pi + \vartheta), 0)(-1)^\kappa . \tag{7}$$

The function $D^k_{\kappa'\kappa}(0, \beta, 0)$ can be found in table D.1 for $k \leqslant 4$. In eq. (6) we utilized the fact that the tensors $\rho^{\mathrm{C}}_{k\kappa}$ are real for even k. The general symmetry properties are collected in table 1. Similar to (6) we may write the angular distribution (5) in the real form

$$\frac{\mathrm{d}^2\sigma}{\mathrm{d}\Omega_{\mathrm{p}}\,\mathrm{d}\Omega_\gamma} = \frac{a^2}{4} \frac{1}{\sin^4(\frac{1}{2}\vartheta)} \frac{1}{\gamma_{\mathrm{f}}} \frac{1}{\sqrt{4\pi}} \sum_{\substack{k\ \mathrm{even}\\ \kappa\geqslant 0}} \rho^{\mathrm{C}}_{k|\kappa|} \sum_{LL'} \delta_L \delta^*_{L'}$$
$$\times\, F_k(LL'I_{\mathrm{ff}}I_{\mathrm{f}})\, Y_{k|\kappa|}(\vartheta^{\mathrm{C}}_\gamma, 0)\{2\cos(\kappa\, \varphi^{\mathrm{C}}_\gamma) - \delta_{\kappa 0}\}. \tag{8}$$

From the expression (8) it is now easy to calculate also the angular distribution of the γ-quanta for a finite opening of the particle or gamma detectors, provided one may neglect the aberration discussed in § X.6. For the case where the particle is detected in a counter which is symmetric around the beam direction (ring counter) the integration over the angle φ leads to the condition $\kappa = 0$ and one finds

$$\frac{\mathrm{d}\sigma}{\mathrm{d}\Omega_\gamma} = \sum_{k\ \mathrm{even}} \beta_k \frac{1}{\gamma_{\mathrm{f}}} \sum_{LL'} \delta_L \delta^*_{L'} F_k(LL'I_{\mathrm{ff}}I_{\mathrm{f}}) P_k(\cos\vartheta_\gamma), \tag{9}$$

where

$$\beta_k = \tfrac{1}{4}a^2\sqrt{2k+1} \int_{\vartheta_1}^{\vartheta_2} \mathrm{d}\vartheta\, \rho^{\mathrm{C}}_{k0}(\vartheta) \frac{\cos(\frac{1}{2}\vartheta)}{\sin^3(\frac{1}{2}\vartheta)} . \tag{10}$$

The opening of the ring counter is indicated by the angles ϑ_1 and ϑ_2.

Formula (9) also applies to the case where the scattered projectile is not observed at all. The scattering angles ϑ_1 and ϑ_2 should then be equal to 0 and π, respectively. It should be noted, however, that the formula only applies if the state $|\mathrm{f}\rangle$ is populated directly by Coulomb excitation and is not fed appreciably by decay processes from higher-lying states.

The angular distribution for finite opening for the γ-detector is a standard problem in angular correlation. The relevant formulae and tables are collected in § X.6.

For the simple case of backwards scattering one sees from table 1 that $\kappa = 0$, and the angular distribution (8) reduces to a form similar to (9). The simplification which appears for backwards scattering is a consequence of the rotation symmetry around the z-axis which implies $M_{\mathrm{i}} = M_{\mathrm{f}}$. As was emphasized in § II.8 the conservation of the magnetic quantum number holds to a good approximation also for other scattering angles in the coordinate system B. In this $\chi(\vartheta)$-approximation we may write the tensors $\rho^{\mathrm{C}}_{k\kappa}$ in the form

$$\rho^{\mathrm{C}}_{k\kappa} \approx \rho^{\mathrm{B}}_{k0} D^{k}_{0\kappa}(0, \tfrac{1}{2}(\pi + \vartheta), 0)(-1)^{\kappa}. \tag{11}$$

A very simple form is obtained if the ground state has spin $I_0 = 0$ in which case the excitation of a given state of spin I is described by one excitation amplitude only, i.e. $a_{I0,00}$. In this case the tensor ρ^{B}_{k0} is given by

$$\rho^{\mathrm{B}}_{k0} = \sqrt{2I+1}(-1)^{I}\begin{pmatrix} I & I & k \\ 0 & 0 & 0 \end{pmatrix} |a_{I0,00}|^2. \tag{12}$$

Inserting the expression (11) and (12) in (6) one finds

$$W^{\mathrm{C}} \approx \sum_{k\,\mathrm{even},\,\kappa} (-1)^{I}\sqrt{2I+1}\begin{pmatrix} I & I & k \\ 0 & 0 & 0 \end{pmatrix} D^{k}_{0\kappa}(0, \tfrac{1}{2}(\pi + \vartheta), 0)(-1)^{\kappa}$$

$$\times \frac{\sqrt{4\pi}}{\gamma_{\mathrm{f}\to\mathrm{ff}}} \sum_{LL'} \delta_L \delta^{*}_{L'} F_k(LL'I_{\mathrm{ff}}I_{\mathrm{f}}) Y_{k\kappa}(\vartheta^{\mathrm{C}}_{\gamma}, \varphi^{\mathrm{C}}_{\gamma}). \tag{13}$$

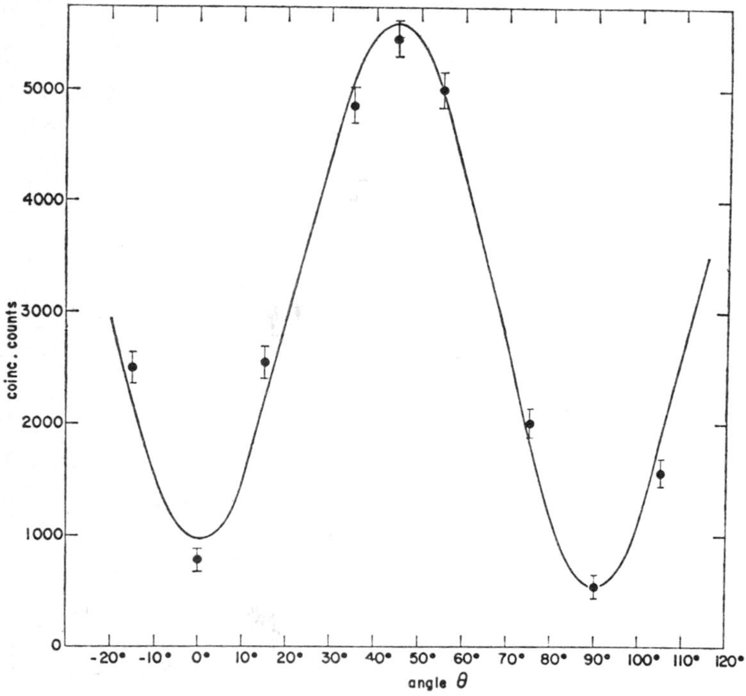

Fig. 3. Angular distribution of the $2^+ \to 0^+$ 300 keV gamma ray of ^{148}Nd excited by backscattered 35 MeV ^{16}O ions and recoiled into a copper catcher. The angular distribution in this case is found to be essentially unperturbed (from Ben Zvi et al., 1967, Nucl. Phys. **A96**, 138).

It is an interesting observation [ALD 62] that this angular distribution is independent of the excitation amplitudes and only depends on the geometry and the δ_L's of the γ-decay. The formula (13) shows that an observation of the γ-angular distribution following Coulomb excitation in even-even nuclei usually gives little information about the nuclear matrix elements. An example of the large anisotropies encountered in such experiments is shown in fig. 3.

§ 4. *Gamma cascades following Coulomb excitation*

In Coulomb excitation, especially with heavy ions, several nuclear states will be populated and the de-excitations will proceed through cascades of gamma rays and conversion electrons. The simple expressions derived in the previous section will apply if the energy of the scattered projectile has been measured in coincidence with the emitted gamma quantum.

We shall generalize this result to the case of a cascade of two gamma rays (see fig. 4) where one measures the triple coincidence between the scattered projectile and the two quanta.

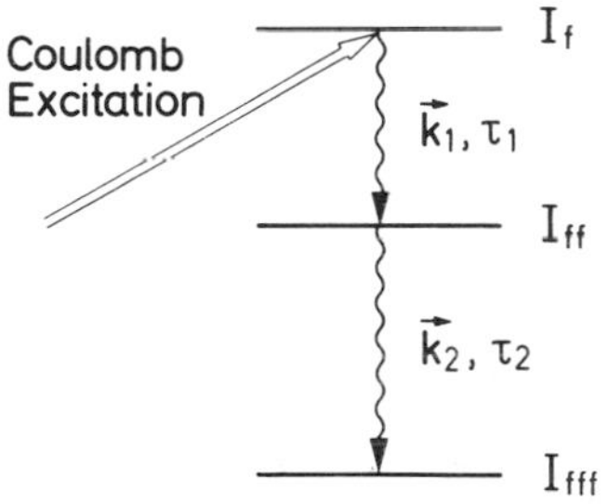

Fig. 4. The figure illustrates the Coulomb excitation of the state of spin I_f and the subsequent detection of the γ-cascade through the state of spin I_{ff} to the state of spin I_{fff}. The detection directions $\boldsymbol{k}_1$ and $\boldsymbol{k}_2$ of the two γ-quanta are indicated as well as the circular polarization quantum numbers τ_1 and τ_2.

With the notation of fig. 4 the probability of observing the first gamma quantum in direction $\boldsymbol{k}_1$ is given by

$$\mathrm{d}P_\gamma(\boldsymbol{k}_1) = \frac{1}{4\pi\gamma_f} \sum_{\substack{k \text{ even} \\ L_1 L_1'}} \delta_{L_1} \delta^*_{L_1'} F_k(L_1 L_1' I_{ff} I_f) D^{k*}_{\kappa 0}(z \to \boldsymbol{k}_1) \times \sqrt{2k+1}\, P_{k\kappa}(I_f)\, \mathrm{d}\Omega_{\gamma_1}, \tag{1}$$

where $P_{k\kappa}(I_f)$ is given by (2.13).

After the detection of the first gamma quantum the density matrix is changed according to (F.23) and the corresponding statistical tensor is given in eq. (G.19)

$$P_{k_2\kappa_2}(I_{\rm ff}) = \frac{1}{4\pi\gamma_{\rm f}} \sum_{\substack{L_1L_1'k \\ \kappa_1\kappa, k \text{ even}}} (-1)^{k+\kappa}\sqrt{\frac{(2k+1)(2k_1+1)}{(2k_2+1)}}\, P_{k\kappa}(I_{\rm f})$$
$$\times\, F_{k_1}^{k_2k}(L_1L_1'I_{\rm ff}I_{\rm f})\delta_{L_1}\delta_{L_1'}^*(D_{\kappa 0}^{k_1}(z\to \boldsymbol{k}_1))^*$$
$$\times \begin{pmatrix} k & k_1 & k_2 \\ -\kappa & \kappa_1 & \kappa_2 \end{pmatrix} \frac{1}{\mathrm{d}P_\gamma(\boldsymbol{k}_1)/\mathrm{d}\Omega_{\gamma_1}}. \qquad (2)$$

This statistical tensor now develops in time according to (F.28) and (F.29) where t is the time since the first gamma quantum was observed.

The probability of observing the second gamma quantum in a direction $\boldsymbol{k}_2$ is obtained by computing the trace of $(\rho\varepsilon)$ over the magnetic quantum number of the state fff. The result is identical to (1) except for an appropriate permutation of the indices, i.e.

$$\mathrm{d}P_{\gamma_2}(\boldsymbol{k}_2) = \frac{1}{4\pi\gamma_{\rm ff}} \sum_{\substack{k_2 \text{ even} \\ L_2L_2'}} \delta_{L_2}\delta_{L_2'}^* F_{k_2}(L_2L_2'I_{\rm fff}I_{\rm ff})\sqrt{2k_2+1}$$
$$\times\, D_{\kappa_2 0}^{k_2 *}(z\to \boldsymbol{k}_2)P_{k_2\kappa_2}(I_{\rm ff})\,\mathrm{d}\Omega_{\gamma_2}, \qquad (3)$$

where we have left out the exponential time factor. The triple coincidence probability is obtained by multiplying the Coulomb excitation probability $P_{\rm f}$ with the probabilities $\mathrm{d}P_{\gamma_1}$ and $\mathrm{d}P_{\gamma_2}$ i.e.

$$P = P_{\rm f}(\vartheta)\,\mathrm{d}P_{\gamma_1}(\boldsymbol{k}_1)\,\mathrm{d}P_{\gamma_2}(\boldsymbol{k}_2). \qquad (4)$$

If the first gamma quantum is detected only up to the time t_1 after the Coulomb excitation and the second gamma quantum is observed only up to the time t_2 after the detection of the first quantum the probability (4) should be multiplied by the factor

$$f(t_1, t_2) = (1 - \exp(-\gamma_{\rm f}t_1))(1 - \exp(-\gamma_{\rm ff}t_2)). \qquad (5)$$

Inserting eqs. (1), (2) and (3) in (4) we may write the triple coincidence probability in the form of a cross section by multiplying (4) with the Rutherford cross section. The result can be written in the form

$$\frac{d^3\sigma}{d\Omega_p\, d\Omega_{\gamma_1}\, d\Omega_{\gamma_2}} = \frac{a^2}{4}\frac{1}{\sin^4(\frac{1}{2}\vartheta)}\frac{1}{\gamma_f\gamma_{ff}}\frac{1}{4\pi}$$

$$\times \sum_{\substack{k_1k_2k\\ k_1,k_2\ \text{even}}} \Big(\sum_{L_1L_1'} F_{k_1}^{k_2k}(L_1L_1'I_{ff}I_f)\delta_{L_1}\delta^*_{L_1'}\Big)\Big(\sum_{L_2L_2'} F_{k_2}(L_2L_2'I_{fff}I_{ff})\delta_{L_2}\delta^*_{L_2'}\Big)$$

$$\times \sqrt{\frac{2k+1}{2k_2+1}}\Big\{\sum_{\kappa_1\kappa_2\kappa}\begin{pmatrix} k & k_1 & k_2\\ \kappa & \kappa_1 & \kappa_2\end{pmatrix}\rho_{k\kappa}(I_f)\,Y_{k_1\kappa_1}(\vartheta_1,\varphi_1)\,Y_{k_2\kappa_2}(\vartheta_2,\varphi_2)\Big\}, \tag{6}$$

where the geometrical coefficient $F_{k_1}^{k_2k}$ is given in eq. (G.21).

In actual Coulomb excitation experiments one rarely performs a complete observation of all emitted γ-quanta and one must then take into account that the state, whose decay is being observed, can be populated not only in the Coulomb excitation itself, but also by decay from higher-lying states through unobserved γ-emission or internal conversion. Let us consider the simplest case illustrated in fig. 5 where one observes the decay of the state ff without detecting the transition f → ff.

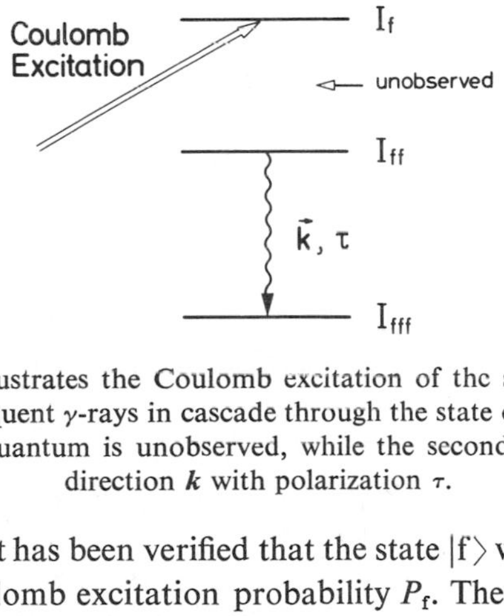

Fig. 5. The figure illustrates the Coulomb excitation of the state of spin I_f and the emission of two subsequent γ-rays in cascade through the state of spin I_{ff} to the state of spin I_{fff}. The first γ-quantum is unobserved, while the second one is detected in the direction $\boldsymbol{k}$ with polarization τ.

We imagine that it has been verified that the state $|f\rangle$ was populated at time $t = 0$ with the Coulomb excitation probability P_f. The statistical tensor for the state $|ff\rangle$ is then according to eq. (G.23) given by

$$P_{k_2\kappa_2}(I_{ff}) = P_{k_2\kappa_2}(I_f)H_{k_2}(I_fI_{ff}), \tag{7}$$

where

$$H_k(I_fI_{ff}) = \frac{\sqrt{(2I_f+1)(2I_{ff}+1)}}{\gamma_f}\sum_L (-1)^{I_f+I_{ff}+L+k}|\delta_L|^2$$

$$\times (1+\alpha(L))\begin{Bmatrix} I_f & I_f & k\\ I_{ff} & I_{ff} & L\end{Bmatrix}, \tag{8}$$

$\alpha(L)$ being the total conversion coefficient for 2^L-pole radiation. If monopole conversion is present one must include a term with $L = 0$ and interpret $|\delta_0|^2(1 + \alpha(0))$ as the monopole conversion intensity.

With this density matrix the probability of observing the γ-quantum of the transition ff $\rightarrow$ fff is given by (3). The probability of observing the excitation of the state $|\mathrm{f}\rangle$ in coincidence with this γ-transition is obtained by multiplying the Coulomb excitation probability P_f with $\mathrm{d}P_{\gamma_2}(\boldsymbol{k}_2)$ since the probability for the transition f $\rightarrow$ ff is unity when the second γ-quantum has been detected

$$P = P_\mathrm{f}\,\mathrm{d}P_{\gamma_2}(\boldsymbol{k}_2). \tag{9}$$

If the γ-quantum is detected only up to time t after the Coulomb excitation this probability should be multiplied by a factor

$$f(t) = \left[1 + \frac{\gamma_\mathrm{f}}{\gamma_\mathrm{ff} - \gamma_\mathrm{f}} \exp(-\gamma_\mathrm{ff} t) - \frac{\gamma_\mathrm{ff}}{\gamma_\mathrm{ff} - \gamma_\mathrm{f}} \exp(-\gamma_\mathrm{f} t)\right]. \tag{10}$$

The differential cross section for the Coulomb excitation populating the state $|\mathrm{f}\rangle$ and the detection of the γ-quantum at any later time may be written in the form

$$\frac{\mathrm{d}^2\sigma}{\mathrm{d}\Omega_\mathrm{p}\,\mathrm{d}\Omega_{\gamma_2}} = \frac{a^2}{4}\frac{1}{\sin^4(\frac{1}{2}\vartheta)} \sum_{\substack{k\ \mathrm{even}\\ \kappa}} \rho^*_{k\kappa}(I_\mathrm{f}) H_k(I_\mathrm{ff} I_\mathrm{f})$$
$$\times \frac{1}{\sqrt{4\pi}\gamma_\mathrm{ff}} \sum_{L_2 L_2'} \delta_{L_2}\delta^*_{L_2'} F_k(L_2 L_2' I_\mathrm{fff} I_\mathrm{ff})\, Y_{k\kappa}(\vartheta_2, \varphi_2). \tag{11}$$

If the decay of the state $|\mathrm{f}\rangle$ proceeds solely through γ-emission the expression (11) can also be obtained from (6) by integration over all angles ϑ_1, φ_1.

Next we may consider the situation indicated in fig. 6 where a state of spin I_4 is populated and one observes the transition $I_2 \rightarrow I_1$, while the decays of the states with spins I_3 and I_4 are unobserved. The cross section for this situation is easily obtained from eq. (11) by introducing an additional factor H_k, i.e.

$$\frac{\mathrm{d}^2\sigma}{\mathrm{d}\Omega_\mathrm{p}\,\mathrm{d}\Omega_\gamma} = \frac{a^2}{4}\frac{1}{\sin^4(\frac{1}{2}\vartheta)} \sum_{\substack{k\ \mathrm{even}\\ \kappa}} \rho^*_{k\kappa}(I_4)\, H_k(I_3 I_4)\, H_k(I_2 I_3)$$
$$\times \frac{1}{\sqrt{4\pi}\gamma_{2\rightarrow 1}} \sum_{LL'} \delta_L \delta^*_{L'} F_k(LL' I_1 I_2)\, Y_{k\kappa}(\vartheta, \varphi), \tag{12}$$

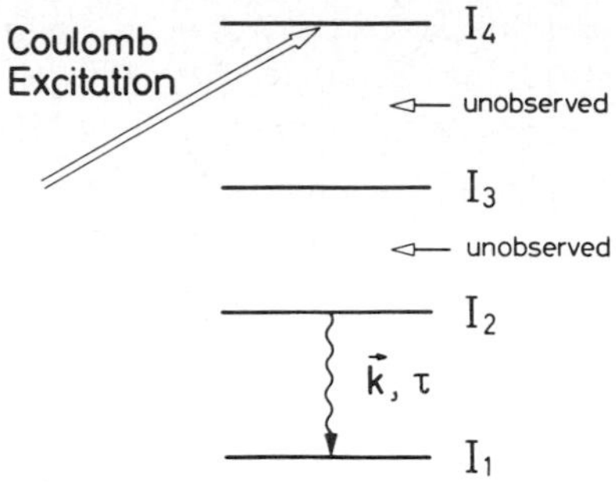

Fig. 6. The figure illustrates the Coulomb excitation of the state with spin I_4 and the emission of three subsequent γ-rays in cascade of which only the last γ-quantum is detected.

where ϑ and φ are the polar angles of the direction of the gamma quantum.

For each additional unobserved γ-transition one thus has to introduce a factor H_k in the cross section. The time dependence of the event is, although elementary, rather complicated.

In many cases of practical interest the energy of the scattered projectile is not detected and the experiment then does not give any information about which state was actually populated in the Coulomb excitation process. In such a situation the evaluation of the angular distribution of the γ-transition $I_2 \rightarrow I_1$ in fig. 7 should be based on the full density matrix (2.5) for Coulomb excitation and one should study the time development of this matrix until

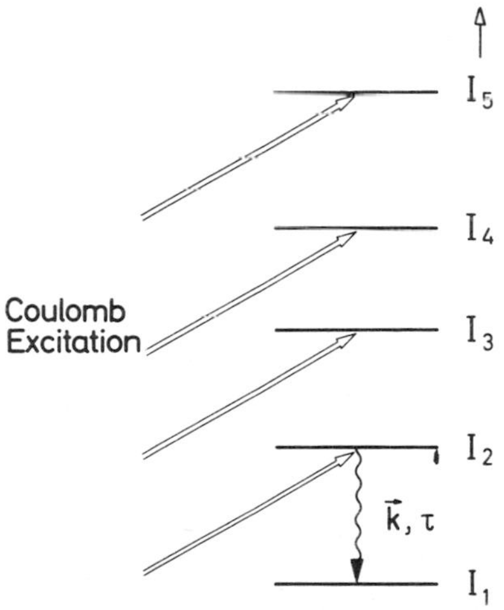

Fig. 7. The figure illustrates the Coulomb excitation into the different nuclear states of spins $I_2, I_3, \ldots$. From the subsequent γ-cascades only the γ-quantum corresponding to the transition $I_2 \rightarrow I_1$ is observed in the direction $\boldsymbol{k}$ with polarization τ.

the time where the quantum is observed. Apart from this time factor the result is given in terms of the above derived formulae for the angular distribution with no, one, two, three etc. preceding unobserved gamma quanta. The cross section for observing the gamma quantum and the projectile without detecting its energy is given simply by adding the results given in formulae (1), (11), (12) etc. and we may write the result in the form

$$\frac{d^2\sigma}{d\Omega_p\, d\Omega_\gamma} = \frac{a^2}{4}\frac{(4\pi)^{-1/2}}{\sin^4(\frac{1}{2}\vartheta)} \sum_{k\kappa} A^*_{k\kappa} \sum_{LL'} \frac{1}{\gamma_{2\to 1}} \delta_L \delta^*_{L'} F_k(LL'I_1I_2)\, Y_{k\kappa}(\vartheta, \varphi), \quad (13)$$

where

$$A_{k\kappa} = \rho_{k\kappa}(I_2) + \rho_{k\kappa}(I_3)\, H_k(I_2I_3) + \rho_{k\kappa}(I_4)\{H_k(I_2I_4) + H_k(I_3I_4)\, H_k(I_2I_3)\} + \cdots. \quad (14)$$

We have here taken into account also that the states I_4, I_5 etc. may decay partly through a cascade, partly directly to the state I_2.

§ 5. *Perturbation of angular distribution*

In an actual Coulomb excitation experiment the target nucleus suffers a considerable recoil motion and will move through the target material until it is stopped or until it leaves the target mostly in a highly ionized state. It is an important practical question whether the state of polarization of the nucleus is disturbed by external perturbations, which may happen during these events. The influence of various types of external fields on the state of polarization has been studied in detail in angular correlation experiments. The theory of attenuation of angular correlation in external fields is reviewed e.g. in [FRA 65, ALD 64, DEV 57]. Sometimes the change in the angular distribution caused e.g. by magnetic fields can be used to measure nuclear magnetic moments [ALD 64] or intrinsic magnetic fields e.g. by the so-called implantation technique [BOE 66, GRO 66].

We consider a nuclear state which at time $t = 0$ is in a state of polarization given by the density matrix $\rho(0)$. At some later time t the density matrix is given by eqs. (F.9) and (F.10), where $H_{\text{int}}(t)$ is the interaction of the nucleus with the external fields.

Reformulating (F.10) in terms of the polarization tensors (F.41) one finds

$$P^*_{k\kappa}(t) = \sum_{k'\kappa'} G^{\kappa'\kappa}_{k'k}(t) P^*_{k'\kappa'}(0) \sqrt{\frac{2k'+1}{2k+1}}, \quad (1)$$

where the perturbation coefficients $G^{\kappa'\kappa}_{k'k}(t)$ are given by

$$G_{k'k}^{\kappa'\kappa}(t) = \sum_{MM'M''M'''} (-1)^{M-M''}\sqrt{(2k+1)(2k'+1)}\begin{pmatrix} I & I & k' \\ M''' & -M'' & \kappa' \end{pmatrix}$$

$$\times \begin{pmatrix} I & I & k \\ M' & -M & \kappa \end{pmatrix} \langle IM|\, U(t,0)\,|IM''\rangle \langle IM'|\, U(t,0)\,|IM'''\rangle^* . \quad (2)$$

The unitary operator $U(t, 0)$ is given by (F.9), i.e.

$$U(t,0) = \mathscr{T}\exp\left\{-\frac{\mathrm{i}}{\hbar}\int_0^t \tilde{H}_{\mathrm{int}}(t)\,\mathrm{d}t\right\}. \quad (3)$$

It is noted that

$$G_{k'k}^{\kappa'\kappa}(0) = \delta_{\kappa\kappa'}\delta_{kk'} . \quad (4)$$

These formulae hold if the state I does not decay. If it decays through γ-emission we may assume that the decay is not influenced by the external fields and the total time development of the statistical tensor is given by the product of decay and attenuation, i.e. we assume

$$\rho_{k\kappa}^{*}(t) = \mathrm{e}^{-\gamma t}\sum_{k'\kappa'} G_{k'k}^{\kappa'\kappa}(t) P_{k'\kappa'}^{*}(0)\sqrt{\frac{2k'+1}{2k+1}} . \quad (5)$$

Through the decay other states will be populated, which in turn may be influenced by external fields.

To obtain the angular distribution of the γ-quanta emitted from the state I, which was populated directly e.g. by Coulomb excitation, one must specify the time t at which the quantum was emitted. The probability of observing the quantum in the time interval $\mathrm{d}t$ and in the direction $\boldsymbol{k}$ is given by

$$\mathrm{d}^2P_\gamma(\boldsymbol{k}) = \frac{\mathrm{d}\Omega_\gamma}{4\pi}\,\mathrm{d}t\,\mathrm{e}^{-\gamma t}\sum_{\substack{k\text{ even}\\ k'\kappa\kappa'}} P_{k'\kappa'}(0) G_{k'k}^{\kappa'\kappa}(t)^*\sqrt{2k'+1}$$

$$\times \sum_{LL'} \delta_L\delta_{L'}^{*} F_k(LL'I_fI) D_{\kappa 0}^{k*}(z\to\boldsymbol{k}) . \quad (6)$$

If the counter is open in a finite time interval, (6) must correspondingly be integrated over time. If the counter is open from time $t = 0$ to infinity it is convenient to define the integrated perturbation factor

$$G_{k'k}^{\kappa'\kappa} = \gamma\int_0^\infty \mathrm{e}^{-\gamma t} G_{k'k}^{\kappa'\kappa}(t)\,\mathrm{d}t \quad (7)$$

and one finds

$$\mathrm{d}P_\gamma(\boldsymbol{k}) = \frac{\mathrm{d}\Omega_\gamma}{4\pi\gamma}\sum_{\substack{k\text{ even}\\ k'\kappa\kappa'}} P_{k'\kappa'}(0)(G_{k'\kappa}^{\kappa'\kappa})^*\sqrt{2k'+1}$$

$$\times \sum_{LL'} \delta_L\delta_{L'}^{*} F_k(LL'I_fI) D_{\kappa 0}^{k*}(z\to\boldsymbol{k}) . \quad (8)$$

The perturbation factors $G^{\kappa'\kappa}_{k'k}(t)$ and $G^{\kappa'\kappa}_{k'k}$ can be found for various cases of interest in [ALD 64] and [FRA 65]. For the important simple case of a constant external magnetic field $\boldsymbol{H}$ along the z-axis one finds

$$G^{\kappa'\kappa}_{k'k}(t) = \delta_{kk'}\delta_{\kappa\kappa'}\exp(\mathrm{i}\kappa\omega_{\mathrm{L}}t) \tag{9}$$

where ω_{L} is the Larmor precession frequency which is given in terms of the magnetic moment μ of the nucleus by

$$\omega_{\mathrm{L}} = \mu H/I\hbar. \tag{10}$$

An example of the implantation technique to determine the hyperfine interaction is given in fig. 8.

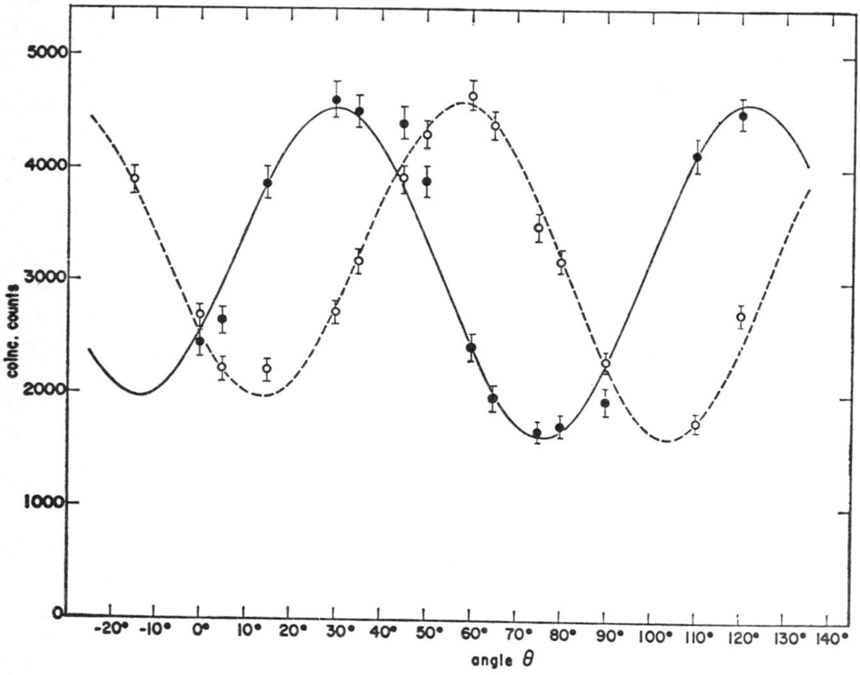

Fig. 8. Gamma-ray angular distribution of the $2^+ \to 0^+$ transition in ^{148}Nd excited by backscattered 35 MeV ^{16}O ions and recoiled into an iron catcher. The iron recoil-catcher was magnetized by an external magnetic field. The solid line shows the computer fit for "field up", the dashed line for "field down". Aside from a rotation the angular distribution is seen to be attenuated as compared to the nearly unperturbed correlation given in fig. 3 (from Ben Zvi et al., 1967, Nucl. Phys. **A96**, 138).

CHAPTER IV

First-Order Perturbation Theory

In the preceding chapter it was shown how one may evaluate cross sections and angular distributions for Coulomb excitation processes from the excitation amplitudes. In this and the following chapters we shall discuss the evaluation of these excitation amplitudes by means of the various approximation methods which were studied in chapter II.

The first-order perturbation treatment is of great interest since this approximation leads to simple and comprehensive expressions for the excitation amplitudes which give an accurate description of Coulomb excitation with light projectiles. At the same time a number of concepts such as orbital integrals, are introduced. These concepts are of importance also for the discussion of more complicated Coulomb excitation processes and this chapter will therefore also serve as a convenient preparation for the systematic study of higher-order processes.

In the present chapter we shall summarize the results of the semiclassical treatment of the first-order perturbation theory. For more details we refer the reader to earlier review articles [ALD 56, BRE 59, BIE 65]. A discussion of the quantum mechanical description is given in chapter IX and we shall occasionally use some of the results derived there.

§ 1. *Excitation amplitudes*

The excitation amplitude of a given nuclear state can only be evaluated by first-order perturbation theory, if the interaction between projectile and target is weak. More specifically, the excitation probability which is computed by the first-order theory should be small compared to unity. As was shown in chapters I and II this condition, although necessary, is not sufficient. To obtain a sufficient condition one must consider all possible transitions from the ground state as well as from the final state. Only if all these transitions are weak the simple first-order theory applies. The strength of transitions is conveniently measured by the matrix elements of the action integral, i.e. by the parameters χ_{nm}. The χ-matrix connecting any pair of nuclear states in a given Coulomb excitation experiment can be classified into different sub-matrices in such a way that all the transitions between states belonging to different sub-matrices are weak ($\chi_{nm} \ll 1$), while the transitions between states within a sub-matrix are strong ($\chi_{nm} \gtrsim 1$). Multiple Coulomb excitation will occur between states in a sub-matrix if any one of these states is populated. The condition for the simple first-order perturbation treatment is therefore that the ground state as well as the final state are not coupled strongly to any other nuclear states. In practice this condition is fulfilled only for projectiles of low charge. The excitation amplitude of the state $|I_f M_f\rangle$ from the ground state $|I_0 M_0\rangle$ is in first-order perturbation theory given by eq. (II.3.13) where we insert for the time-development operator the first two terms in the expansion (II.3.8), i.e.

$$a_{I_f M_f, I_0 M_0} = \frac{1}{i\hbar} \int_{-\infty}^{+\infty} \langle I_f M_f | \, V(t) \, | I_0 M_0 \rangle \exp\left\{\frac{i}{\hbar}(E_f - E_0)t\right\} dt, \tag{1}$$

where E_f and E_0 are the energies of the final and initial states, respectively. The result (1) can also be obtained from the coupled equations (II.4.5) by assuming that $a_m(t)$ on the right-hand side is given by

$$a_m(t) = a_{IM}(t) = \delta_{II_0}\delta_{MM_0}. \tag{2}$$

If we consider electric excitation only we find from eqs. (II.1.11-12) the following result

$$a_{I_f M_f, I_0 M_0} = \frac{4\pi Z_1 e}{i\hbar} \sum_{\lambda\mu} \frac{(-1)^\mu}{2\lambda + 1} \langle I_f M_f | \, \mathscr{M}(E\lambda, -\mu) \, | I_0 M_0 \rangle \, S_{E\lambda\mu}, \tag{3}$$

where we have introduced the notation

$$S_{E\lambda\mu} = \int_{-\infty}^{+\infty} \bar{S}_{E\lambda\mu}(t) \, e^{i\omega t} \, dt = \int_{-\infty}^{+\infty} Y_{\lambda\mu}[\theta(t), \phi(t)]/[r(t)]^{\lambda+1} \, e^{i\omega t} \, dt \tag{4}$$

with

$$\omega = (E_f - E_0)/\hbar. \tag{5}$$

In expression (3) the nuclear properties enter only through the electric multipole matrix elements and all dependences on the parameters of the collision are contained in the orbital integral $S_{E\lambda\mu}$. In the following section we shall discuss the properties of these orbital integrals in more detail. Since with the definition (II.1.3)

$$(-1)^{\mu}\mathcal{M}(\mathrm{E}\lambda, -\mu) = \mathcal{M}(\mathrm{E}\lambda, \mu)^{\dagger}, \tag{6}$$

we may write eq. (3) in the form

$$a_{I_f M_f, I_0 M_0} = \frac{4\pi Z_1 e}{i\hbar} \sum_{\lambda\mu} \frac{1}{2\lambda + 1} \langle I_0 M_0 | \mathcal{M}(\mathrm{E}\lambda, \mu) | I_f M_f \rangle^* S_{\mathrm{E}\lambda\mu}. \tag{7}$$

While the orbital integrals are in general complex numbers the multipole matrix element can be chosen to be real. A common convention for the phases of the nuclear state vectors was discussed in § II.8 and in appendix E. According to this convention the multipole matrix elements in (3) and (7) are real numbers.

The dependence of the excitation amplitude on the magnetic quantum numbers can be exhibited by expressing the nuclear matrix element as a product of a 3-j symbol and a reduced matrix element according to the definition

$$\langle I_0 M_0 | \mathcal{M}(\mathrm{E}\lambda, \mu) | I_f M_f \rangle = (-1)^{I_0 - M_0} \begin{pmatrix} I_0 & \lambda & I_f \\ -M_0 & \mu & M_f \end{pmatrix} \langle I_0 \| \mathcal{M}(\mathrm{E}\lambda) \| I_f \rangle. \tag{8}$$

Introducing this expression into eq. (7) one obtains the following result for the first-order excitation amplitude

$$a_{I_f M_f, I_0 M_0} = \frac{4\pi Z_1 e}{i\hbar} \sum_{\lambda\mu} \frac{1}{2\lambda + 1} (-1)^{I_0 - M_0} \begin{pmatrix} I_0 & \lambda & I_f \\ -M_0 & \mu & M_f \end{pmatrix} \times \langle I_0 \| \mathcal{M}(\mathrm{E}\lambda) \| I_f \rangle S_{\mathrm{E}\lambda\mu}. \tag{9}$$

The result shows that in first order it is possible to separate the excitation amplitude in a factor which depends only on the nuclear properties, through the reduced multipole matrix element, a factor which depends on the angular momentum geometry through a 3-j symbol, and a factor which depends on the kinematics of the hyperbolic motion through the orbital integrals. The quantity λ indicates the total angular momentum transfer to the nucleus from the orbital motion. The magnetic quantum number μ is the negative value of the z-component of this angular momentum, i.e.

$$M_f - M_0 = -\mu. \tag{10}$$

§ 2. *Classical orbital integrals*

The orbital integrals defined by eq. (1.4) play an important role for the discussion of Coulomb excitation processes and we shall, therefore, consider their properties in some detail. As is seen from their definition, they depend on the energy loss and on the parameters describing the hyperbolic orbit. The orbital integrals transform under rotation of the reference system like spherical tensors. Under complex conjugation they fulfill the following relation

$$S^{*}_{\mathrm{E}\lambda\mu}(\omega) = (-1)^{\mu} S_{\mathrm{E}\lambda-\mu}(-\omega), \tag{1}$$

where we specified the dependence on the energy parameter (1.5). It is seen that $S_{\mathrm{E}\lambda\mu}$ is a non-Hermitian spherical tensor.

While the orbital integrals $S_{\mathrm{E}\lambda\mu}$ depend on the dimension as well as on the form of the hyperbolic orbit, it is possible through the introduction of the dimensionless collision function $Q_{\lambda\mu}(w)$ in (II.9.18) to separate the dependence on the size of the orbit. Introducing (II.9.10) in (1.4) we find

$$S_{\mathrm{E}\lambda\mu} = \frac{1}{a^{\lambda}v} \frac{(\lambda-1)!}{(2\lambda-1)!!} \sqrt{\frac{2\lambda+1}{\pi}}\, R_{\lambda\mu}(\vartheta, \xi), \tag{2}$$

where the orbital integral $R_{\lambda\mu}(\vartheta, \xi)$ is defined by

$$R_{\lambda\mu}(\vartheta, \xi) = \int_{-\infty}^{+\infty} Q_{\lambda\mu}(\varepsilon, w) \exp\{\mathrm{i}\xi[\varepsilon \sinh w + w]\}\, \mathrm{d}w. \tag{3}$$

These integrals depend on the scattering angle ϑ which is connected with the excentricity of the hyperbola by

$$\varepsilon = 1/\sin(\tfrac{1}{2}\vartheta) \tag{4}$$

and on the excitation energy through the parameter ξ that is defined by*

$$\xi = \frac{a\omega}{v} = \frac{a}{v}\frac{E_{\mathrm{f}} - E_{0}}{\hbar}. \tag{5}$$

For the explicit evaluation of the orbital integrals one must use a definite coordinate system and it is convenient here to use the parameterization of the

* It should be noted that for the quantities v and a appearing in eq. (5) one should use symmetrized expressions as described in § 7 and § VI.8 below.

TABLE 1

Symmetry properties of the Coulomb excitation functions $I_{\lambda\mu}$ and the orbital integrals $R_{\lambda\mu}$ in the coordinate systems A and B.

	μ-dependence	$\xi \to -\xi$
	$I_{\lambda\mu}(\vartheta, \xi) \neq 0$ for all μ	$I_{\lambda\mu}(\vartheta, \xi) = I_{\lambda-\mu}(\vartheta, -\xi)$
A	$R^{A}_{\lambda\mu}(\vartheta, \xi) = 0$ for $\lambda + \mu$ odd	$R^{A}_{\lambda\mu}(\vartheta, \xi) = (-1)^{\mu} R^{A}_{\lambda-\mu}(\vartheta, -\xi)$
B	$R^{B}_{\lambda\mu}(\vartheta, \xi) = R^{B}_{\lambda-\mu}(\vartheta, \xi)$	$R^{B}_{\lambda\mu}(\vartheta, \xi) = (-1)^{\mu} R^{B}_{\lambda\mu}(\vartheta, -\xi)$

	$\xi = 0$	$\vartheta = \pi$
	$I_{\lambda\mu}(\vartheta, 0) = I_{\lambda-\mu}(\vartheta, 0)$	$I_{\lambda\mu}(\pi, \xi)$ independent of μ
A	$R^{A}_{\lambda\mu}(\vartheta, 0) = (-1)^{\mu} R^{A}_{\lambda-\mu}(\vartheta, 0)$	$R^{A}_{\lambda\mu}(\pi, \xi) = Y_{\lambda\mu}(\frac{1}{2}\pi, 0) I_{\lambda\lambda}(\pi, \xi)$
B	$R^{B}_{\lambda\mu}(\vartheta, 0) = 0$ for μ odd	$R^{B}_{\lambda\mu}(\pi, \xi) = 0$ for $\mu \neq 0$

The columns show the characteristic properties of these quantities as a function of μ, under inversion of the sign of ξ, for $\xi = 0$ and for backwards scattering. Note that all the quantities given in the table are real.

hyperbolic orbit as indicated in § II.9. It follows from the general symmetry properties of § II.8 that the orbital integrals have especially simple properties in the coordinate systems A and B. While in general they are complex functions satisfying eq. (1) above, they are in coordinate systems A and B according to eq. (II.8.23) real numbers. Other properties which follow from the symmetries discussed in § II.8 are collected in table 1.

According to the definitions in § II.9 one finds explicitly in the coordinate system A

$$R^{A}_{\lambda\mu}(\vartheta, \xi) = \frac{(2\lambda-1)!!}{(\lambda-1)!} \sqrt{\frac{\pi}{2\lambda+1}}\, Y_{\lambda\mu}(\tfrac{1}{2}\pi, 0) I_{\lambda\mu}(\vartheta, \xi), \tag{6}$$

where

$$I_{\lambda\mu}(\vartheta, \xi) = \int_{-\infty}^{+\infty} \frac{[\cosh w + \varepsilon + \mathrm{i}\sqrt{\varepsilon^2 - 1}\sinh w]^{\mu}}{[\varepsilon \cosh w + 1]^{\lambda+\mu}} \exp\{\mathrm{i}\xi[\varepsilon \sinh w + w]\}\, \mathrm{d}w. \tag{7}$$

These functions, which we shall refer to as Coulomb excitation functions, can, in general, only be expressed in terms of complicated hypergeometric functions of two variables. They have been evaluated numerically for $\lambda \leqslant 4$ for a wide range of parameters ξ and ϑ [ALD 56a].

In appendix H we have summarized a number of mathematical properties of the Coulomb excitation functions.

For $\lambda = 1$ the Coulomb excitation functions can be expressed in terms of modified Hankel functions (see appendix H)

$$I_{1\pm1}(\vartheta, \xi) = -2\xi \exp(-\tfrac{1}{2}\pi\xi)\left[K'_{i\xi}(\xi\varepsilon) \pm \frac{\sqrt{\varepsilon^2-1}}{\varepsilon} K_{i\xi}(\xi\varepsilon)\right], \tag{8}$$

where K' represents the derivative of the function K with respect to the argument.

We notice furthermore that the Coulomb excitation functions for $\xi = 0$ can be expressed in terms of elementary functions [ALD 56a], i.e. (see appendix H)

$$I_{\lambda\mu}(\vartheta, 0) = (-1)^{\mu} \frac{2(\lambda-1)!}{(\lambda+\mu-1)!} (\mathrm{i} \tan \tfrac{1}{2}\vartheta)^{\lambda} Q^{\mu}_{\lambda-1}(\mathrm{i} \tan \tfrac{1}{2}\vartheta), \tag{9}$$

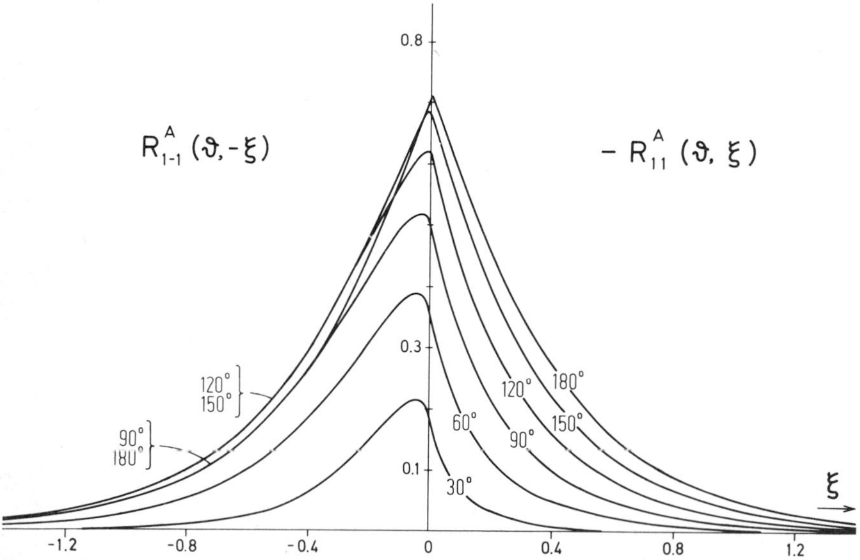

Fig. 1. The orbital integral $-R^A_{11}(\vartheta, \xi)$ in the coordinate system A as a function of ξ for different values of ϑ. The values of $R^A_{1-1}(\vartheta, \xi)$ follow from the relation $R^A_{1-1}(\vartheta, \xi) = -R^A_{11}(\vartheta, -\xi)$.

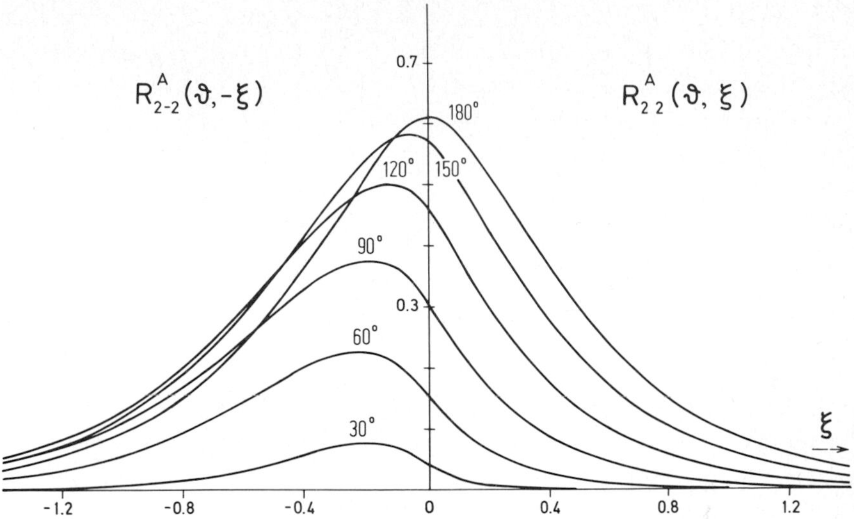

Fig. 2. The orbital integral $R^{A}_{22}(\vartheta, \xi)$ in the coordinate system A as a function of ξ for different values of ϑ. The values of $R^{A}_{2-2}(\vartheta, \xi)$ follow from the relation $R^{A}_{2-2}(\vartheta, \xi) = R^{A}_{22}(\vartheta, -\xi)$.

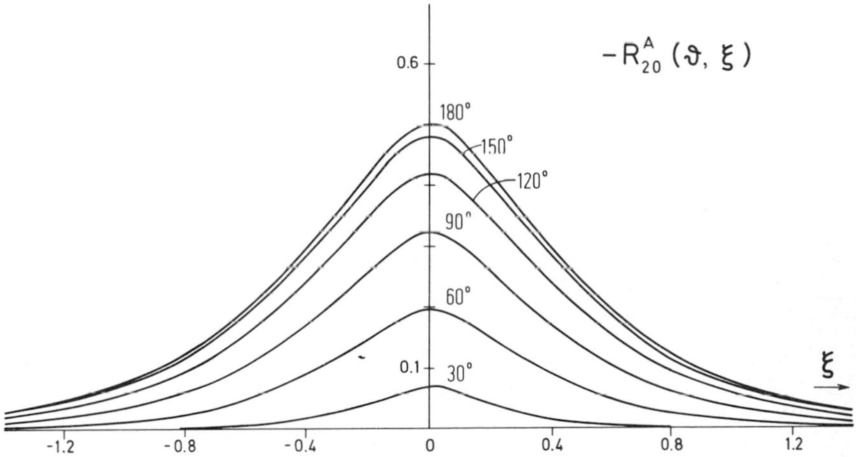

Fig. 3. The orbital integral $-R^{A}_{20}(\vartheta, \xi)$ in the coordinate system A as a function of ξ for different values of ϑ.

where Q_n^m are the associated Legendre functions of the second kind. Explicit expressions of $I_{\lambda\mu}(\vartheta, 0)$ in terms of elementary functions are given in table H.1. For the special case of $\vartheta = \pi$ we have the simple result

$$I_{\lambda\mu}(\pi, 0) = 2(\lambda - 1)!/(2\lambda - 1)!!. \qquad (10)$$

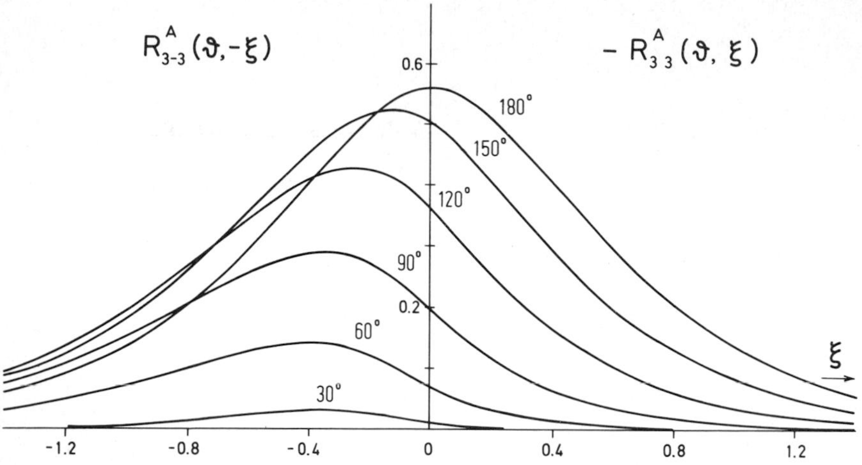

Fig. 4. The orbital integral $-R^A_{33}(\vartheta, \xi)$ in the coordinate system A as a function of ξ for different values of ϑ. The values of $R^A_{3\ -3}(\vartheta, \xi)$ follow from the relation $R^A_{3\ -3}(\vartheta, \xi) = -R^A_{33}(\vartheta, -\xi)$.

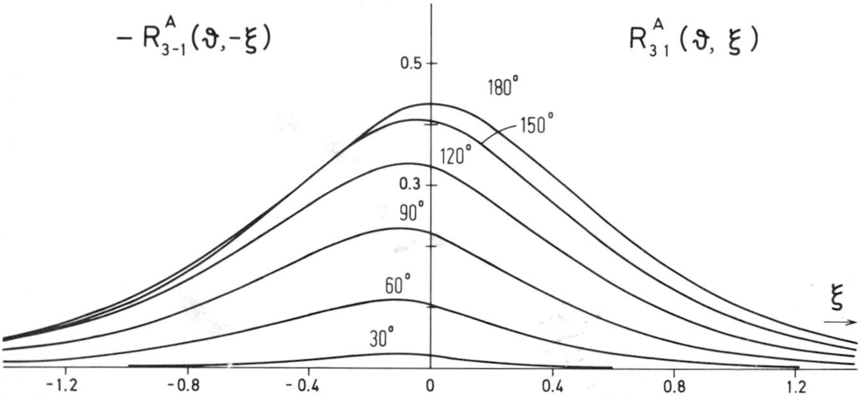

Fig. 5. The orbital integral $R^A_{31}(\vartheta, \xi)$ in the coordinate system A as a function of ξ for different values of ϑ. The values of $R^A_{3\ -1}(\vartheta, \xi)$ follow from the relation $R^A_{3\ -1}(\vartheta, \xi) = -R^A_{31}(\vartheta, -\xi)$.

For backwards scattering and $\xi \neq 0$ one finds, in accordance with table I, that $I_{\lambda\mu}(\pi, \xi)$ is independent of μ. A characteristic feature of the orbital integrals is the fact that they vanish exponentially for large values of the parameter $\varepsilon\xi$ or $\xi(\vartheta)$. As was discussed in chapter I this can be related to the adiabatic behaviour of the collision process in this limit.

The orbital integrals $R^A_{\lambda\mu}$ are illustrated in figs. 1–5 for $\lambda = 1$, 2 and 3 and for $\vartheta = 180$, 150, 120, 90, 60 and 30 degrees as functions of the parameter ξ.

It is seen from the result (10) that they are normalized such that

$$\sum_{\mu} |R^{\mathrm{A}}_{\lambda\mu}(\pi, 0)|^2 = 1. \tag{11}$$

While the orbital integrals in the coordinate system A take a relatively simple form, the coordinate system B is often more convenient, especially if one considers large scattering angles. Since the orbital integrals $S_{\lambda\mu}$ or $R_{\lambda\mu}$ transform like spherical tensors under rotation one may also, in this coordinate system, express them in terms of the Coulomb excitation functions $I_{\lambda\mu}$, i.e. according to eq. (D.11),

$$S^{\mathrm{B}}_{\mathrm{E}\lambda\mu} = \sum_{\nu} S^{\mathrm{A}}_{\mathrm{E}\lambda\nu} D^{\lambda}_{\mu\nu}(0, -\tfrac{1}{2}\pi, 0). \tag{12}$$

Inserting the definitions (2) and (6) one obtains

$$R^{\mathrm{B}}_{\lambda\mu}(\vartheta, \xi) = \frac{(2\lambda-1)!!}{(\lambda-1)!}\sqrt{\frac{\pi}{2\lambda+1}}\sum_{\nu} Y_{\lambda\nu}(\tfrac{1}{2}\pi, 0) D^{\lambda}_{\mu\nu}(0, -\tfrac{1}{2}\pi, 0) I_{\lambda\nu}(\vartheta, \xi). \tag{13}$$

In table 2 we have listed explicit relations between the integrals $R^{\mathrm{B}}_{\lambda\mu}$ and $I_{\lambda\mu}$ for $\lambda = 1, 2, 3$ and 4. The integrals $R^{\mathrm{B}}_{\lambda\mu}$ are illustrated in figs. 6–8 as functions

TABLE 2

The explicit relationship between the orbital integrals in the coordinate system B and the Coulomb excitation function for $\lambda = 1, 2, 3$ and 4.

$R^{\mathrm{B}}_{11} = \frac{1}{4}\sqrt{\frac{1}{2}}(I_{1\,-1} - I_{11})$
$R^{\mathrm{B}}_{10} = \frac{1}{4}(I_{1\,-1} + I_{11})$
$R^{\mathrm{B}}_{22} = \frac{3}{16}\sqrt{\frac{3}{2}}\{I_{2\,-2} + I_{22} - 2I_{20}\}$
$R^{\mathrm{B}}_{21} = \frac{3}{8}\sqrt{\frac{3}{2}}(I_{2\,-2} - I_{22})$
$R^{\mathrm{B}}_{20} = \frac{3}{16}\{3(I_{2\,-2} + I_{22}) + 2I_{20}\}$
$R^{\mathrm{B}}_{33} = \frac{15}{128}\sqrt{5}\{(I_{3\,-3} - I_{33}) - 3(I_{3\,-1} - I_{31})\}$
$R^{\mathrm{B}}_{32} = \frac{15}{64}\sqrt{\frac{15}{2}}\{I_{3\,-3} + I_{33} - I_{3\,-1} - I_{31}\}$
$R^{\mathrm{B}}_{31} = \frac{15}{128}\sqrt{3}\{5(I_{3\,-3} - I_{33}) + (I_{3\,-1} - I_{31})\}$
$R^{\mathrm{B}}_{30} = \frac{15}{64}\{5(I_{3\,-3} + I_{33}) + 3(I_{3\,-1} + I_{31})\}$
$R^{\mathrm{B}}_{44} = \frac{35}{512}\sqrt{\frac{35}{2}}\{I_{4\,-4} + I_{44} - 4(I_{4\,-2} + I_{42}) + 6I_{40}\}$
$R^{\mathrm{B}}_{43} = \frac{35}{256}\sqrt{35}\{I_{4\,-4} - I_{44} - 2(I_{4\,-2} - I_{42})\}$
$R^{\mathrm{B}}_{42} = \frac{35}{256}\sqrt{\frac{5}{2}}\{7(I_{4\,-4} + I_{44}) - 4(I_{4\,-2} + I_{42}) - 6I_{40}\}$
$R^{\mathrm{B}}_{41} = \frac{35}{256}\sqrt{5}\{7(I_{4\,-4} - I_{44}) - 2(I_{4\,-2} - I_{42})\}$
$R^{\mathrm{B}}_{40} = \frac{35}{512}\{35(I_{4\,-4} + I_{44}) + 20(I_{4\,-2} + I_{42}) + 18I_{40}\}$

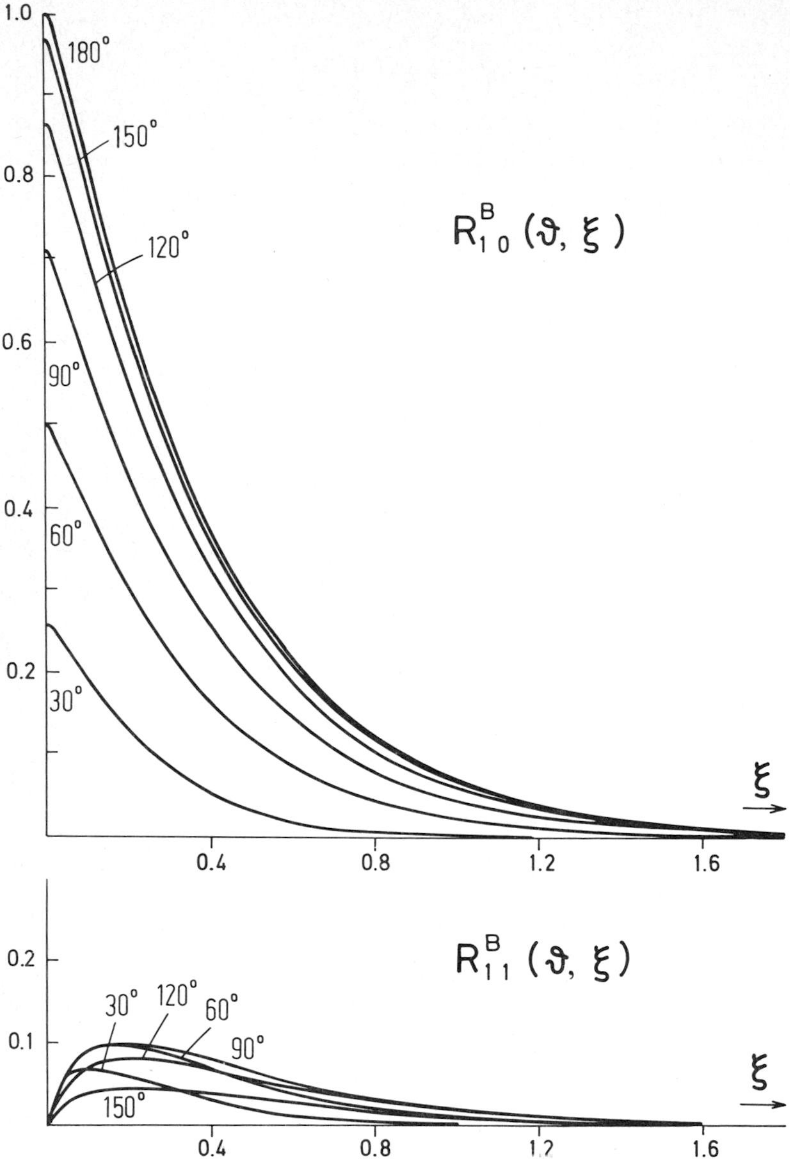

Fig. 6. The orbital integrals $R^B_{1\mu}(\vartheta, \xi)$ in the coordinate system B as functions of ξ for different values of ϑ. The function R^B_{10} is symmetric while the function R^B_{11} is anti-symmetric in ξ.

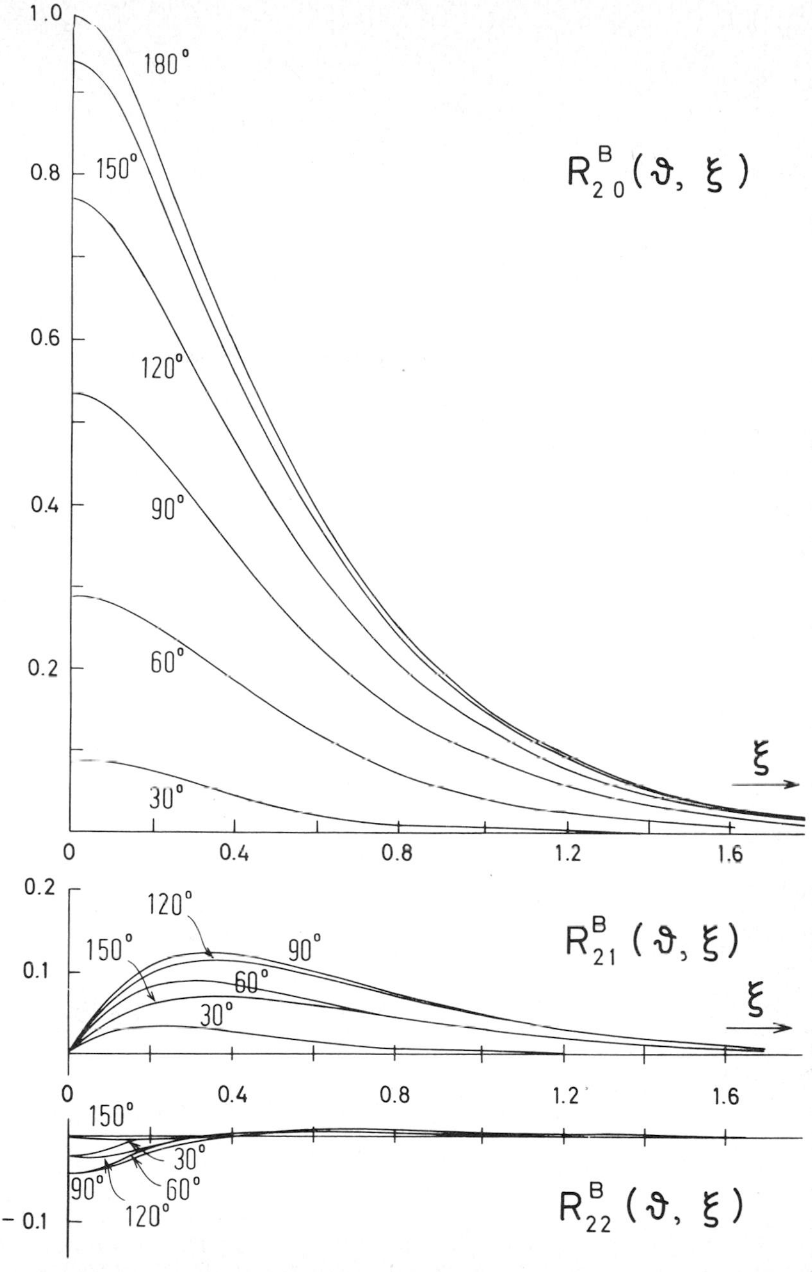

Fig. 7. The orbital integrals $R^{B}_{2\mu}(\vartheta, \xi)$ in the coordinate system B as functions of ξ for different values of ϑ. The functions R^{B}_{20} and R^{B}_{22} are symmetric in ξ while the function R^{B}_{21} is antisymmetric in ξ.

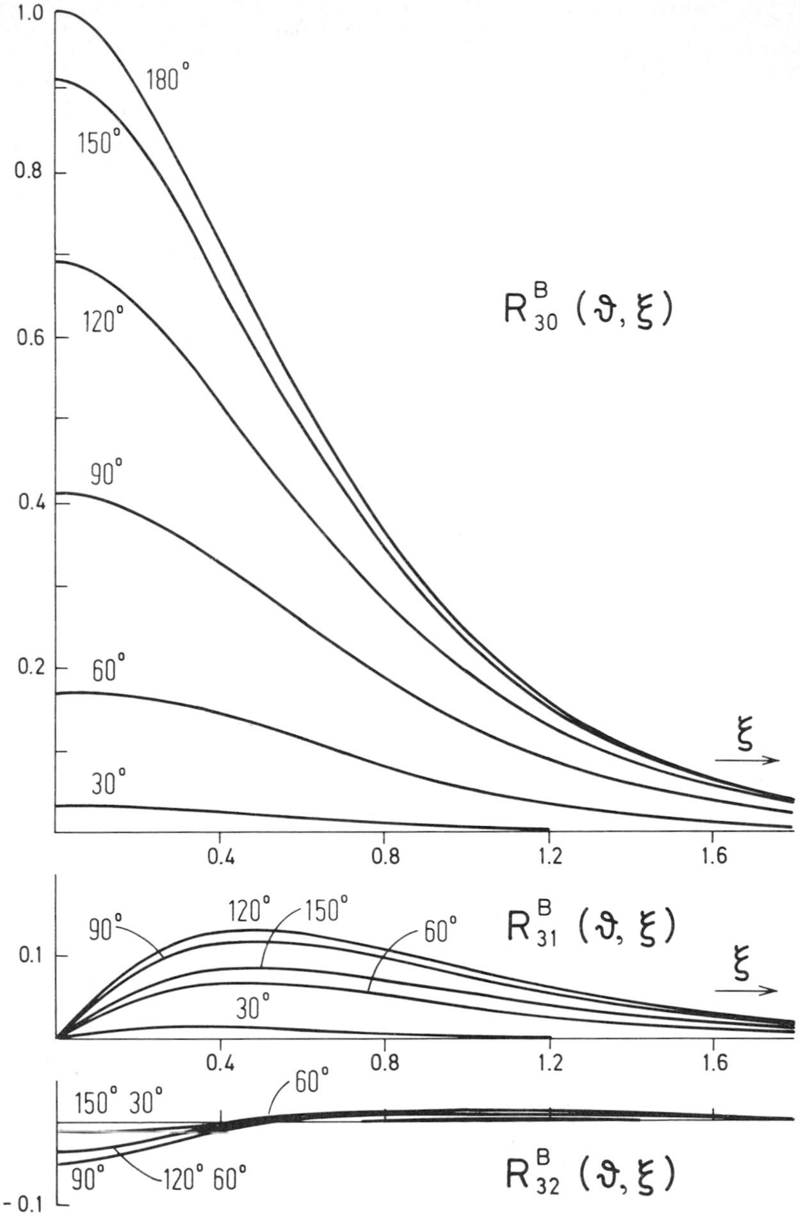

Fig. 8. The orbital integrals $R^B_{3\mu}(\vartheta, \xi)$ in the coordinate system B as functions of ξ for different values of ϑ. The functions R^B_{30} and R^B_{32} are symmetric in ξ while the functions R^B_{33} and R^B_{31} are antisymmetric in ξ. The functions R^B_{33} are left out because they do not deviate sufficiently from zero to show up in the drawing with the present scale.

of the parameter ξ for $\lambda = 1$, 2 and 3 for $\vartheta = 180$, 150, 120, 90, 60 and 30 degrees.

The maximum value of $R^{\mathrm{B}}_{\lambda 0}$ is obtained for $\vartheta = \pi$ and $\xi = 0$, where $R^{\mathrm{B}}_{\lambda\mu}$, in accordance with table 1, vanishes for $\mu \neq 0$. The magnitude of $R^{\mathrm{B}}_{\lambda 0}$ can be found from the normalization (11) which holds in any coordinate system. One thus finds

$$R^{\mathrm{B}}_{\lambda\mu}(\pi, 0) = \delta_{\mu 0}. \tag{14}$$

It is noted that especially for small values of ξ and for angles larger than 90 degrees the integral $R^{\mathrm{B}}_{\lambda 0}$ with $\mu = 0$ gives the dominating contribution, by far.

§ 3. *Excitation probabilities*

As was seen in the last section the rather complicated dependence of the orbital integrals $S_{\mathrm{E}\lambda\mu}$ on the kinematics of the hyperbolic motion can be isolated into a dimensionless integral $R_{\lambda\mu}$ which depends only on the scattering angle ϑ and the parameter ξ. The magnitude of the orbital integrals is especially simple to overlook in the coordinate system B where the approximate conservation of the angular momentum along the symmetry axis leads to the dominance of the integrals $R^{\mathrm{B}}_{\lambda\mu}$ for $\mu = 0$. For other purposes the coordinate system A may be more convenient. In terms of the orbital integrals the excitation amplitudes may be written, according to eqs. (1.9) and (2.2)

$$a_{I_{\mathrm{f}}M_{\mathrm{f}},I_0M_0} = \frac{4\pi Z_1 e}{i\hbar v} \sum_{\lambda\mu} \frac{1}{a^{\lambda}} \frac{(\lambda-1)!}{(2\lambda+1)!!} \sqrt{\frac{2\lambda+1}{\pi}} (-1)^{I_0 - M_0}$$
$$\times \begin{pmatrix} I_0 & \lambda & I_{\mathrm{f}} \\ -M_0 & \mu & M_{\mathrm{f}} \end{pmatrix} \langle I_0 \| \mathscr{M}(\mathrm{E}\lambda) \| I_{\mathrm{f}} \rangle \, R_{\lambda\mu}(\vartheta, \xi). \tag{1}$$

From this expression one may derive the excitation probability (III.1.2) of the state of spin I_{f}. Utilizing the orthogonality properties of the 3-j symbols one obtains

$$P_{\mathrm{f}} = \sum_{\lambda} |\chi^{(\lambda)}_{0\to\mathrm{f}}|^2 R^2_{\lambda}(\vartheta, \xi), \tag{2}$$

where we have separated the dependence on ϑ, ξ through the definition

$$R^2_{\lambda}(\vartheta, \xi) = \sum_{\mu} |R_{\lambda\mu}(\vartheta, \xi)|^2. \tag{3}$$

This function which is independent of the coordinate system can be expressed in terms of the Coulomb excitation functions by using eq. (2.6), i.e.

$$R_\lambda^2(\vartheta, \xi) = \left|\frac{(2\lambda-1)!!}{(\lambda-1)!}\right|^2 \frac{\pi}{2\lambda+1} \sum_\mu |Y_{\lambda\mu}(\tfrac{1}{2}\pi, 0) I_{\lambda\mu}(\vartheta, \xi)|^2 . \tag{4}$$

It is seen from eq. (2.11) that $R_\lambda^2(\vartheta, \xi)$ is normalized such that

$$R_\lambda^2(\pi, 0) = 1 . \tag{5}$$

Equation (2) implies an accurate definition of the strength parameter $\chi_{\mathrm{f}0}^{(\lambda)}$ which we have used several times earlier for more qualitative discussions. The exact definition is*

$$\chi_{\mathrm{f}0}^{(\lambda)} = \chi_{0\to\mathrm{f}}^{(\lambda)} = \frac{\sqrt{16\pi}(\lambda-1)!}{(2\lambda+1)!!} \frac{Z_1 e}{\hbar v} \frac{\langle I_0 \| \mathscr{M}(\mathrm{E}\lambda) \| I_\mathrm{f}\rangle}{a^\lambda \sqrt{2I_0+1}} . \tag{6}$$

The square of the strength parameter $|\chi_{0\to\mathrm{f}}^{(\lambda)}|^2$ measures the λ-pole excitation probability of the state $|\mathrm{f}\rangle$ for $\vartheta = \pi$ and $\xi = 0$ in accordance with the definitions in (I.27). The quantity $R_\lambda^2(\vartheta, \xi)$ thus measures the excitation probability for multipole order λ relative to the case of $\vartheta = \pi$ and $\xi = 0$. These relative probabilities are tabulated in table 3.

In the following we shall always assume that the phase convention (II.8.5) is used and that therefore the multipole matrix elements and the strength parameters are real quantities. The phase convention, however, does not determine the sign of the matrix element. As was mentioned in appendix E we shall often assume that the matrix element of the lowest multipolarity connecting a given state with the ground state is positive.

Inserting eq. (6) in (1) we may write the excitation amplitude in the following form

$$a_{I_\mathrm{f}M_\mathrm{f}, I_0 M_0} = -\mathrm{i} \sum_\lambda (-1)^{I_0 - M_0} \sqrt{2I_0+1}\sqrt{2\lambda+1} \begin{pmatrix} I_0 & \lambda & I_\mathrm{f} \\ -M_0 & \mu & M_\mathrm{f} \end{pmatrix}$$

$$\times\ \chi_{0\to\mathrm{f}}^{(\lambda)} R_{\lambda\mu}(\vartheta, \xi) . \tag{7}$$

While the quantity $\chi_{0\to\mathrm{f}}^{(\lambda)}$ measures the λ-pole strength of the coupling of the states $|0\rangle$ and $|\mathrm{f}\rangle$ in a collision of $\vartheta = \pi$ (and $\xi = 0$), the coupling for other scattering angles is measured by the parameter $\chi_{0\to\mathrm{f}}^{(\lambda)}(\vartheta)$ defined by

$$\chi_{0\to\mathrm{f}}^{(\lambda)}(\vartheta) = \chi_{0\to\mathrm{f}}^{(\lambda)} R_\lambda(\vartheta, 0) \tag{8}$$

in accordance with eqs. (I.25–26). One may formally generalize the χ-

* It should be noted that for the quantities v and a appearing in eq. (6) one should use symmetrized expressions as described in § 7 and § VI.8 below.

parameter to include the dependence on ξ by the definition

$$\chi^{(\lambda)}_{0\to f}(\vartheta, \xi) = \chi^{(\lambda)}_{0\to f} R_\lambda(\vartheta, \xi), \tag{9}$$

but must realize that the decrease of this parameter for increasing ξ does not correspond to a decrease in the strength of the interaction but rather to a gradual transition to an adiabatic situation. Another generalization of the strength parameter $\chi^{(\lambda)}$ is to consider its dependence on the magnetic quantum

TABLE 3

The relative excitation probabilities $R^2_\lambda(\vartheta, \xi)$ for $\lambda = 1, 2, 3$ and 4 as functions of ϑ and ξ. (These quantities are independent of the choice of the coordinate system.)

$R^2_1(\vartheta, \xi)$

ξ \ ϑ	180	160	140	120	100	80	60	40	20
0.0	1.0000	0.9698	0.8830	0.7500	0.5868	0.4132	0.2500	0.1170	0.0302
0.05	0.8232	0.7989	0.7291	0.6218	0.4895	0.3477	0.2129	0.1011	0.0258
0.1	0.6573	0.6387	0.5849	0.5017	0.3979	0.2850	0.1754	0.0822	0.0182
0.2	0.4024	0.3921	0.3617	0.3137	0.2518	0.1815	0.1102	0.0475	0.0068
0.4	0.1394	0.1365	0.1277	0.1126	0.0913	0.0649	0.0364	0.0121	0.0006
0.6	0.0458	0.0451	0.0427	0.0380	0.0308	0.0211	0.0106	0.0026	—
0.8	0.0146	0.0144	0.0138	0.0124	0.0099	0.0065	0.0029	0.0005	—
1.0	0.0046	0.0045	0.0044	0.0039	0.0031	0.0019	0.0007	0.0001	—
1.2	0.0014	0.0014	0.0014	0.0012	0.0010	0.0006	0.0002	—	—
1.4	0.0004	0.0004	0.0004	0.0004	0.0003	0.0002	—	—	—
1.6	0.0001	0.0001	0.0001	0.0001	0.0001	—	—	—	—

$R^2_2(\vartheta, \xi)$

ξ \ ϑ	180	160	140	120	100	80	60	40	20
0.0	1.0000	0.9464	0.7998	0.5974	0.3864	0.2080	0.0860	0.0224	0.0019
0.05	0.9650	0.9141	0.7744	0.5810	0.3782	0.2054	0.0859	0.0228	0.0020
0.1	0.8897	0.8443	0.7193	0.5448	0.3593	0.1985	0.0848	0.0231	0.0021
0.2	0.6959	0.6639	0.5745	0.4460	0.3040	0.1743	0.0772	0.0214	0.0017
0.4	0.3516	0.3397	0.3046	0.2492	0.1801	0.1085	0.0486	0.0120	0.0004
0.6	0.1539	0.1507	0.1400	0.1198	0.0900	0.0550	0.0234	0.0046	0.0001
0.8	0.0620	0.0615	0.0590	0.0523	0.0402	0.0243	0.0096	0.0015	—
1.0	0.0236	0.0237	0.0234	0.0214	0.0166	0.0098	0.0035	0.0004	—
1.2	0.0086	0.0088	0.0089	0.0083	0.0065	0.0037	0.0012	0.0001	—
1.4	0.0031	0.0032	0.0033	0.0031	0.0024	0.0013	0.0004	—	—
1.6	0.0011	0.0011	0.0012	0.0011	0.0009	0.0005	0.0001	—	—
1.8	0.0004	0.0004	0.0004	0.0004	0.0003	0.0002	—	—	—
2.0	0.0001	0.0001	0.0001	0.0001	0.0001	—	—	—	—

Continued overleaf.

TABLE 3 (*continued*)

$R_3^2(\vartheta, \xi)$

ξ \ ϑ	180	160	140	120	100	80	60	40	20
0.0	1.0000	0.9234	0.7273	0.4810	0.2604	0.1094	0.0318	0.0048	0.0001
0.05	0.9862	0.9122	0.7190	0.4769	0.2592	0.1094	0.0320	0.0048	0.0001
0.1	0.9483	0.8786	0.6960	0.4653	0.2556	0.1092	0.0323	0.0050	0.0002
0.2	0.8239	0.7676	0.6181	0.4242	0.2409	0.1069	0.0329	0.0053	0.0002
0.4	0.5185	0.4911	0.4143	0.3044	0.1872	0.0899	0.0294	0.0047	0.0001
0.6	0.2775	0.2679	0.2379	0.1873	0.1234	0.0624	0.0206	0.0029	—
0.8	0.1334	0.1314	0.1227	0.1026	0.0711	0.0368	0.0117	0.0014	—
1.0	0.0595	0.0598	0.0585	0.0515	0.0370	0.0192	0.0058	0.0005	—
1.2	0.0250	0.0257	0.0262	0.0241	0.0177	0.0091	0.0025	0.0002	—
1.4	0.0101	0.0105	0.0112	0.0107	0.0080	0.0040	0.0010	—	—
1.6	0.0039	0.0042	0.0046	0.0045	0.0034	0.0017	0.0004	—	—
1.8	0.0015	0.0016	0.0018	0.0019	0.0014	0.0007	0.0001	—	—
2.0	0.0006	0.0006	0.0007	0.0007	0.0006	0.0003	—	—	—

$R_4^2(\vartheta, \xi)$

ξ \ ϑ	180	160	140	120	100	80	60	40	20
0.0	1.0000	0.9031	0.6625	0.3893	0.1776	0.0589	0.0122	0.0011	—
0.05	0.9919	0.8962	0.6584	0.3877	0.1774	0.0590	0.0123	0.0011	—
0.1	0.9682	0.8761	0.6464	0.3831	0.1776	0.0593	0.0124	0.0011	—
0.2	0.8825	0.8029	0.6016	0.3650	0.1732	0.0597	0.0128	0.0012	—
0.4	0.6316	0.5849	0.4606	0.3004	0.1547	0.0578	0.0133	0.0013	—
0.6	0.3883	0.3679	0.3079	0.2179	0.1221	0.0491	0.0118	0.0011	—
0.8	0.2135	0.2074	0.1850	0.1414	0.0851	0.0360	0.0087	0.0007	—
1.0	0.1079	0.1076	0.1020	0.0836	0.0532	0.0232	0.0055	0.0004	—
1.2	0.0511	0.0523	0.0525	0.0457	0.0304	0.0134	0.0030	0.0002	—
1.4	0.0229	0.0241	0.0256	0.0234	0.0161	0.0071	0.0015	—	—
1.6	0.0099	0.0106	0.0119	0.0114	0.0080	0.0035	0.0007	—	—
1.8	0.0041	0.0045	0.0053	0.0053	0.0038	0.0016	0.0003	—	—
2.0	0.0016	0.0019	0.0023	0.0023	0.0017	0.0007	0.0001	—	—

number. The quantity

$$\chi_{0\to f}^{(\lambda\mu)}(\vartheta) = \chi_{0\to f}^{(\lambda)} R_{\lambda\mu}(\vartheta, 0) \tag{10}$$

measures the interaction strength where an angular momentum $\lambda\hbar$ with z-component $-\mu\hbar$ is transferred. In the coordinate system B it is seen from figs. 6–8 that except for very forward scattering the couplings with $\mu \neq 0$ are much smaller than those for $\mu = 0$. This approximate conservation of the magnetic quantum number for small ξ is an important qualitative feature of the Coulomb excitation process (see also eqs. (II.8.24) and (III.3.11)).

§ 4. *Cross section and angular distribution*

From the excitation amplitudes (3.1) and (3.7) one may easily compute the cross sections and the angular distributions of de-excitation γ-quanta. The differential cross section (III.1.1) for electric excitations is conveniently written in the form

$$\mathrm{d}\sigma = \sum_{\lambda} \mathrm{d}\sigma_{\mathrm{E}\lambda}, \tag{1}$$

where

$$\mathrm{d}\sigma_{\mathrm{E}\lambda} = \tfrac{1}{4}a^2|\chi_{0\to\mathrm{f}}^{(\lambda)}|^2 R_{\lambda}^2(\vartheta, \xi)\, \mathrm{d}\Omega/\sin^4(\tfrac{1}{2}\vartheta). \tag{2}$$

The cross section may alternatively [ALD 56] be written*

$$\mathrm{d}\sigma_{\mathrm{E}\lambda} = \left(\frac{Z_1 e}{\hbar v}\right)^2 a^{-2\lambda+2} B(\mathrm{E}\lambda, I_0 \to I_\mathrm{f})\, \mathrm{d}f_{\mathrm{E}\lambda}(\vartheta, \xi). \tag{3}$$

In this expression $B(\mathrm{E}\lambda, I_0 \to I_\mathrm{f})$ is the reduced transition probability associated with the radiative transition of multipole order Eλ. It is related to the nuclear matrix elements of the electric multipole operator by the formula

$$\begin{aligned} B(\mathrm{E}\lambda, I_0 \to I_\mathrm{f}) &= \sum_{\mu M_\mathrm{f}} |\langle I_0 M_0|\, \mathcal{M}(\mathrm{E}\lambda, \mu)\, |I_\mathrm{f} M_\mathrm{f}\rangle|^2 \\ &= \frac{1}{2I_0 + 1} |\langle I_0 \|\, \mathcal{M}(\mathrm{E}\lambda)\, \| I_\mathrm{f}\rangle|^2. \end{aligned} \tag{4}$$

The differential cross section function $\mathrm{d}f_{\mathrm{E}\lambda}$ is thus defined by

$$\mathrm{d}f_{\mathrm{E}\lambda}(\vartheta,\xi) = 4\pi \left|\frac{(\lambda-1)!}{(2\lambda+1)!!}\right|^2 R_{\lambda}^2(\vartheta, \xi) \sin^{-4}(\tfrac{1}{2}\vartheta)\, \mathrm{d}\Omega. \tag{5}$$

The functions $\mathrm{d}f_{\mathrm{E}\lambda}/\mathrm{d}\Omega$ have been extensively tabulated in [ALD 56a] and the results are illustrated in fig. 9. One notices how the forward scattering cross section decreases for increasing values of ξ. This can be seen explicitly for the case of dipole excitations where there exists an analytic expression for the differential cross section in terms of modified Bessel functions. This expression is obtained by inserting eq. (2.8) into eq. (5), i.e.

$$\mathrm{d}f_{\mathrm{E}1}(\vartheta,\xi) = \tfrac{4}{9}\pi\xi^2\varepsilon^4\, \mathrm{e}^{-\pi\xi}\left[|K'_{\mathrm{i}\xi}(\varepsilon\xi)|^2 + \frac{\varepsilon^2-1}{\varepsilon^2}|K_{\mathrm{i}\xi}(\varepsilon\xi)|^2\right]\mathrm{d}\Omega \tag{6}$$

* It should be noted that for the quantities v and a, appearing in eq. (3), one should use symmetrized expressions as discussed in § IV.7 and § 8 below.

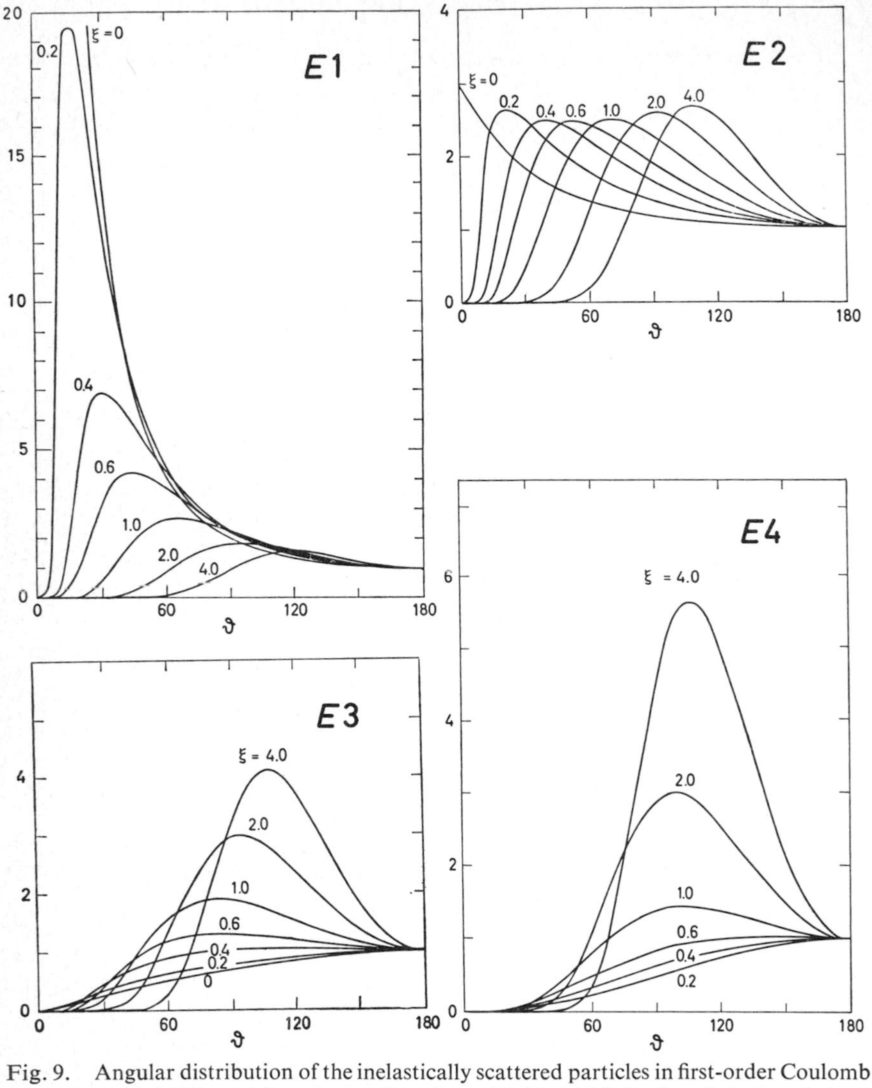

Fig. 9. Angular distribution of the inelastically scattered particles in first-order Coulomb excitation. The differential excitation functions $\mathrm{d}f_{\mathrm{E}\lambda}(\vartheta, \xi)/\mathrm{d}\Omega$ for $\lambda = 1, 2, 3$ and 4 are plotted as functions of ϑ for different values of ξ. The curves have been normalized to unity at $\vartheta = 180°$. The absolute values can be obtained from the table below.

	$\xi = 0$	$\xi = 0.2$	$\xi = 0.4$	$\xi = 0.6$	$\xi = 1.0$	$\xi = 2.0$	$\xi = 4.0$
$\mathrm{d}f_{\mathrm{E1}}(\pi, \xi)/\mathrm{d}\Omega$	1.40(0)	5.62(−1)	1.95(−1)	6.40(−2)	6.39(−3)	1.66(−5)	8.41(−11)
$\mathrm{d}f_{\mathrm{E2}}(\pi, \xi)/\mathrm{d}\Omega$	5.58(−2)	3.89(−2)	1.96(−2)	8.60(−3)	1.32(−3)	6.76(−6)	7.37(−11)
$\mathrm{d}f_{\mathrm{E3}}(\pi, \xi)/\mathrm{d}\Omega$	4.56(−3)	3.76(−3)	2.36(−3)	1.26(−3)	2.71(−4)	2.51(−6)	5.58(−11)
$\mathrm{d}f_{\mathrm{E4}}(\pi, \xi)/\mathrm{d}\Omega$	5.16(−4)	4.47(−4)	3.20(−4)	1.97(−4)	5.46(−5)	8.39(−7)	3.57(−11)

The entry is given by a number and the power of ten by which it should be multiplied.

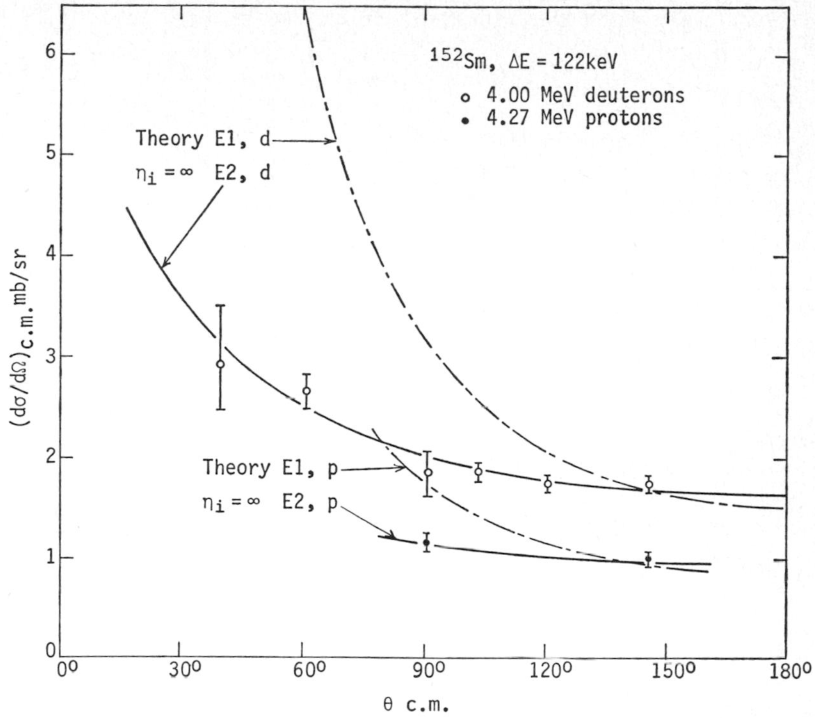

Fig. 10. Angular distributions of 4.00 MeV deuterons and 4.27 MeV protons inelastically scattered from ^{152}Sm. The theoretical E2 deuteron curve is a least-squares fit to the experimental points. The theoretical E2 proton curve is an absolute prediction of the theory calculated from the deuteron results. To illustrate the sensitiveness of the angular distribution to the multipole order of the excitation, the theoretical E1 curves are also shown. These are both normalized to the deuteron cross section at the back angle (from Bernstein, E. M. and E. Z. Skurnik, 1961, Phys. Rev. **121**, 841).

from which it follows that

$$\mathrm{d}f_{\mathrm{E1}}\,(\vartheta,\xi) \approx \tfrac{4}{9}\pi^2\xi\varepsilon^3 \exp\{-2\xi(\varepsilon + \tfrac{1}{2}\pi)\}\,\mathrm{d}\Omega \qquad \text{for } \varepsilon \gg 1. \tag{7}$$

An example of an experimental verification of (5) is given in fig. 10.

The total electric excitation cross section for a state $|\mathrm{f}\rangle$ may, in analogy to eq. (3) be written in the form

$$\sigma = \sum_{\lambda} \sigma_{\mathrm{E}\lambda}, \tag{8}$$

with

$$\sigma_{\mathrm{E}\lambda} = \left(\frac{Z_1 e}{\hbar v}\right)^2 a^{-2\lambda+2} B(\mathrm{E}\lambda, I_0 \to I_{\mathrm{f}}) f_{\mathrm{E}\lambda}(\xi), \tag{9}$$

where the total cross section function $f_{E\lambda}(\xi)$ is given by

$$f_{E\lambda}(\xi) = \int_{\Omega} df_{E\lambda}(\vartheta, \xi). \tag{10}$$

These functions have also been extensively tabulated [ALD 56a] and the results are illustrated in fig. 11 as a function of ξ in a logarithmic scale.

For the case of $\lambda = 1$, the function $f_{E1}(\xi)$ can again be evaluated in terms of modified Bessel functions. It thus follows from (6) that

$$f_{E1}(\xi) = -\tfrac{32}{9}\pi^2 \, e^{-\pi\xi} \xi K_{i\xi}(\xi) K'_{i\xi}(\xi). \tag{11}$$

This explicit expression may e.g. be used to study the typical behaviour of the Coulomb excitation cross section for large values of ξ. Thus one finds [ALD 56]

$$f_{E1}(\xi) \approx \frac{32\pi^3}{9\sqrt{3}} e^{-2\pi\xi}. \tag{12}$$

We may express the cross sections $\sigma_{E\lambda}$ or $d\sigma_{E\lambda}$ in terms of the parameter χ and the cross section functions $f_{E\lambda}$ or $df_{E\lambda}$, i.e.

$$\sigma_{E\lambda} = a^2 |\chi_{0\to f}^{(\lambda)}|^2 \frac{[(2\lambda+1)!!]^2}{16\pi[(\lambda-1)!]^2} f_{E\lambda}(\xi). \tag{13}$$

Comparing with fig. 11 it is seen that the order of magnitude of $\sigma_{E\lambda}$ for small ξ is $a^2|\chi^{(\lambda)}|^2$.

The different energy dependences of the total cross sections (13) for E1 and E2 excitations are illustrated in figs. 12 and 13.

In a similar way as the multipolarity of the excitation process can be determined from the energy dependence of the yield, one may determine the excitation energy. This can be a useful tool in assigning the position of a de-excitation gamma line in a level scheme, as is illustrated in figs. 14 a-c. To obtain thick-target yields, the cross sections have been integrated (see § X.5).

For the evaluation of the angular distribution of de-excitation γ-quanta it is practical first to compute the statistical tensors $\rho_{k\kappa}$ introduced in chapter III. These quantities take an especially simple form in the first-order perturbation theory. Introducing the excitation amplitudes (3.7) into the expression (III.2.11) for the statistical tensor one obtains

$$\rho_{k\kappa} = \sqrt{2I_f+1}(-1)^{I_0+I_f} \sum_{\lambda\lambda'} \chi_{0\to f}^{(\lambda)} \chi_{0\to f}^{(\lambda')} \begin{Bmatrix} \lambda & \lambda' & k \\ I_f & I_f & I_0 \end{Bmatrix}$$
$$\times \sqrt{(2\lambda+1)(2\lambda'+1)} \sum_{\mu\mu'} (-1)^{\mu'} \begin{pmatrix} \lambda & \lambda' & k \\ \mu & -\mu' & \kappa \end{pmatrix} R^*_{\lambda\mu} R_{\lambda'\mu'}. \tag{14}$$

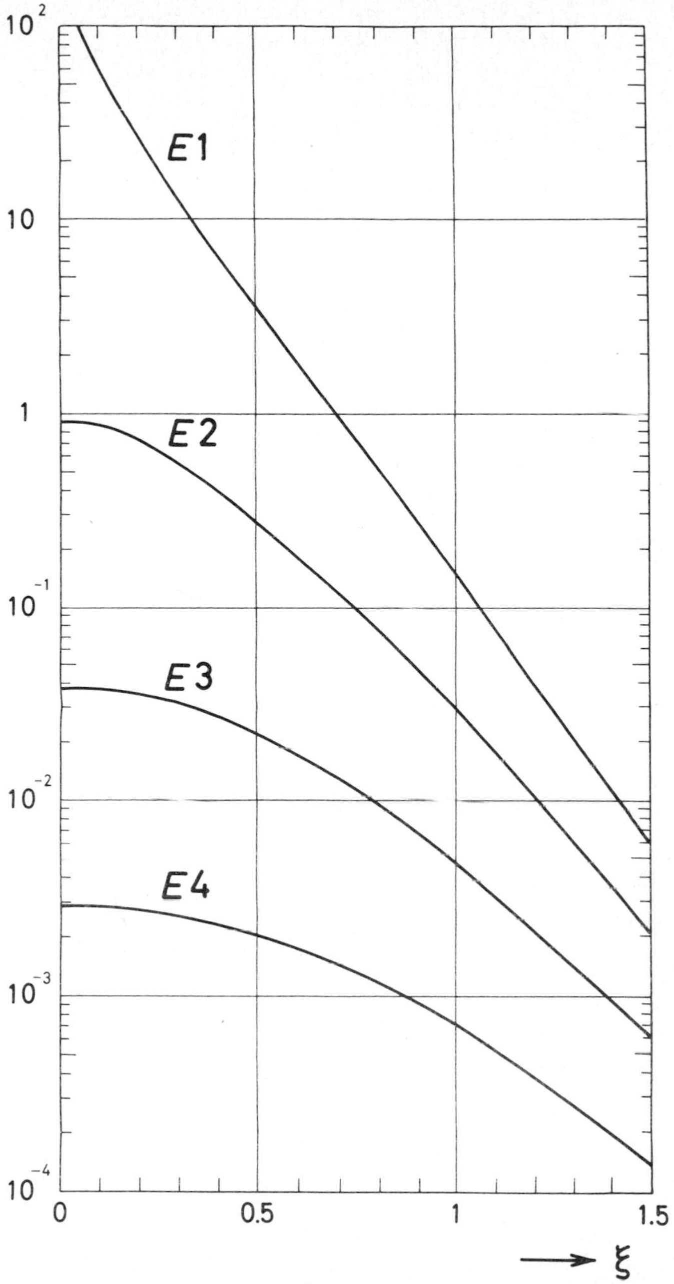

Fig. 11. The total cross section functions $f_{E\lambda}(\xi)$ for $\lambda = 1, 2, 3$ and 4 as functions of ξ in a logarithmic scale.

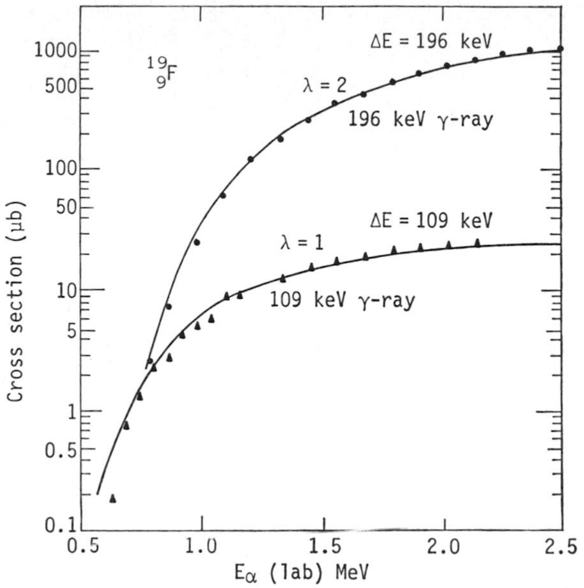

Fig. 12. Excitation function for levels in ^{19}F. The figure shows the measured excitation cross sections for the 109 keV and 196 keV γ-rays observed in α bombardment of a thin target of CaF_2 (Sherr, Li and Christy, 1954, Phys. Rev. **96**, 1258). The theoretical excitation functions given by the full drawn curves are obtained by assuming E1 excitation with $\Delta E = 109$ keV for the 109 keV γ-ray. The excitation functions are not sensitive to the multipole order, but the assumed values of λ are those indicated by other experimental evidence. The theoretical curves are normalized to the experimental cross sections at $E_\alpha = 1.55$ MeV.

We have in the usual way reduced the sum over the magnetic quantum numbers of the product of three 3-j symbols to a 6-j symbol. It is seen that $\rho_{k\kappa}$ is a real number in the coordinate systems A and B where $R_{\lambda\mu}$ is real. It follows furthermore from the symmetry relations quoted in table III.1 that in coordinate system B

$$\rho^{\mathrm{B}}_{k\kappa} = 0 \qquad \text{for } k + \kappa \text{ odd}. \tag{15}$$

In the coordinate system A the statistical tensors $\rho_{k\kappa}$ fulfill the relation

$$\rho^{\mathrm{A}}_{k\kappa} = 0 \qquad \text{for } \kappa \text{ odd}. \tag{16}$$

From the spherical tensors (14) we may compute directly the vector polarization of the nucleus after the excitation. According to eq. (III.2.18) one finds from either expression (15) or (16) that the polarization is perpendicular to

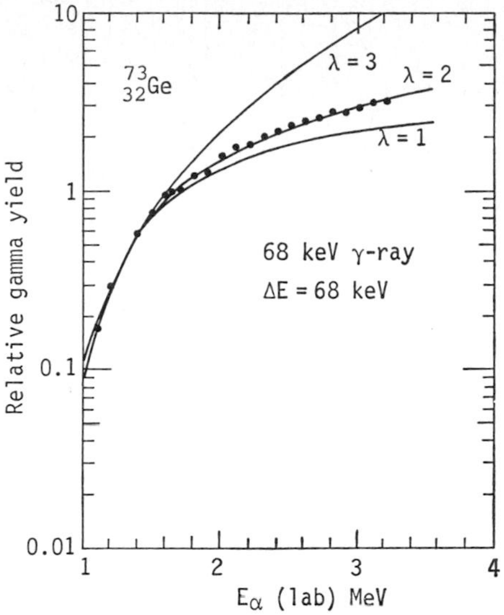

Fig. 13. Dependence of excitation function on multipole order. The figure shows the thin target yield of the 68 keV γ-ray observed in a bombardment of ^{73}Ge (Temmer, G. M. and N. P. Heydenburg, 1954, Phys. Rev. **96**, 426). The full drawn curves give the theoretical excitation functions for E1, E2 and E3 Coulomb excitation, assuming $\Delta E = 68$ keV. The curves are normalized to the experimental value at 1.4 MeV. The possibility of distinguishing in the present case between E1 and E2 excitation on the basis of the yield function is associated with the rather small ξ values for the excitation ($\xi = 0.14$ for $E_\alpha = 3$ MeV).

the plane of the orbit. For the magnitude we find in coordinate system A

$$P_z = \sqrt{(I_f + 1)/I_f}\, \rho_{10}/\rho_{00}, \tag{17}$$

which according to eqs. (14) and (2.6) for a pure multipole excitation may be written explicitly in terms of the Coulomb excitation functions as

$$P_z = \sqrt{\frac{(2I_f + 1)(I_f + 1)}{I_f}}\,(-1)^{I_0 + I_f + \lambda + 1}(2\lambda + 1) \times \begin{Bmatrix} \lambda & \lambda & 1 \\ I_f & I_f & I_0 \end{Bmatrix} \begin{pmatrix} \lambda & \lambda & 1 \\ \lambda & -\lambda & 0 \end{pmatrix} \Pi_{E\lambda}(\vartheta, \xi), \tag{18}$$

where

$$\Pi_{E\lambda}(\vartheta, \xi) = \frac{\sum_\mu (-1)^{\lambda+\mu+1} \begin{pmatrix} \lambda & \lambda & 1 \\ \mu & -\mu & 0 \end{pmatrix} |Y_{\lambda\mu}(\tfrac{1}{2}\pi, 0) I_{\lambda\mu}(\vartheta, \xi)|^2}{\begin{pmatrix} \lambda & \lambda & 1 \\ \lambda & -\lambda & 0 \end{pmatrix} \sum_\mu |Y_{\lambda\mu}(\tfrac{1}{2}\pi, 0) I_{\lambda\mu}(\vartheta, \xi)|^2}. \tag{19}$$

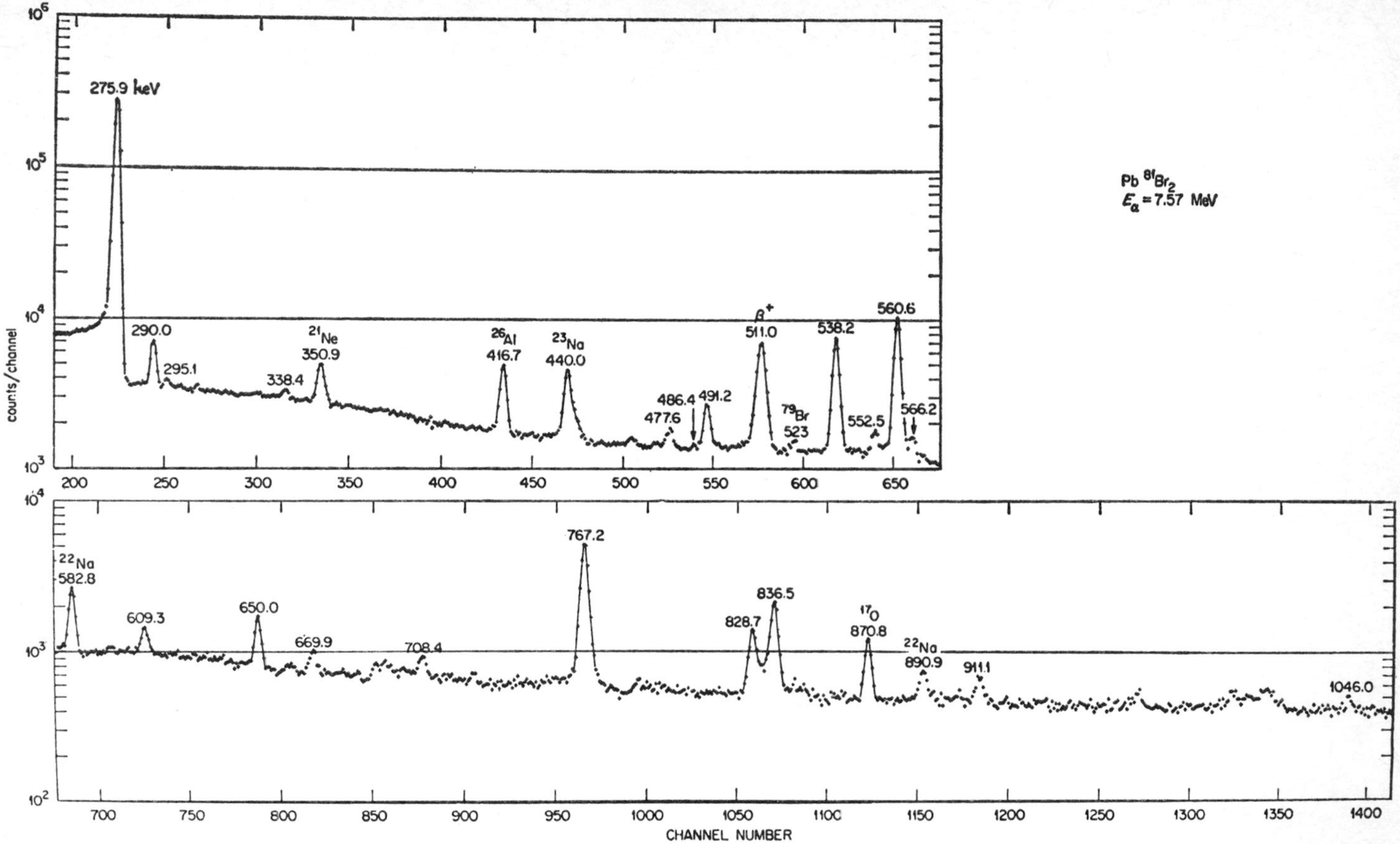

Fig. 14a. Gamma-ray spectrum of ^{81}Br resulting from bombardment of a thick target with 7.57 MeV α-particles (from Robinson, R. L. et al., 1972, Nucl. Phys. **A193**, 14).

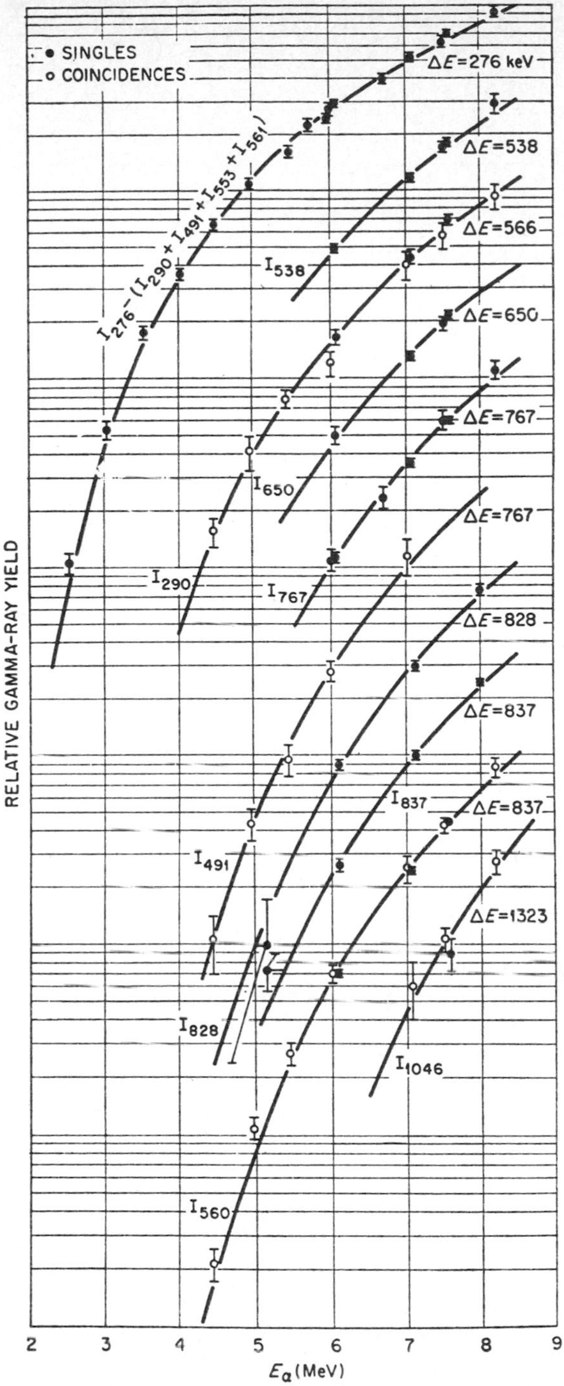

Fig. 14b. Relative yields of the γ-rays deduced from singles and coincident spectra as a function of the α-particle energy. The experimental points are the yields I_E for the γ-ray of energy E and the solid curves are theoretical predictions for E2 Coulomb excitation for states of energy ΔE (from Robinson, R. L. et al., 1972, Nucl. Phys. **A193**, 14).

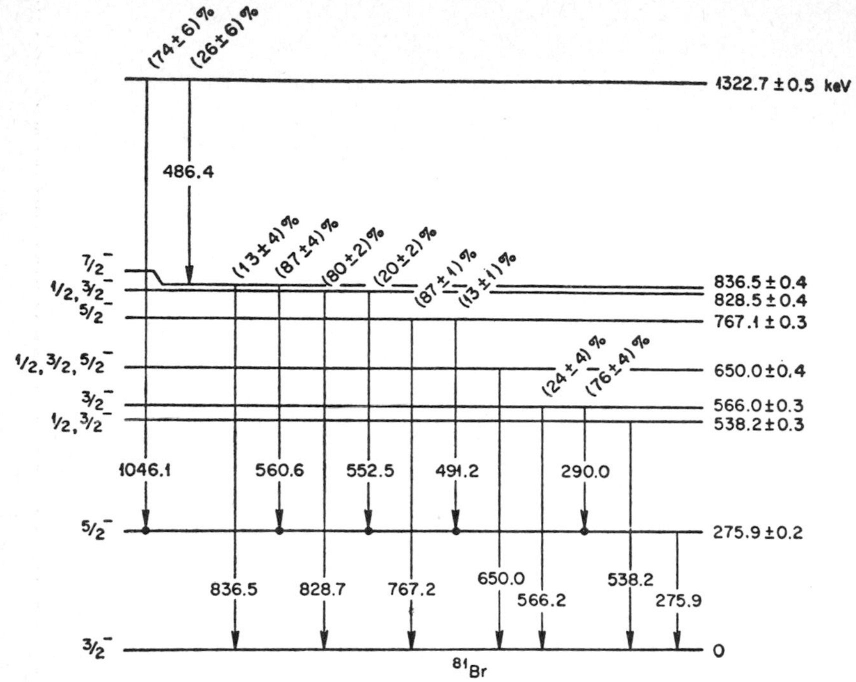

Fig. 14c. Levels Coulomb excited in ^{81}Br. The position of the γ-rays terminating at a dot were verified with coincident studies. The numbers above the γ-rays give relative branching ratios obtained in the work (from Robinson, R. L. et al., 1972, Nucl. Phys. **A193**, 14).

The polarization function $\Pi_{E\lambda}(\vartheta, \xi)$ is illustrated in fig. 15 for $\lambda = 1$ and 2 as a function of ϑ for a few values of ξ. It is normalized in such a way that for forwards scattering, where $I_{\lambda-\lambda}$ dominates (see e.g. appendix H), one finds

$$\lim_{\vartheta \to 0} \Pi_{E\lambda}(\vartheta, \xi) = 1 . \tag{20}$$

It is noted that for even-even nuclei with $I_0 = 0$ and $I_f = \lambda$ one finds that

$$P_z = \Pi_{E\lambda}(\vartheta, \xi) . \tag{21}$$

In this case there is, according to (20), a complete polarization for forward scattering. The direction of the polarization is in the positive z-direction as is to be expected from a simple classical picture.

As was seen in chapter III the statistical tensors $\rho_{2\kappa}$ can be illustrated geometrically in terms of the cartesian tensor polarization P_{ij}. The components of P_{ij} can be evaluated explicitly from eq. (III.2.22). It is seen that

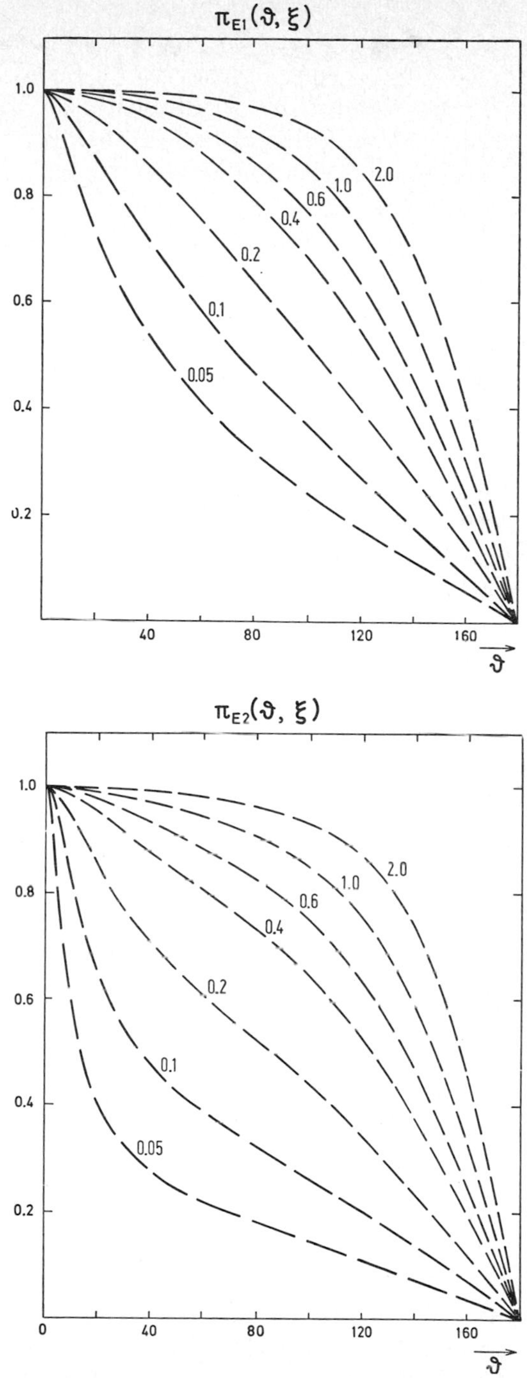

Fig. 15. The polarization function $\Pi_{E\lambda}(\vartheta, \xi)$ defined in eq. (4.19) for $\lambda = 1$ and 2. The polarizations are plotted as functions of ϑ for different values of ξ. In the limit $\xi = 0$ they approach the step function.

in both coordinate systems, A and B, the elements P_{xy}, P_{yz} and P_{xz} vanish. This means that the polarization ellipsoid in first-order perturbation theory has its three main axes along the coordinate axes. We shall not indicate any explicit expression, but note that in the limit for backwards scattering the ellipsoid has rotational symmetry around the z-axis in coordinate system B. Usually this main axis is much larger than the two others. Actually, for ground state spin $I_0 = 0$ the ellipsoid degenerates into a cylinder corresponding to the fact that for backwards scattering $M = 0$, i.e. $P_{zz} = 0$ in coordinate system B. For decreasing scattering angle the main axis of the ellipsoid along the y-axis increases, while the main axis along the symmetry axis of the hyperbolic orbit decreases. Finally, for very small scattering angles, these two axes become equal, corresponding to the fact that in this limit one has a complete vector polarization perpendicular to the scattering plane. Roughly speaking, the polarization ellipsoid thus changes from a cigar for backwards scattering to a pancake for forwards scattering.

In the perturbation theory the spherical tensor takes a form which is similar to the spherical tensor after a γ-decay (see appendix G). It is in such cases convenient to write the angular distribution in terms of a γ-γ correlation modified by a particle parameter. In the coordinate system C where the z-axis is chosen along the direction of the incoming projectile the result (III.3.6) may thus be written

$$W^{\mathrm{C}} = \sum_{\substack{k\,\mathrm{even}\\ \kappa}} \left\{ \sum_{\lambda\lambda'} F_k(\lambda\lambda' I_0 I_{\mathrm{f}}) \chi^{(\lambda)}_{0\to\mathrm{f}} \chi^{(\lambda')}_{0\to\mathrm{f}} b^{\lambda,\lambda'}_{k\kappa} \right\}$$

$$\times \left\{ \sum_{LL'} F_k(LL' I_{\mathrm{ff}} I_{\mathrm{f}}) \delta_L \delta^*_{L'} \right\} Y_{k\kappa}(\vartheta^{\mathrm{C}}_\gamma, \varphi^{\mathrm{C}}_\gamma) \tag{22}$$

where the particle parameter $b^{\lambda;\lambda'}_{k\kappa}$ for mixed λ, λ' excitations is given by

$$b^{\lambda,\lambda'}_{k\kappa} = -(2k+1)^{-1/2} \begin{pmatrix} \lambda & \lambda' & k \\ 1 & -1 & 0 \end{pmatrix}^{-1} \sum_{\substack{\mu\mu' \\ \kappa'\,\mathrm{even}}} (-1)^{\mu'} \begin{pmatrix} \lambda & \lambda' & k \\ \mu & -\mu' & \kappa' \end{pmatrix}$$

$$\times R^{\mathrm{B}}_{\lambda\mu} R^{\mathrm{B}}_{\lambda'\mu'} D^k_{\kappa'\kappa} \left(\tfrac{1}{2}\pi, \tfrac{1}{2}(\pi+\vartheta), \pi\right) \tag{23}$$

in terms of the orbital integrals $R^{\mathrm{B}}_{\lambda\mu}$ in coordinate system B. If one wants to express the particle parameter in terms of the orbital integrals $R^{\mathrm{A}}_{\lambda\mu}$ in coordinate system A, the same formula (23) applies, except that the Eulerian angles in the D-function should be replaced by $\frac{1}{2}\pi + \frac{1}{2}\vartheta$, $\frac{1}{2}\pi$, $-\frac{1}{2}\pi$.

In most cases of practical interest only one multipole order will be of importance and the sum over λ and λ' reduces to one term. For this case the particle parameter b has been evaluated in terms of the Coulomb excitation functions $I_{\lambda\mu}$ in [ALD 56].

In the case where one uses a ring counter for the detection of the scattered particles or where one does not observe the scattered particles at all, an integration over $\varphi_\gamma^{\mathrm{C}}$ should be made, which leads to $\kappa = 0$. It is then convenient to write the angular distribution (22) in the normalized form

$$W^{\mathrm{C}}(\vartheta, \theta_\gamma) = \sum_{k \text{ even}} F_k(\lambda\lambda I_0 I_{\mathrm{f}}) a_{k0}^{\mathrm{E}\lambda}(\vartheta, \xi) \times \frac{1}{\gamma_{\mathrm{f}\to\mathrm{ff}}} \left\{ \sum_{LL'} F_k(LL'I_{\mathrm{ff}}I_{\mathrm{f}})\delta_L \delta_{L'}^* \right\} P_k(\cos\theta_\gamma), \tag{24}$$

where the normalized particle parameters are given by

$$a_{k0}^{\mathrm{E}\lambda}(\vartheta, \xi) = b_{k0}^{(\lambda,\lambda)}(\vartheta, \xi)/b_{00}^{(\lambda,\lambda)}(\vartheta, \xi). \tag{25}$$

In figs. 16 and 17 we have illustrated the coefficients $a_{k0}^{\mathrm{E}\lambda}(\vartheta, \xi)$ for a pure dipole and quadrupole excitation as a function of the deflection angle ϑ in a ring counter for various values of ξ.

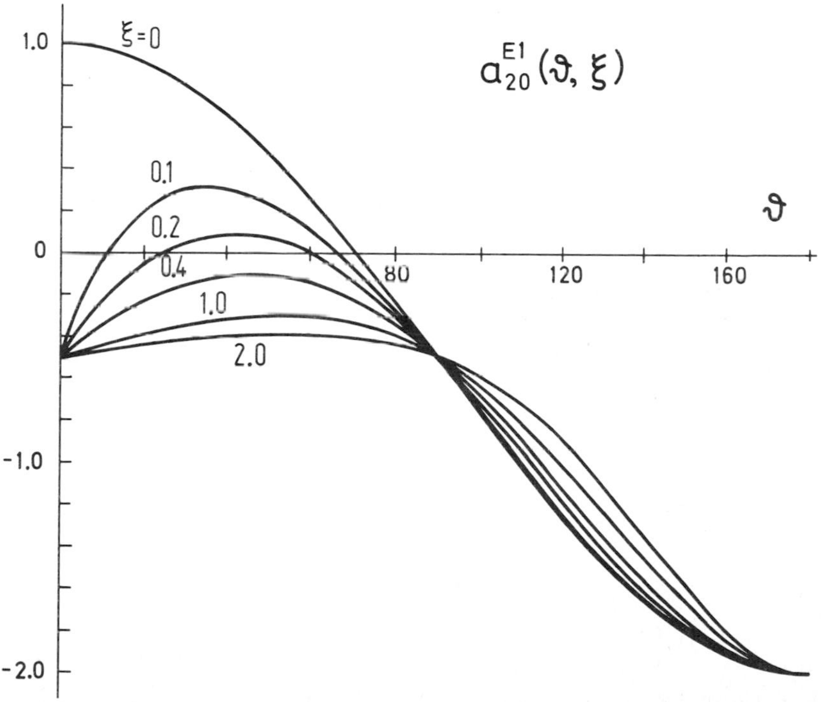

Fig. 16. The normalized particle parameter $a_{20}^{\mathrm{E}1}(\vartheta, \xi)$ for dipole excitation. The parameter which is defined in eq. (4.25) is given as a function of ϑ for different values of ξ.

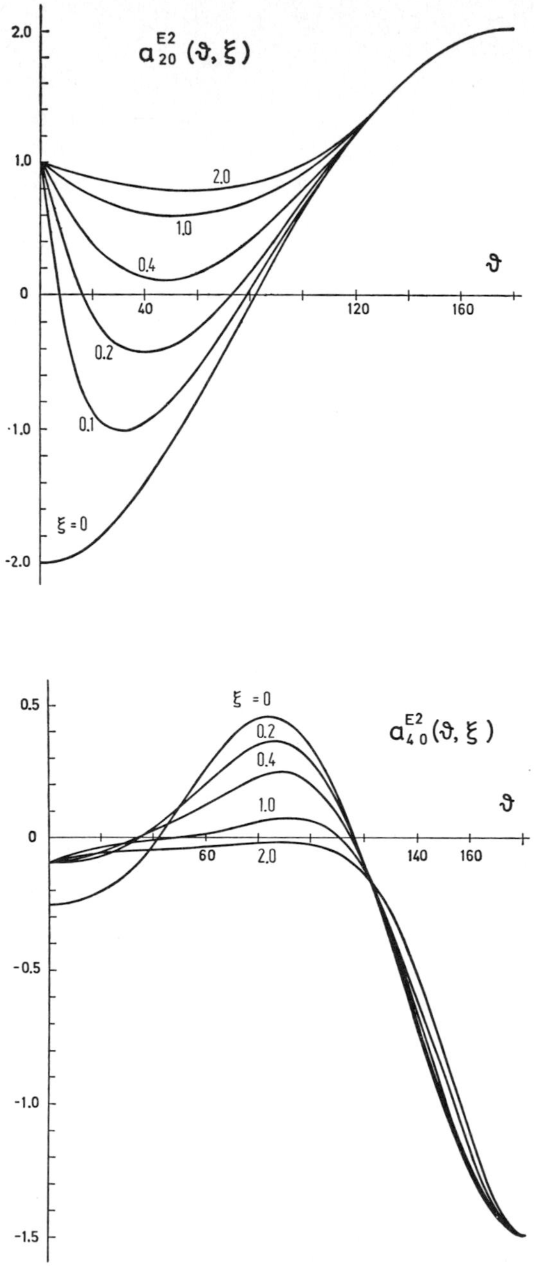

Fig. 17. The normalized particle parameters $a^{E2}_{k0}(\vartheta, \xi)$ for quadrupole excitation. The parameters which are defined in eq. (4.25) are given as functions of ϑ for different values of ξ.

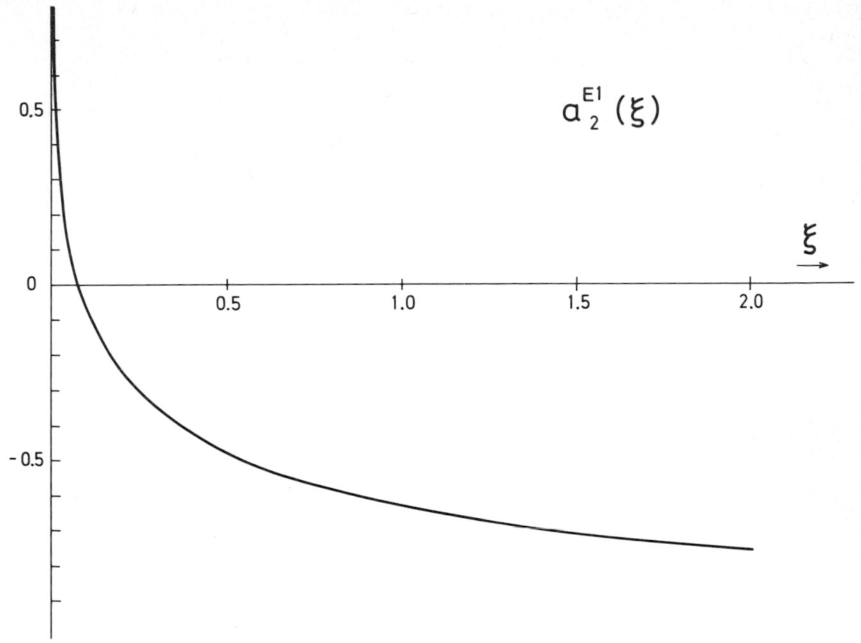

Fig. 18. The normalized particle parameter $a_2^{E1}(\xi)$ for dipole excitation. The parameter which is defined in eqs. (4.27–28) is plotted as a function of ξ.

For a finite opening of the particle counter one must integrate the expression (22) together with the Rutherford cross section. We shall write down explicitly the result for the angular distribution of de-excitation γ-rays following an inelastic scattering where the scattered projectile is not observed. In first-order perturbation theory this angular distribution, according to eq. (III.3.9) for a pure multipole excitation, may be written in the form

$$W(\theta_\gamma) = \sum_{k\,\text{even}} F_k(\lambda\lambda I_0 I_f) a_k^{E\lambda}(\xi) \times \frac{1}{\gamma_{f\to ff}} \left\{ \sum_{LL'} F_k(LL'I_{ff}I_f)\delta_L \delta_{L'}^* \right\} P_k(\cos\theta_\gamma), \tag{26}$$

where the normalized particle parameters $a_k^{E\lambda}(\xi)$ are given by

$$a_k^{E\lambda}(\xi) = b_k^{(\lambda)}(\xi)/b_0^{(\lambda)}(\xi) \tag{27}$$

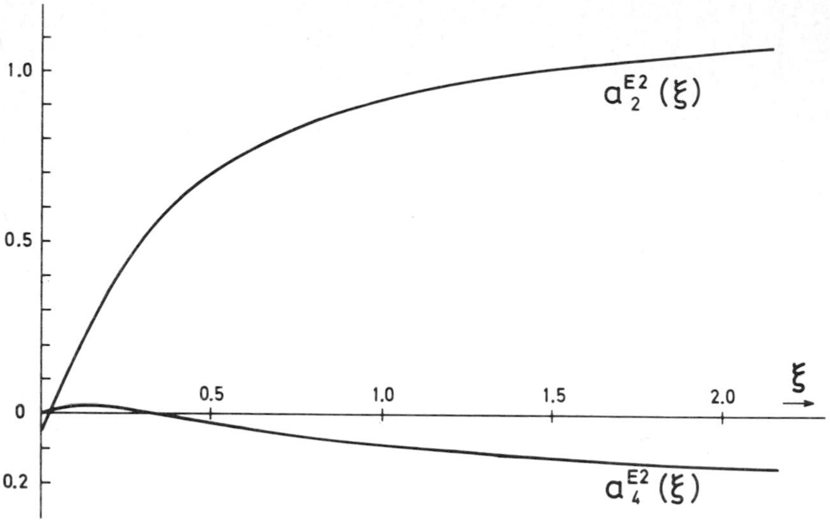

Fig. 19. The normalized particle parameters $a_k^{E2}(\xi)$ for quadrupole excitation. The parameters which are defined in eqs. (4.27–28) are plotted as functions of ξ.

with

$$b_k^{(\lambda)}(\xi) = \begin{pmatrix} \lambda & \lambda & k \\ 1 & -1 & 0 \end{pmatrix}^{-1} \int_0^{\pi} \frac{d\vartheta \cos(\frac{1}{2}\vartheta)}{\sin^3(\frac{1}{2}\vartheta)}$$

$$\times \sum_{\substack{\mu\mu' \\ \kappa \text{ even}}} (-1)^{\mu'+\kappa/2} D_{\kappa 0}^{k}(0, \tfrac{1}{2}(\pi+\vartheta), 0) R_{\lambda\mu}^{B} R_{\lambda\mu'}^{B} \begin{pmatrix} \lambda & \lambda & k \\ \mu & -\mu' & \kappa \end{pmatrix}. \tag{28}$$

These parameters $a_k^{E\lambda}$ have been evaluated numerically in [ALD 56] and the results for $\lambda = 1$ and 2 are illustrated in figs. 18 and 19 as functions of ξ.

§ 5. *Magnetic excitation*

In a Coulomb excitation process the excitation is mainly caused by the interaction between the electric monopole moment of the projectile and the electric multipole moments of the target nucleus. In this section we shall consider the much weaker excitation that is caused by the magnetic field created by the moving charge of the projectile, while the subsequent section will contain a discussion of the excitation that may be caused by higher multipole terms in eq. (II.1.7). The order of magnitude of the amplitudes, e.g. for a magnetic dipole excitation, is v/c times the amplitude for electric quadrupole excitation, if the corresponding nuclear matrix elements are both of the order of the single particle value. Under special conditions the effect of the magnetic excitation may therefore be detected, especially in experiments where the interference between electric and magnetic excitations can be studied.

The contribution of the magnetic interaction to the excitation amplitude is in first-order perturbation theory given by

$$a_{I_f M_f, I_0 M_0} = \frac{4\pi Z_1 e}{i\hbar} \sum_{\lambda\mu} \frac{(-1)^\mu}{2\lambda+1} \langle I_f M_f | \mathscr{M}(\mathrm{M}\lambda, -\mu) | I_0 M_0 \rangle S_{\mathrm{M}\lambda\mu}, \tag{1}$$

where the orbital integral $S_{\mathrm{M}\lambda\mu}$ is defined by

$$S_{\mathrm{M}\lambda\mu} = \int_{-\infty}^{+\infty} \bar{S}_{\mathrm{M}\lambda\mu}(t)\, e^{i\omega t}\, dt. \tag{2}$$

The collision function $\bar{S}_{\mathrm{M}\lambda\mu}(t)$ is given by the expressions (II.2.18) or (II.9.23–24).

The magnetic multipole matrix elements occuring in (1) are real numbers if we use the phase convention for the nuclear state vectors that was discussed earlier (see appendix E). Furthermore, since

$$(-1)^\mu \mathscr{M}(\mathrm{M}\lambda, -\mu) = \mathscr{M}(\mathrm{M}\lambda, \mu)^\dagger \tag{3}$$

we find

$$a_{I_f M_f, I_0 M_0} = \frac{4\pi Z_1 e}{i\hbar} \sum_{\lambda\mu} \frac{1}{2\lambda+1} \langle I_0 M_0 | \mathscr{M}(\mathrm{M}\lambda, \mu) | I_f M_f \rangle S_{\mathrm{M}\lambda\mu}. \tag{4}$$

Because, in general, the magnetic excitation amplitude (4) interferes with the corresponding electric excitation amplitude (1.7) it is convenient to introduce a notation which is analogous to (3.7). We thus define a dimensionless strength parameter $\chi_{0\to f}^{(\mathrm{M}\lambda)}$ and dimensionless orbital integrals $R_{\mathrm{M}\lambda\mu}$ in the following way

$$\chi_{f0}^{(\mathrm{M}\lambda)} = \chi_{0\to f}^{(\mathrm{M}\lambda)} = \frac{\sqrt{16\pi} Z_1 e}{\hbar c} \frac{(\lambda-1)!}{(2\lambda+1)!!} \frac{\langle I_0 \| \mathscr{M}(\mathrm{M}\lambda) \| I_f \rangle}{a^\lambda \sqrt{2I_0+1}}, \tag{5}$$

and

$$S_{\mathrm{M}\lambda\mu} = \frac{1}{a^\lambda c} \frac{(\lambda-1)!}{(2\lambda-1)!!} \sqrt{\frac{2\lambda+1}{\pi}}\, R_{\mathrm{M}\lambda\mu}(\vartheta, \xi), \tag{6}$$

where in comparison with the corresponding definitions in eqs. (2.2) and (3.6) for the electric excitations, the velocity v has been substituted by c.

In § II.9 we introduced a dimensionless collision function $Q_{\mathrm{M}\lambda\mu}(w)$.

According to eqs. (II.9.27–28) one finds from (6)

$$R_{M\lambda\mu}(\vartheta, \xi) = \int_{-\infty}^{+\infty} Q_{M\lambda\mu}(\varepsilon, w) \exp\{i\xi[\varepsilon \sinh w + w]\} \, dw. \tag{7}$$

Introducing the reduced matrix element of the magnetic multipole operator (see (1.8)) and the definitions (5) and (6) into (4) the magnetic excitation amplitude takes the form

$$a_{I_f M_f, I_0 M_0} = -i \sum_{\lambda} (-1)^{I_0 - M_0} \sqrt{(2I_0 + 1)(2\lambda + 1)} \times \begin{pmatrix} I_0 & \lambda & I_f \\ -M_0 & \mu & M_f \end{pmatrix} \chi_{0\to f}^{(M\lambda)} R_{M\lambda\mu}(\vartheta, \xi). \tag{8}$$

For the explicit evaluation of the orbital integrals $R_{M\lambda\mu}$ it is convenient to use the coordinate system A where they can be easily expressed in terms of the Coulomb excitation functions. According to eqs. (6), (2.2) and (II.9.26)

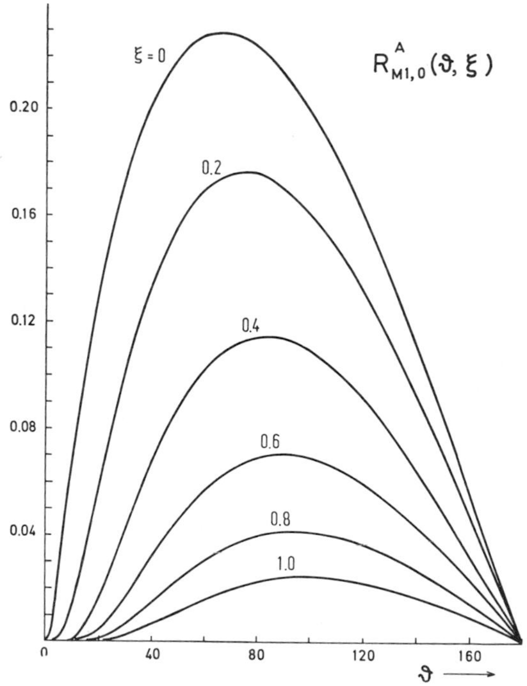

Fig. 20. The orbital integral $R^{A}_{M10}(\vartheta, \xi)$ for magnetic dipole excitation in the coordinate system A. The integral is plotted as a function of ϑ for different values of ξ.

one finds

$$R^{\mathrm{A}}_{\mathrm{M}\lambda\mu}(\vartheta, \xi) = \frac{(2\lambda-1)!!}{\lambda!}\sqrt{\frac{\pi}{2\lambda+3}}\,[(\lambda+1)^2-\mu^2]^{1/2}$$

$$\times\ Y_{\lambda+1,\mu}(\tfrac{1}{2}\pi, 0) I_{\lambda+1,\mu}(\vartheta, \xi) \cot(\tfrac{1}{2}\vartheta). \tag{9}$$

Since the orbital integrals $R_{\mathrm{M}\lambda\mu}$ transform like spherical tensors of order λ, μ they can also be easily evaluated in coordinate system B. Similar to eq. (2.12) one finds

$$R^{\mathrm{B}}_{\mathrm{M}\lambda\mu} = \sum_{\nu} R^{\mathrm{A}}_{\mathrm{M}\lambda\nu} D^{\lambda}_{\mu\nu}(0, -\tfrac{1}{2}\pi, 0). \tag{10}$$

In the coordinate systems A and B the orbital integrals are, according to (9) and (10), real quantities. It is noted that in the coordinate system A, $R_{\mathrm{M}\lambda\mu}$ vanishes for even values of $\lambda+\mu$, which means that for dipole excitation only the term with $\mu = 0$ contributes. The orbital integral $R^{\mathrm{A}}_{\mathrm{M}10}$ is illustrated in fig. 20 as a function of ϑ for some values of ξ.

In complete analogy to (3.2) the magnetic excitation probability may be written

$$P_{\mathrm{f}} = \sum_{\lambda} |\chi^{(\mathrm{M}\lambda)}_{0\to\mathrm{f}}|^2 R^2_{\mathrm{M}\lambda}(\vartheta, \xi), \tag{11}$$

where the relative excitation probabilities are defined by

$$R^2_{\mathrm{M}\lambda}(\vartheta, \xi) = \sum_{\mu} |R_{\mathrm{M}\lambda\mu}(\vartheta, \xi)|^2. \tag{12}$$

This quantity is independent of the choice of the coordinate system and one finds from (9) the following explicit expression in terms of the Coulomb excitation functions

$$R^2_{\mathrm{M}\lambda}(\vartheta, \xi) = \frac{\pi}{2\lambda+3}\left(\frac{(2\lambda-1)!!}{\lambda!}\right)^2 \sum_{\mu} [(\lambda+1)^2-\mu^2]$$

$$\times\ |Y_{\lambda+1,\mu}(\tfrac{1}{2}\pi, 0)|^2 |I_{\lambda+1,\mu}(\vartheta, \xi)|^2 \cot^2(\tfrac{1}{2}\vartheta). \tag{13}$$

The relative magnetic probabilities are normalized differently from the corresponding electric ones, since they vanish for backwards scattering. Instead one finds according to (2.10) that

$$\lim_{\vartheta\to\pi} \frac{R^2_{\mathrm{M}\lambda}(\vartheta, 0)}{\cot^2(\frac{1}{2}\vartheta)} = \frac{\lambda(\lambda+1)}{2(2\lambda+1)^2}. \tag{14}$$

The functions $R^2_{\mathrm{M}\lambda}$ are given in table 4 for $\lambda = 1$ and 2 for a few scattering

TABLE 4

The relative excitation probabilities $R^2_{M\lambda}(\vartheta, \xi)$ for magnetic dipole and quadrupole excitation as functions of ϑ and ξ. (These quantities are independent of the choice of the coordinate system.)

ξ \ ϑ	180	160	140	120	100	80	60	40	20
					$R^2_{M1}(\vartheta, \xi)$				
0.0	0.0000	0.0034	0.0127	0.0260	0.0401	0.0505	0.0521	0.0409	0.0177
0.2	0.0000	0.0023	0.0087	0.0175	0.0261	0.0310	0.0287	0.0177	0.0033
0.4	0.0000	0.0012	0.0043	0.0084	0.0120	0.0131	0.0104	0.0049	0.0003
0.6	0.0000	0.0005	0.0018	0.0035	0.0047	0.0048	0.0032	0.0010	—
0.8	0.0000	0.0002	0.0007	0.0013	0.0017	0.0016	0.0009	0.0002	—
1.0	0.0000	0.0001	0.0003	0.0005	0.0006	0.0005	0.0003	—	—
1.2	0.0000	0.0000	0.0001	0.0002	0.0002	0.0002	0.0001	—	—
1.4	0.0000	0.0000	0.0000	0.0001	0.0001	—	—	—	—
					$R^2_{M2}(\vartheta, \xi)$				
0.0	0.0000	0.0035	0.0126	0.0232	0.0302	0.0297	0.0211	0.0091	0.0012
0.2	0.0000	0.0029	0.0104	0.0190	0.0247	0.0240	0.0166	0.0066	0.0006
0.4	0.0000	0.0018	0.0065	0.0119	0.0153	0.0144	0.0093	0.0031	0.0001
0.6	0.0000	0.0010	0.0035	0.0063	0.0080	0.0072	0.0043	0.0011	—
0.8	0.0000	0.0005	0.0017	0.0030	0.0037	0.0032	0.0017	0.0003	—
1.0	0.0000	0.0002	0.0007	0.0013	0.0016	0.0013	0.0006	0.0001	—
1.2	0.0000	0.0001	0.0003	0.0006	0.0007	0.0005	0.0002	—	—
1.4	0.0000	0.0000	0.0001	0.0002	0.0003	0.0002	0.0001	—	—
1.6	0.0000	0.0000	0.0000	0.0001	0.0001	0.0001	—	—	—

angles and a few values of ξ. As is seen from the table the maximum values of $R^2_{M\lambda}$ are obtained for $\vartheta \approx 90$ degrees where $R^2_{M\lambda}$ attains approximately the value 0.05. The excitation probability always vanishes for backwards scattering.

Before we enter into a discussion of the cross sections and angular distributions following magnetic excitations, it should be emphasized that the magnetic excitations and the electric excitations, in general, occur simultaneously and, hence, they will interfere. The mixed excitation amplitude for electromagnetic excitation is according to (3.7) and (8) given by

$$a_{I_f M_f, I_0 M_0} = -\mathrm{i} \sum_{\lambda} (-1)^{I_0 - M_0} \sqrt{(2I_0 + 1)(2\lambda + 1)}$$
$$\times \begin{pmatrix} I_0 & \lambda & I_f \\ -M_0 & \mu & M_f \end{pmatrix} \{\chi^{(\lambda)}_{0\to f} R_{\lambda\mu}(\vartheta, \xi) + \chi^{(M\lambda)}_{0\to f} R_{M\lambda\mu}(\vartheta, \xi)\}. \quad (15)$$

The complete symmetry (15) between electric and magnetic excitations makes

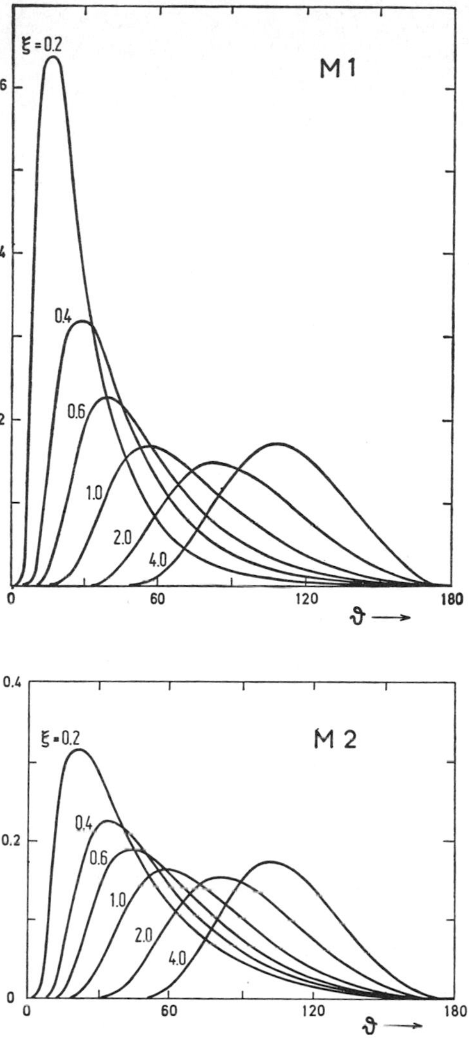

Fig. 21. Angular distribution of the inelastically scattered particles after magnetic dipole and quadrupole excitation. The differential excitation functions are plotted as a function of ϑ for different values of ξ. The curves have been normalized to give the same total cross section. The absolute values can be obtained from the table below.

	$\xi = 0.0$	$\xi = 0.2$	$\xi = 0.4$	$\xi = 0.6$	$\xi = 1.0$	$\xi = 2.0$	$\xi = 4.0$
$df_{M1}(90°, \xi)/d\Omega$	2.58(−1)	1.63(−1)	7.23(−2)	2.76(−2)	3.25(−3)	8.97(−6)	3.16(−11)
$df_{M2}(90°, \xi)/d\Omega$	6.94(−3)	5.64(−3)	3.45(−3)	1.77(−3)	3.44(−4)	2.30(−6)	2.36(−11)

The entry is given by a number and power of ten by which it should be multiplied.

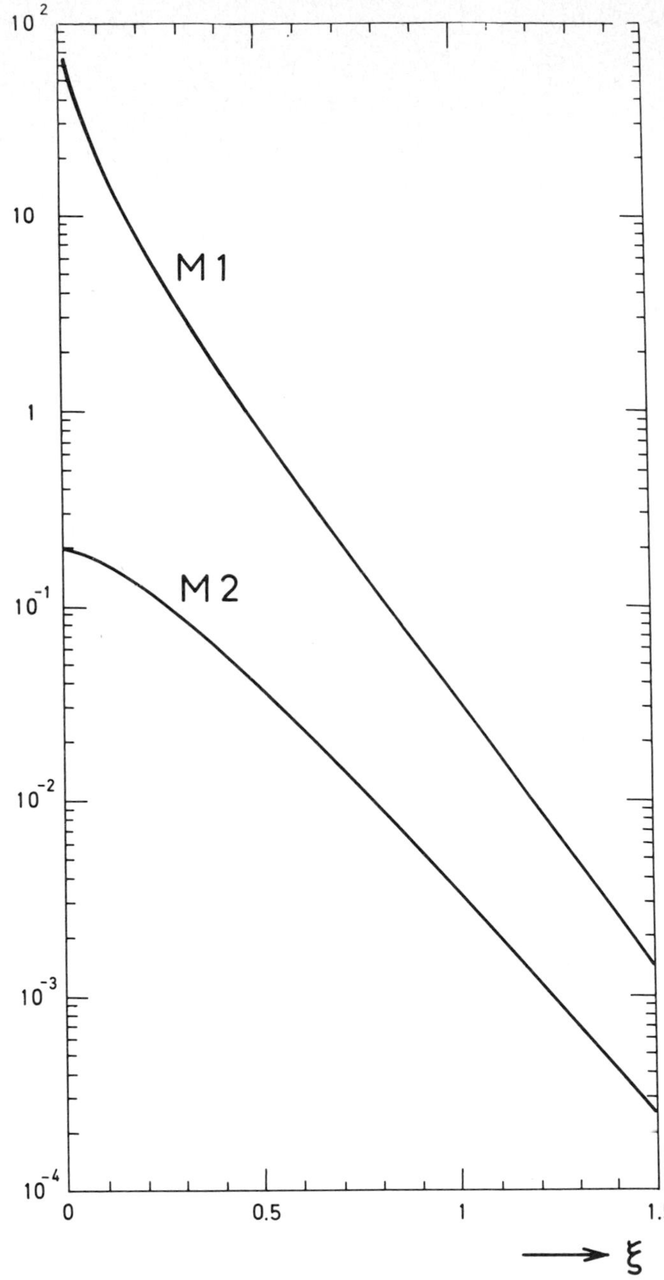

Fig. 22. The total cross section function $f_{M\lambda}(\xi)$ for magnetic dipole and quadrupole excitations in a logarithmic scale.

it very easy to evaluate cross sections and angular distribution, since one may take over the formulae given in § 4 simply by performing the substitution

$$\chi_{0\to f}^{(\lambda)} R_{\lambda\mu} \to \chi_{0\to f}^{(\lambda)} R_{\lambda\mu} + \chi_{0\to f}^{(M\lambda)} R_{M\lambda\mu}. \tag{16}$$

It is noted that since two given states can never be connected by electric and magnetic multipole matrix elements of the same multipolarity, the excitation probabilities and cross sections do not contain interference terms if the initial state is unpolarized. Hence, one finds for the differential cross section

$$\mathrm{d}\sigma = \sum_{\lambda} (\mathrm{d}\sigma_{E\lambda} + \mathrm{d}\sigma_{M\lambda}), \tag{17}$$

where $\mathrm{d}\sigma_{E\lambda}$ is given by eqs. (4.2)–(4.3) and $\mathrm{d}\sigma_{M\lambda}$ by

$$\mathrm{d}\sigma_{M\lambda} = \tfrac{1}{4}a^2 |\chi_{0\to f}^{(M\lambda)}|^2 R_{M\lambda}^2(\vartheta, \xi)\, \mathrm{d}\Omega/\sin^4(\tfrac{1}{2}\vartheta). \tag{18}$$

It may be convenient to write (18) in a form similar to (4.3), i.e.

$$\mathrm{d}\sigma_{M\lambda} = \left(\frac{Z_1 e}{\hbar c}\right)^2 a^{-2\lambda+2} B(M\lambda, I_0 \to I_f)\, \mathrm{d}f_{M\lambda}(\vartheta, \xi), \tag{19}$$

where $B(M\lambda, I_0 \to I_f)$ is the reduced transition probability for the magnetic transition defined analogous to (4.4). The differential cross section function is defined similarly to (4.5) and is illustrated in fig. 21 for $\lambda = 1$ and 2.

The total magnetic cross section is conveniently written

$$\sigma = \sum_{\lambda} \sigma_{M\lambda} = \left(\frac{Z_1 e}{\hbar c}\right)^2 \sum_{\lambda} a^{-2\lambda+2} B(M\lambda, I_0 \to I_f) f_{M\lambda}(\xi). \tag{20}$$

The function $f_{M\lambda}(\xi)$ is illustrated in fig. 22 as a function of ξ for $\lambda = 1$ and 2 on a logarithmic scale.

In the angular distribution of de-excitation γ-rays there will appear interference terms between electric and magnetic excitations. We shall here not state any explicit formulae for the particle parameters. They can be obtained directly from eqs. (4.22–26) by the substitution (16).

§ 6. *Multipole-multipole excitation*

In the collision of two heavy ions, where both of them may have large deformations, one might expect that the deviation from the pure point charge field can give rise to significant corrections to the Coulomb excitation amplitudes as evaluated from a simple monopole-multipole field. In this section we shall study the effect of the multipole-multipole interactions in the semiclassical picture and show that, in most experiments, the effects are of the order $1/\eta^2$, which means that they are indeed negligibly small. In principle, there are also corrections to the magnetic excitations arising e.g. from an interaction between the magnetic moment of the projectile and the magnetic multipole moments of the targets. These corrections will be mentioned at the end of the section.

The electric multipole-multipole interaction is given by eq. (II.1.7), i.e.

$$W_{\mathrm{E}}(1,2;t) = \sum_{\lambda_1\lambda_2\mu_1\mu_2} (4\pi)^{3/2}\left[\frac{(2\lambda_1+2\lambda_2)!}{(2\lambda_1+1)!\,(2\lambda_2+1)!}\right]^{1/2}\begin{pmatrix}\lambda_1 & \lambda_2 & \lambda_1+\lambda_2\\ \mu_1 & \mu_2 & \mu\end{pmatrix}$$

$$\times (-1)^{\lambda_2}\mathscr{M}_1(\mathrm{E}\lambda_1,\mu_1)\mathscr{M}_2(\mathrm{E}\lambda_2,\mu_2)\frac{1}{r(t)^{\lambda_1+\lambda_2+1}}Y_{\lambda_1+\lambda_2,\mu}[\theta(t),\phi(t)]. \quad (1)$$

This time-dependent interaction may cause a simultaneous excitation of both target and projectile and we must, therefore, specify the initial and final states of both particles.

We denote the initial (ground) state by $|0\rangle = |I_0^1M_0^1\rangle|I_0^2M_0^2\rangle$ and the final state by $|\mathrm{f}\rangle = |I_\mathrm{f}^1M_\mathrm{f}^1\rangle\,|I_\mathrm{f}^2M_\mathrm{f}^2\rangle$ where the indices 1 and 2 refer to projectile and target nucleus, respectively. The symbols I and M denote as usual the spin and magnetic quantum numbers.

In the first-order perturbation treatment one finds in analogy to (1.1)

$$a_{0\to\mathrm{f}} = \frac{1}{\mathrm{i}\hbar}\int_{-\infty}^{+\infty}\mathrm{d}t\langle I_\mathrm{f}^1M_\mathrm{f}^1I_\mathrm{f}^2M_\mathrm{f}^2|\;W_\mathrm{E}(1,2;t)\;|I_0^1M_0^1I_0^2M_0^2\rangle$$

$$\times\exp\left\{\frac{\mathrm{i}}{\hbar}t[E_\mathrm{f}^1+E_\mathrm{f}^2-E_0^1-E_0^2]\right\}, \quad (2)$$

where E^1 and E^2 denote the energies of projectile and target, respectively. It is seen directly from eq. (1) that the time integral leads to the orbital integrals $S_{\mathrm{E}(\lambda_1+\lambda_2),\mu}$ which were discussed in § 2. The matrix elements of the multipole operators can again be incorporated into the parameters $\chi_{0\to\mathrm{f}}^{(\lambda)}$ defined by eq. (3.6). Using the normalized orbital integrals defined by (2.2) the excitation amplitude may thus be written

$$a_{0\to f} = -\frac{i}{\eta} \sum_{\lambda_1\lambda_2\mu_1\mu_2} (-1)^{I_0{}^1+I_0{}^2-M_0{}^1-M_0{}^2+\lambda_2+\mu} \sqrt{(2I_0^1+1)(2I_0^2+1)}$$

$$\times \begin{pmatrix} I_0^1 & \lambda_1 & I_f^1 \\ -M_0^1 & \mu_1 & M_f^1 \end{pmatrix} \begin{pmatrix} I_0^2 & \lambda_2 & I_f^2 \\ -M_0^2 & \mu_2 & M_f^2 \end{pmatrix} \tfrac{1}{2}\lambda_1\lambda_2\, d(\lambda_1, \lambda_2)$$

$$\times \begin{pmatrix} \lambda_1 & \lambda_2 & \lambda_1+\lambda_2 \\ -\mu_1 & -\mu_2 & \mu \end{pmatrix} \chi^{(\lambda_1)}_{I_0{}^1\to I_f{}^1} \chi^{(\lambda_2)}_{I_0{}^2\to I_f{}^2} R_{\lambda_1+\lambda_2,\mu}(\vartheta, \xi_1+\xi_2), \tag{3}$$

where

$$d(\lambda_1, \lambda_2) = \tfrac{1}{2}\sqrt{(2\lambda_1+2\lambda_2)(2\lambda_1+2\lambda_2+1)}$$

$$\times \sqrt{\frac{(2\lambda_1+1)!\,(2\lambda_2+1)!}{(2\lambda_1+2\lambda_2-1)!}} \left(\frac{(\lambda_1+\lambda_2-1)!}{\lambda_1!\,\lambda_2!}\right)^2. \tag{4}$$

The parameter ξ_1 is defined by

$$\xi_1 = \frac{E_f^1 - E_0^1}{\hbar} \frac{a}{v} \tag{5}$$

and ξ_2 is defined correspondingly.

The formula (3) includes the excitation amplitude for monopole-multipole excitation (3.7) when $\lambda_1 = 0$. In this case we have

$$d(0, \lambda) = (2\lambda + 1)/\lambda. \tag{6}$$

As was discussed in chapter I (cf. eq. (1.23)) the χ-parameter for $\lambda = 0$ is proportional to η. With the normalization which we have used for χ in (3.6) one should interpret $\chi^{(0)}$ in the following way

$$\tfrac{1}{2}\lambda\chi^{(\lambda)} = \eta, \qquad \text{for } \lambda = 0. \tag{7}$$

It is seen that (3) becomes identical to (3.7) for $\lambda_1 = 0$ (or $\lambda_2 = 0$).

In comparison with the monopole-multipole excitation the multipole-multipole excitation amplitude is reduced by a factor of the order χ/η which for heavy ions is usually a small number. Because of the orthogonality properties of the 3-j symbols appearing in eq. (3) there will be no interference between the multipole-multipole and the monopole-multipole excitations in most experiments. Thus one finds for the mutual excitation probability of target and projectile

$$P^{\text{mut}}(I_0^1 \to I_f^1; I_0^2 \to I_f^2) = \frac{1}{4\eta^2} \sum_{\lambda_1\lambda_2} \frac{\lambda_1^2\lambda_2^2 d^2(\lambda_1, \lambda_2)}{(2\lambda_1+1)(2\lambda_2+1)(2\lambda_1+2\lambda_2+1)}$$

$$\times |\chi^{(\lambda_1)}_{I_0{}^1\to I_f{}^1}|^2 |\chi^{(\lambda_2)}_{I_0{}^2\to I_f{}^2}|^2 R^2_{\lambda_1+\lambda_2}(\vartheta, \xi_1+\xi_2), \tag{8}$$

where $R^2_\lambda(\vartheta, \xi)$ is defined in eqs. (3.3-4).

The state $|f\rangle$ can also be populated by an accidental simultaneous excitation of both target and projectile through the monopole-multipole interaction. The probability of this process is given by

$$P^{\text{sim}}(I_0^1 \to I_f^1; I_0^2 \to I_f^2) = P(I_0^1 \to I_f^1)P(I_0^2 \to I_f^2)$$
$$= \sum_{\lambda_1\lambda_2} |\chi^{(\lambda_1)}_{I_0{}^1 \to I_f{}^1}|^2 |\chi^{(\lambda)_2}_{I_0{}^2 \to I_f{}^2}|^2 R^2_{\lambda_1}(\vartheta, \xi_1) R^2_{\lambda_2}(\vartheta, \xi_2), \quad (9)$$

where we have used the expression (3.2) for the excitation probabilities P. If only one value of each λ_1 and λ_2 contributes, the ratio between the mutual and the simultaneous excitation is of the order of magnitude $1/\eta^2$.

As another example we may consider the correction to the usual Coulomb excitation probability from the quadrupole moments of the projectile (static as well as transition moments), for the case where the state of excitation of the projectile is not detected. One finds then from eq. (8)

$$P(I_0^2 \to I_f^2) = \sum_\lambda |\chi^{(\lambda)}_{I_0{}^2 \to I_f{}^2}|^2 \Big\{ R^2_\lambda(\vartheta, \xi_2) + \sum_{I_f{}^1} \frac{3}{8} \frac{\lambda^2(\lambda+1)^3(\lambda+2)}{(2\lambda+1)(2\lambda+3)} \frac{1}{\eta^2}$$
$$\times |\chi^{(2)}_{I_0{}^1 \to I_f{}^1}|^2 R^2_{\lambda+2}(\vartheta, \xi_1(I_0^1 \to I_f^1) + \xi_2) \Big\}. \quad (10)$$

A summation over the possible final states I_f^1 of the projectile after the quadrupole excitation appears in the correction term. From an order of magnitude estimate one finds that the correction term is at most about 1%.

An interference term between the monopole-multipole and the multipole-multipole excitation occurs if polarized projectiles or targets are used or if one studies the angular distribution of subsequent γ-rays.

According to § II.1, the magnetic multipole-multipole excitations are described by the same expressions as the electric multipole-multipole excitations, if one substitutes the electric multipole moments by the corresponding magnetic moments. The most important term is the excitation caused by the magnetic dipole moment of the projectile. This excitation has been considered by [BIE 65].

§ 7. *Quantum mechanical effects*

In the present treatment of the Coulomb excitation process we have used the semiclassical approximation which is valid provided the parameter η is large compared to unity. Actual values of η for different projectiles and targets are indicated in fig. X.4, and it is seen that for projectiles of low charge the parameter η can be as small as 2 or 3. Still, one finds, even in such cases, in first-order perturbation theory that the semiclassical theory is quite accurate. Since, however, the Coulomb excitation process can be measured and calculated to a high degree of accuracy one should consider the quantum mechanical corrections.

In chapter IX we formulate a complete quantum mechanical treatment of the Coulomb excitation and shall limit the discussion in the present section to a summary of the results of a first-order perturbation treatment.

While in the complete semiclassical treatment one does not distinguish between the initial and the final velocity the cross sections which have been computed in this chapter depend rather sensitively on the projectile velocity and one might find rather different results when one inserts the initial velocity v_i or the final velocity v_f for v in the formulae for the cross section. This is basically a quantum mechanical effect which cannot be incorporated in a classical treatment since the moment at which an energy loss occurs is undetermined.

It is, however, a general consequence of time-reflection invariance (detailed balance) that the Coulomb excitation cross section in first-order perturbation theory is of the form

$$d\sigma = \frac{v_f}{v_i} F(v_i, v_f), \tag{1}$$

where the function F is symmetrical under the interchange of v_i and v_f. From this argument it may be conjectured that an improvement of the classical picture can be obtained if one interprets the velocity v as a suitable average of initial and final velocities and corrects the expressions for the cross sections by a factor v_f/v_i.

Thus, one should introduce instead of half the distance of closest approach defined in eq. (II.2.3) a parameter a_{if} defined by

$$a_{if} = Z_1 Z_2 e^2 / m_0 v_i v_f, \tag{2}$$

which is symmetric in the initial and final velocities. The main dependence of the cross section on the velocity enters through the parameter ξ. If one uses the parameter (2) in the expression (2.5) one finds the following symmetrized expression

$$\xi = \frac{\Delta E}{\hbar} \frac{a_{if}}{(v_i + v_f)/2} = \frac{Z_1 Z_2 e^2}{\hbar} \left(\frac{1}{v_f} - \frac{1}{v_i} \right) = \eta_f - \eta_i, \tag{3}$$

where we have used that $\Delta E = \frac{1}{2} m_0 (v_i^2 - v_f^2)$. Finally, the cross sections are proportional to the factor $(Z_1 e/\hbar v)^2$ which should be replaced by $(Z_1 e/\hbar \sqrt{v_i v_f})^2$.

According to eq. (1) the cross section (4.3) should therefore be replaced by

$$d\sigma_{E\lambda} = \left(\frac{Z_1 e}{\hbar v_i}\right)^2 a_{if}^{-2\lambda+2} B(E\lambda, I_0 \rightarrow I_f)\, df_{E\lambda}(\vartheta, \xi). \tag{4}$$

The improvement which is obtained in the description of the Coulomb excitation process through the symmetrization can only be judged on the basis of a more complete quantum mechanical treatment. In perturbation theory this treatment is the distorted wave Born approximation which for the cases of $\lambda = 1$ and 2 have been carried out completely. The results which will be discussed in chapter IX show that the symmetrized classical expressions for the total cross section are accurate to better than 5% for $\eta > 3$, while a corresponding accuracy in the angular distribution is only obtained for $\eta > 10$. One may give arguments for more refined symmetrizations [BIE 65], which for small values of η give somewhat better results than eq. (4).

CHAPTER V

Higher-Order Perturbation Theory

The validity of the first-order perturbation theory depends on the smallness of the parameters χ describing the strength of the Coulomb excitation. As was discussed in chapter I these quantities may, especially in heavy ion bombardments, become quite large. If the χ-parameters are larger than or comparable to unity one must treat the Coulomb excitation process by solving directly the time-dependent Schrödinger equation for the nucleus. In a number of cases of practical interest the strength of the interaction is such that the deviation from the first-order theory can be described by considering second-order corrections. This second-order perturbation theory also gives an insight into the mechanism of Coulomb excitation which provides a useful picture for the discussion of the more complicated multiple excitations. We shall in the present chapter consider the semiclassical description of the Coulomb excitation process in a higher-(mainly second-) order perturbation theory utilizing the results of the first-order theory.

§ 1. *Excitation amplitudes*

In the following we consider the cases where the deviations from the first-order theory are small. We thus assume that all matrix elements of the χ-matrix (cf. § IV.1) are relatively small, say, at most of the order of 0.5. The perturbation expansion (II.3.8) is essentially a series expansion in the χ-parameters and, with the above limitation, it is expected that the series shows a reasonable convergence. The measurable quantities M in Coulomb excitation like excitation probabilities, cross sections, angular distributions etc. are quadratic expressions in the amplitudes, and they will therefore be of the form

$$M = a_2\chi^2 + a_3\chi^3 + a_4\chi^4 + \cdots \tag{1}$$

where we use the symbol χ to indicate typical values of χ. The coefficients a_2, a_3 etc. will depend on the scattering angle and on the ξ's and will, in general, form a series of somewhat decreasing numbers. The first term corresponds to first-order perturbation theory, the second term is an interference between first- and second-order, while the third term is partly due to second-order perturbation theory, partly due to an interference between first- and third-order etc.

From this qualitative picture one sees that the relative accuracy of the first-order theory is of the order $a_3\chi/a_2$ and second-order corrections to the results of the last chapter may, in general, be expected to be of the order of 5% already for χ of the order 0.1. This estimate is in many cases too high because the coefficient a_3 may be small. In the limit of the sudden approximation one thus finds $a_3 = 0$ (see § 2 below).

The accuracy of the second-order theory may be estimated from the magnitude of the third term in eq. (1) which is uncertain due to third-order corrections. If we again demand a 5% relative accuracy, the second-order theory is expected to be valid for $\chi < (0.05a_2/a_4)^{1/2}$ which may be of the order of 0.4. Similar estimates for higher-order perturbation treatments show that the range of χ-values, where higher-order theory is valid, only increases slowly with the order. It is thus hardly of practical interest to push the perturbation expansion beyond second order.

In some cases it may happen that the first-order term vanishes or is very small because of special selection rules. The series will then start with the term a_4, and the second-order treatment is then appropriate in the range $\chi < 0.05a_4/a_5$. Although the above considerations show the rather limited applicability of the second-order theory, it gives a useful estimate of the deviations from the first-order theory and we shall in the following carry through the second order theory for electric excitations in all details.

To second order the excitation amplitude is according to eq. (II.3.8) of the form

$$a_n = a_n^{(1)} + a_n^{(2)}, \tag{2}$$

where $a^{(1)}$ is the first-order amplitude (IV.1.1) computed in the last chapter, while $a^{(2)}$ is given by

$$a_n^{(2)} = \sum_z \left(\frac{-\mathrm{i}}{\hbar}\right)^2 \int_{-\infty}^{+\infty} \mathrm{d}t \,\langle n|\, V_{\mathrm{E}}(t)\,|z\rangle \exp\left\{\mathrm{i}\,\frac{E_n - E_z}{\hbar}\,t\right\}$$
$$\times \int_{-\infty}^{t} \mathrm{d}t' \,\langle z|\, V_{\mathrm{E}}(t')\,|0\rangle \exp\left\{\mathrm{i}\,\frac{E_z - E_0}{\hbar}\,t'\right\}. \tag{3}$$

We have introduced a summation over a complete set of nuclear states $|z\rangle$ which includes the states $|0\rangle$ and $|n\rangle$. Using the multipole expansion of the interaction energy (II.2.16) and specifying the nuclear states by their spin quantum numbers the second-order amplitude may be written

$$a^{(2)}_{I_fM_f,I_0M_0} = -\frac{16\pi^2 Z_1^2 e^2}{\hbar^2} \sum_{\substack{\lambda\lambda' \\ \mu\mu' z}} \frac{1}{(2\lambda+1)(2\lambda'+1)}$$
$$\times \langle I_0M_0|\, \mathcal{M}(\mathrm{E}\lambda, \mu)\, |I_zM_z\rangle^* \langle I_zM_z|\, \mathcal{M}(\mathrm{E}\lambda', \mu')\, |I_fM_f\rangle^*$$
$$\times \int_{-\infty}^{+\infty} \mathrm{d}t\, \bar{S}_{\mathrm{E}\lambda'\mu'}(t) \exp(\mathrm{i}\omega_{fz}t) \int_{-\infty}^{t} \mathrm{d}t'\, \bar{S}_{\mathrm{E}\lambda\mu}(t') \exp(\mathrm{i}\omega_{z0}t'), \qquad (4)$$

where ω_{fz} and ω_{z0} are given by eq. (IV.1.5) for the transitions $z \to \mathrm{f}$ and $0 \to z$, respectively.

We note that if the second integral in eq. (4) would have the upper limit $+\infty$ the second-order amplitude would be the product of the first-order amplitudes. In some cases one is entitled to extend the second integral to infinity. This is the case if the product of the two integrals is a symmetric function in t and t', which occurs when $\lambda = \lambda'$, $\mu = \mu'$ and $\omega_{fz} = \omega_{z0}$. Under these circumstances the double integral is just half the product of the two first-order integrals, i.e.

$$a^{(2)}_{I_fM_f,I_0M_0} = \tfrac{1}{2} \sum_{I_zM_z} a^{(1)}_{I_fM_f,I_zM_z} a^{(1)}_{I_zM_z,I_0M_0}. \qquad (5)$$

More systematically one may introduce the step function (see eq. (II.3.15)) defined by

$$\theta(t-t') = \tfrac{1}{2}[1+\varepsilon(t-t')] = \begin{cases} 1, & t > t' \\ 0, & t < t' \end{cases} \qquad (6)$$

in the double integral of eq. (4). The function $\varepsilon(t-t')$ has the convenient integral representation

$$\varepsilon(t-t') = \frac{\mathrm{i}}{\pi} \mathcal{P} \int_{-\infty}^{+\infty} \frac{\exp\{-\mathrm{i}(t-t')q\}}{q}\, \mathrm{d}q, \qquad (7)$$

where $\mathcal{P}$ denotes the principal part of the integral. Introducing eqs. (6) and (7) into eq. (4) one finds

$$a^{(2)}_{I_fM_f,I_0M_0} = \tfrac{1}{2} \sum_{I_zM_z} a^{(1)}_{I_fM_f,I_zM_z}(\omega_{fz}) a^{(1)}_{I_zM_z,I_0M_0}(\omega_{z0})$$
$$+ \frac{\mathrm{i}}{2\pi} \sum_{I_zM_z} \mathcal{P} \int_{-\infty}^{+\infty} \frac{\mathrm{d}q}{q}\, a^{(1)}_{I_fM_f,I_zM_z}(\omega_{fz}-q) a^{(1)}_{I_zM_z,I_0M_0}(\omega_{z0}+q), \qquad (8)$$

where we have introduced a notation for the first-order amplitudes specifying explicitly the frequency ω. The first term in eq. (8) is the product of the first-order amplitudes discussed above and identical to eq. (5).

It is noted that the second term also disappears in the limit of the sudden approximation ($\omega_{fz} = \omega_{z0} = 0$), independent of the multipolarity. This is most easily seen in coordinate system B where $M_f - M_i$ must be even (see eq. (II.8.20)). Since it follows furthermore from expression (IV.3.7) and table IV.1 that

$$a^{(1)}_{I_fM_f,I_0M_0}(q) = (-1)^{M_f-M_0}a^{(1)}_{I_fM_f,I_0M_0}(-q) \tag{9}$$

the integrand in eq. (8) is an odd function in q.

In chapter IV it was shown that in coordinate systems A and B the first-order amplitudes are purely imaginary. In the excitation probability, which is the numerical square of the amplitude (2), the first term in eq. (8), which is purely real, will interfere neither with the second term in eq. (8) nor with the first-order amplitude.

It is interesting to observe that the second-order perturbation treatment can also be obtained from eq. (II.3.36) if one expands the exponential function in the S-matrix, i.e.

$$\exp(2\mathrm{i}Q) = \exp\{\mathrm{i}[R + G]\} \approx 1 + \mathrm{i}R - \tfrac{1}{2}R^2 + \mathrm{i}G + \cdots. \tag{10}$$

With the expressions (II.3.27) and (II.3.28) for the quantities R and G one finds that the matrix element of $\mathrm{i}R$ gives rise to the first-order amplitude $a^{(1)}$, while the two last terms give rise to the second-order amplitude $a^{(2)}$. The contributions of $-\frac{1}{2}R^2$ and of $\mathrm{i}G$ are identical to the first and second term in eq. (8), respectively.

For the evaluation of the second-order amplitude one introduces the result (IV.3.7) into eq. (8). One hereby obtains

$$\begin{aligned} a^{(2)}_{I_fM_f,I_0M_0} = -\tfrac{1}{2} \sum_{\substack{\lambda\lambda'\mu\mu' \\ I_zM_z}} & \sqrt{(2I_0+1)(2I_z+1)(2\lambda+1)(2\lambda'+1)} \\ & \times (-1)^{I_0-M_0+I_z-M_z} \begin{pmatrix} I_0 & \lambda & I_z \\ -M_0 & \mu & M_z \end{pmatrix} \begin{pmatrix} I_z & \lambda' & I_f \\ -M_z & \mu' & M_f \end{pmatrix} \chi^{(\lambda)}_{0\to z}\chi^{(\lambda')}_{z\to f} \\ & \times \Big\{ R_{\lambda\mu}(\vartheta, \xi_{z0}) R_{\lambda'\mu'}(\vartheta, \xi_{fz}) \\ & \qquad + \frac{\mathrm{i}}{\pi}\mathscr{P}\int_{-\infty}^{+\infty} \frac{\mathrm{d}\xi}{\xi} R_{\lambda\mu}(\vartheta, \xi_{z0}+\xi) R_{\lambda'\mu'}(\vartheta, \xi_{fz}-\xi) \Big\}. \end{aligned} \tag{11}$$

Under rotations of the reference system the single terms in the second-order amplitude transform like a product of two tensor operators of orders (λ, μ) and (λ', μ'). It is convenient, especially for the evaluation of angular distributions, to decompose this product into a sum of tensor operators. Such tensor operators are conveniently defined by

$$R_{(\lambda\lambda')k\kappa}(\vartheta, \xi_{z0}, \xi_{fz}) = \sum_{\mu\mu'} (-1)^{\kappa} \begin{pmatrix} \lambda & \lambda' & k \\ \mu & \mu' & -\kappa \end{pmatrix} R_{\lambda\mu}(\vartheta, \xi_{z0}) R_{\lambda'\mu'}(\vartheta, \xi_{fz}), \tag{12}$$

and

$$G_{(\lambda\lambda')k\kappa}(\vartheta, \xi_{z0}, \xi_{fz}) = \sum_{\mu\mu'} (-1)^{\kappa} \begin{pmatrix} \lambda & \lambda' & k \\ \mu & \mu' & -\kappa \end{pmatrix} G^{\lambda\lambda'}_{\mu\mu'}(\vartheta, \xi_{z0}, \xi_{fz}), \tag{13}$$

where the double integral G is defined by

$$G^{\lambda\lambda'}_{\mu\mu'}(\vartheta, \xi, \xi') = \frac{1}{\pi} \mathscr{P} \int_{-\infty}^{+\infty} \frac{dq}{q} R_{\lambda\mu}(\vartheta, \xi + q)\, R_{\lambda'\mu'}(\vartheta, \xi' - q). \tag{14}$$

Utilizing the orthogonality relation of the 3-j symbol to express the product of the orbital integrals by means of the spherical tensors $R_{(\lambda\lambda')k\kappa}$ and $G_{(\lambda\lambda')k\kappa}$ and performing the summation over μ, μ' and M_z, one obtains the following result for the second-order excitation amplitude

$$\begin{aligned} a^{(2)}_{I_f M_f, I_0 M_0} = -\tfrac{1}{2} \sum_{\substack{\lambda\lambda' k\kappa \\ I_z}} & (2k+1)\sqrt{(2I_0+1)(2I_z+1)(2\lambda+1)(2\lambda'+1)} \\ & \times (-1)^{I_0+I_f+k+\lambda+\lambda'} \begin{Bmatrix} I_0 & I_f & k \\ \lambda' & \lambda & I_f \end{Bmatrix} (-1)^{I_0-M_0} \begin{pmatrix} I_0 & k & I_f \\ -M_0 & \kappa & M_f \end{pmatrix} \\ & \times \chi^{(\lambda)}_{0\to z} \chi^{(\lambda')}_{z\to f} \{R_{(\lambda\lambda')k\kappa}(\vartheta, \xi_{z0}, \xi_{fz}) + \mathrm{i}G_{(\lambda\lambda')k\kappa}(\vartheta, \xi_{z0}, \xi_{fz})\}. \end{aligned} \tag{15}$$

The expressions (11) and (15) offer in the coordinate systems A and B a decomposition of the second-order amplitude in real and imaginary parts. The expression (15) is formally very similar to eq. (IV.3.7) for the first-order amplitude, if one replaces λ and μ by k and κ. This is of some help in the discussion of angular distributions as we shall see later.

§ 2. *Double integrals*

The evaluation of the second-order amplitude was reduced in the last section to the computation of the quantities $R_{(\lambda\lambda')k\kappa}$ and $G_{(\lambda\lambda')k\kappa}$. They can be evaluated from the knowledge of the first-order orbital integrals $R_{\lambda\mu}$ (see eqs. (1.12–14)). In this section we shall discuss some of their properties.

The second-order integrals $R_{(\lambda\lambda')k\kappa}$ and $G_{(\lambda\lambda')k\kappa}$ depend not only on the multipole orders λ, λ', k and κ, but also on the scattering angle ϑ and on the two adiabaticity parameters ξ and ξ'. In general, they are complex quantities which, according to eq. (IV.2.1) satisfy the relations

$$\begin{aligned} & R^{*}_{(\lambda\lambda')k\kappa}(\vartheta, \xi, \xi') = (-1)^{\lambda+\lambda'+k+\kappa} R_{(\lambda\lambda')k-\kappa}(\vartheta, -\xi, -\xi') \\ \text{and} \\ & G^{*}_{(\lambda\lambda')k\kappa}(\vartheta, \xi, \xi') = (-1)^{\lambda+\lambda'+k+\kappa+1} G_{(\lambda\lambda')k-\kappa}(\vartheta, -\xi, -\xi'). \end{aligned} \tag{1}$$

It also follows from their definition (1.12) and (1.13) that in the coordinate systems A and B they are real numbers. They show a number of further symmetry relations that can be obtained from the symmetry properties of the functions $R_{\lambda\mu}$ which are summarized in table IV.1. For the coordinate systems A and B the symmetry properties of the tensors $R_{(\lambda\lambda')k\kappa}$ and $G_{(\lambda\lambda')k\kappa}$ are collected in table 1. As an example of the use of this table we note that for $\lambda = \lambda'$, $\mu = \mu'$ and $\xi = \xi'$ one finds in both coordinate systems

$$G_{\mu\mu}^{\lambda\lambda}(\vartheta, \xi, \xi) = 0 \tag{2}$$

in agreement with the remarks in connection with eq. (1.4). In the sudden approximation where $\xi = \xi' = 0$ one finds from table 1 in coordinate system B

$$G_{(\lambda\lambda')k\kappa}^{\mathrm{B}}(\vartheta, 0, 0) = 0 \qquad \text{for } \kappa \text{ even}. \tag{3}$$

The fact that the principal part integral in (1.11) or (1.15) vanishes in the sudden approximation also for odd values of κ can only be seen explicitly when the summation over the intermediate set of states $|z\rangle$ is performed leading to terms with the same nuclear matrix elements but interchanged values of λ and λ'.

The orbital integrals have been evaluated numerically in coordinate system A and the results are given for $\lambda = \lambda' = 2$ in [DOU 62]. The quantities which are tabulated in this reference are defined in [ALD 56] and are related to the orbital integrals in coordinate system A by

$$\alpha_{k\kappa} = \frac{(-1)^{\kappa}(\lambda-1)!\,(\lambda'-1)!}{(2\lambda-1)!!\,(2\lambda'-1)!!}\,\frac{\sqrt{(2\lambda+1)(2\lambda'+1)}}{\pi}\,R_{(\lambda\lambda')k-\kappa}^{\mathrm{A}}$$

and

$$\beta_{k\kappa} = \frac{(-1)^{\kappa}(\lambda-1)!\,(\lambda'-1)!}{(2\lambda-1)!!\,(2\lambda'-1)!!}\,\frac{\sqrt{(2\lambda+1)(2\lambda'+1)}}{\pi}\,G_{(\lambda\lambda')k-\kappa}^{\mathrm{A}}. \tag{4}$$

For the numerical evaluation it is convenient to rewrite the principal part integral (1.14) in the form

$$G_{\mu\mu'}^{\lambda\lambda'}(\vartheta, \xi, \xi') = \frac{1}{\pi}\int_0^{\infty}\frac{\mathrm{d}q}{q}\{R_{\lambda\mu}(\vartheta, \xi+q)\,R_{\lambda'\mu'}(\vartheta, \xi'-q) - R_{\lambda\mu}(\vartheta, \xi-q)\,R_{\lambda'\mu'}(\vartheta, \xi'+q)\}. \tag{5}$$

It is seen from this expression that the main contribution arises from the region of q-values of the order of $\frac{1}{2}|\xi' - \xi|$. It is noted that for large values of $|\xi + \xi'|$, G vanishes exponentially. Thus, in a situation where both ξ and ξ' are large and positive, which corresponds to a double excitation of a high energy state, both the real and the imaginary part of the second-order amplitude thus vanish exponentially with the total energy of the final state.

TABLE 1

Symmetry properties of the second-order integrals. The table shows the symmetry properties of the double integrals (1.14) and the quantity $R^{\lambda\lambda'}_{\mu\mu'}(\vartheta, \xi, \xi') = R_{\lambda\mu}(\vartheta, \xi) R_{\lambda'\mu'}(\vartheta, \xi')$ as well as of the tensors (1.12) and (1.13) in the coordinate systems A and B. (These properties follow from the definitions and from the symmetry properties of the orbital integrals summarized in table IV.1.)

	coordinate system A	coordinate system B
μ-dependence	$R^{\lambda\lambda'}_{\mu\mu'}(\vartheta, \xi, \xi') = 0$ for $\lambda + \mu$ or $\lambda' + \mu'$ odd	$R^{\lambda\lambda'}_{\mu\mu'}(\vartheta, \xi, \xi') = R^{\lambda\lambda'}_{\lvert\mu\rvert\lvert\mu'\rvert}(\vartheta, \xi, \xi')$
interchange	$R^{\lambda\lambda'}_{\mu\mu'}(\vartheta, \xi, \xi') = R^{\lambda'\lambda}_{\mu'\mu}(\vartheta, \xi', \xi)$	$R^{\lambda\lambda'}_{\mu\mu'}(\vartheta, \xi, \xi') = R^{\lambda'\lambda}_{\mu'\mu}(\vartheta, \xi', \xi)$
$\xi \to -\xi$	$R^{\lambda\lambda'}_{\mu\mu'}(\vartheta, \xi, \xi') = (-1)^{\mu'} R^{\lambda\lambda'}_{\mu-\mu'}(\vartheta, \xi, -\xi')$	$R^{\lambda\lambda'}_{\mu\mu'}(\vartheta, \xi, \xi') = (-1)^{\mu} R^{\lambda\lambda'}_{\mu\mu'}(\vartheta, -\xi, \xi')$
	$= (-1)^{\mu} R^{\lambda\lambda'}_{-\mu'\mu}(\vartheta, -\xi, \xi') = (-1)^{\mu+\mu'} R^{\lambda\lambda'}_{-\mu-\mu'}(\vartheta, -\xi, -\xi')$	$= (-1)^{\mu'} R^{\lambda\lambda'}_{\mu\mu'}(\vartheta, \xi, -\xi') = (-1)^{\mu+\mu'} R^{\lambda\lambda'}_{\mu\mu'}(\vartheta, -\xi, -\xi')$
$\xi = 0$	$R^{\lambda\lambda'}_{\mu\mu'}(\vartheta, \xi, 0) = (-1)^{\mu} R^{\lambda\lambda'}_{\mu-\mu'}(\vartheta, \xi, 0)$	$R^{\lambda\lambda'}_{\mu\mu'}(\vartheta, 0, \xi') = 0$ for μ odd
	$R^{\lambda\lambda'}_{\mu\mu'}(\vartheta, 0, \xi') = (-1)^{\mu} R^{\lambda\lambda'}_{-\mu\mu'}(\vartheta, 0, \xi')$	$R^{\lambda\lambda'}_{\mu\mu'}(\vartheta, \xi, 0) = 0$ for μ' odd
$\vartheta = \pi$	$R^{\lambda\lambda'}_{\mu\mu'}(\pi, \xi, \xi')$ only geometrically dependent on μ and μ'	$R^{\lambda\lambda'}_{\mu\mu'}(\pi, \xi, \xi') = 0$ for $\mu \neq 0$ or $\mu' \neq 0$
μ-dependence	$G^{\lambda\lambda'}_{\mu\mu'}(\vartheta, \xi, \xi') = 0$ for $\lambda + \mu$ or $\lambda' + \mu'$ odd	$G^{\lambda\lambda'}_{\mu\mu'}(\vartheta, \xi, \xi') = G^{\lambda\lambda'}_{\lvert\mu\rvert\lvert\mu'\rvert}(\vartheta, \xi, \xi')$
interchange	$G^{\lambda\lambda'}_{\mu\mu'}(\vartheta, \xi, \xi') = -G^{\lambda'\lambda}_{\mu'\mu}(\vartheta, \xi', \xi)$	$G^{\lambda\lambda'}_{\mu\mu'}(\vartheta, \xi, \xi') = -G^{\lambda'\lambda}_{\mu'\mu}(\vartheta, \xi', \xi)$
$\xi \to -\xi$	$G^{\lambda\lambda'}_{\mu\mu'}(\vartheta, \xi, \xi') = (-1)^{\mu+\mu'+1} G^{\lambda\lambda'}_{-\mu-\mu'}(\vartheta, -\xi, -\xi')$	$G^{\lambda\lambda'}_{\mu\mu}(\vartheta, \xi, \xi') = (-1)^{\mu+\mu'+1} G^{\lambda\lambda'}_{\mu\mu'}(\vartheta, -\xi, -\xi')$
$\xi = \xi' = 0$	$G^{\lambda\lambda'}_{\mu\mu'}(\vartheta, 0, 0) = (-1)^{\mu+\mu'+1} G^{\lambda\lambda'}_{-\mu-\mu'}(\vartheta, 0, 0)$	$G^{\lambda\lambda'}_{\mu\mu'}(\vartheta, 0, 0) = 0$ if $\mu + \mu'$ even
$\vartheta = \pi$	$G^{\lambda\lambda'}_{\mu\mu'}(\pi, \xi, \xi')$ only geometrically dependent on μ and μ'	$G^{\lambda\lambda'}_{\mu\mu'}(\pi, \xi, \xi') = 0$ for $\mu \neq 0$ or $\mu' \neq 0$
μ-dependence	$R_{(\lambda\lambda')k\kappa}(\vartheta, \xi, \xi') = 0$ if $\lambda + \lambda' + \kappa$ odd	$R_{(\lambda\lambda')k\kappa}(\vartheta, \xi, \xi') = (-1)^{\lambda+\lambda'+k} R_{(\lambda\lambda')k-\kappa}(\vartheta, \xi, \xi')$
interchange	$R_{(\lambda\lambda')k\kappa}(\vartheta, \xi, \xi') = (-1)^{\lambda+\lambda'+k} R_{(\lambda'\lambda)k\kappa}(\vartheta, \xi', \xi)$	$R_{(\lambda\lambda')k\kappa}(\vartheta, \xi, \xi') = (-1)^{\lambda+\lambda'+k} R_{(\lambda'\lambda)k\kappa}(\vartheta, \xi', \xi)$
$\xi \to -\xi$	$R_{(\lambda\lambda')k\kappa}(\vartheta, \xi, \xi') = (-1)^{\lambda+\lambda'+k-\kappa} R_{(\lambda\lambda')k-\kappa}(\vartheta, -\xi, -\xi')$	$R_{(\lambda\lambda')k\kappa}(\vartheta, \xi, \xi') = (-1)^{\kappa} R_{(\lambda\lambda')k\kappa}(\vartheta, -\xi, -\xi')$
$\xi = \xi' = 0$	$R_{(\lambda\lambda')k\kappa}(\vartheta, 0, 0) = (-1)^{\lambda+\lambda'+k+\kappa} R_{(\lambda\lambda')k-\kappa}(\vartheta, 0, 0)$	$R_{(\lambda\lambda')k\kappa}(\vartheta, 0, 0) = 0$ for κ odd
$\vartheta = \pi$	$R_{(\lambda\lambda')k\kappa}(\pi, \xi, \xi')$ only geometrically dependent on κ	$R_{(\lambda\lambda')k\kappa}(\pi, \xi, \xi') = 0$ for $\kappa \neq 0$
μ-dependence	$G_{(\lambda\lambda')k\kappa}(\vartheta, \xi, \xi') = 0$ for $\lambda + \lambda' + \kappa$ odd	$G_{(\lambda\lambda')k\kappa}(\vartheta, \xi, \xi') = (-1)^{\lambda+\lambda'+k} G_{(\lambda\lambda')k-\kappa}(\vartheta, \xi, \xi')$
interchange	$G_{(\lambda\lambda')k\kappa}(\vartheta, \xi, \xi') = (-1)^{\lambda+\lambda'+k+1} G_{(\lambda'\lambda)k\kappa}(\vartheta, \xi', \xi)$	$G_{(\lambda\lambda')k\kappa}(\vartheta, \xi, \xi') = (-1)^{\lambda+\lambda'+k+1} G_{(\lambda'\lambda)k\kappa}(\vartheta, \xi', \xi)$
$\xi \to -\xi$	$G_{(\lambda\lambda')k\kappa}(\vartheta, \xi, \xi') = (-1)^{\lambda+\lambda'+k+\kappa+1} G_{(\lambda\lambda')k-\kappa}(\vartheta, -\xi, -\xi')$	$G_{(\lambda\lambda')k\kappa}(\vartheta, \xi, \xi') = (-1)^{\kappa+1} G_{(\lambda\lambda')k\kappa}(\vartheta, -\xi, -\xi')$
$\xi = \xi' = 0$	$G_{(\lambda\lambda')k\kappa}(\vartheta, 0, 0) = (-1)^{\lambda+\lambda'+k+\kappa+1} G_{(\lambda\lambda')k-\kappa}(\vartheta, 0, 0)$	$G_{(\lambda\lambda')k\kappa}(\vartheta, 0, 0) = 0$ for κ even
$\vartheta = \pi$	$G_{(\lambda\lambda')k\kappa}(\pi, \xi, \xi')$ only geometrically dependent on κ	$G_{(\lambda\lambda')k\kappa}(\pi, \xi, \xi') = 0$ for $\kappa \neq 0$

Another limiting case is the one in which a low excited state is excited through an intermediate state of very high energy, i.e.

$$\xi' \approx -\xi. \tag{6}$$

One finds again from (1.12) that $R_{(\lambda\lambda')k\kappa}$ vanishes exponentially with ξ. The integral $G_{(\lambda\lambda')k\kappa}$, however, receiving the main contribution from the region around $q \approx \frac{1}{2}(\xi' - \xi)$ vanishes only as $1/\xi$.

We can obtain an approximate expression for $G_{(\lambda\lambda')k\kappa}$ in this limit by taking the nearly constant factor $1/q \approx 2/(\xi' - \xi)$ outside the integral in (1.14). Using the notation

$$\bar{\xi} = \tfrac{1}{2}(\xi - \xi') \tag{7}$$

and

$$\tilde{\xi} = \xi + \xi' \tag{8}$$

we find from eqs. (1.13) and (1.14)

$$G_{(\lambda\lambda')k\kappa}(\vartheta, \xi, \xi') \approx -\sum_{\mu\mu'}(-1)^{\kappa}\begin{pmatrix}\lambda & \lambda' & k\\ \mu & \mu' & -\kappa\end{pmatrix}\frac{1}{\pi}\frac{1}{\bar{\xi}}$$

$$\times \int_{-\infty}^{+\infty} \mathrm{d}q\, R_{\lambda\mu}(\xi + q) R_{\lambda'\mu'}(\xi' - q). \tag{9}$$

Using the definitions (IV.1.4) and (IV.2.2) of the orbital integral $R_{\lambda\mu}$ in terms of an integral over the time t the integration over q leads to a delta function $\delta(t - t')$. The expression (9) is thus reduced to a single orbital integral over a product of two spherical harmonics $Y_{\lambda\mu}(\theta, \phi)\; Y_{\lambda'\mu'}(\theta, \phi)$ taken at the same time. The summation over μ and μ' may now be performed leading to the following result

$$G_{(\lambda\lambda')k\kappa}(\vartheta, \xi, \xi')$$

$$\approx -\frac{1}{\bar{\xi}}\begin{pmatrix}\lambda & \lambda' & k\\ 0 & 0 & 0\end{pmatrix}\frac{(2\lambda-1)!!\,(2\lambda'-1)!!\,(k-1)!}{(\lambda-1)!\,(\lambda'-1)!\,(2k-1)!!}P_{(\lambda\lambda')k\kappa}(\vartheta, \tilde{\xi}), \tag{10}$$

where

$$P_{(\lambda\lambda')k\kappa}(\vartheta, \tilde{\xi}) = \sqrt{\frac{\pi}{2k+1}}\,\frac{(2k-1)!!}{(k-1)!}\,a^{\lambda+\lambda'+1}v$$

$$\times \int_{-\infty}^{+\infty} \frac{1}{r^{\lambda+\lambda'+2}}\, Y_{k\kappa}(\theta, \phi) \exp\left(\mathrm{i}\frac{v}{a}\tilde{\xi}t\right) \mathrm{d}t. \tag{11}$$

Since the orbital integral is independent of the energy of the intermediate state, the tensor $P_{(\lambda\lambda')k\kappa}$ vanishes only proportional to $1/\bar{\xi}$ or $1/(E_z - E_0)$ for large values of the intermediate energy E_z.

The orbital integrals appearing in eq. (11) can be expressed in terms of the Coulomb excitation functions $I_{\lambda\mu}$. In the coordinate system A one finds

$$P^{A}_{(\lambda\lambda')k\kappa} = \sqrt{\frac{\pi}{2k+1}}\, Y_{k\kappa}(\tfrac{1}{2}\pi, 0)\, \frac{(2k-1)!!}{(k-1)!}\, I_{\lambda+\lambda'+1,\kappa}(\vartheta, \tilde{\xi}). \tag{12}$$

Due to the selection rules of the 3-j symbol and eq. (II.9.16) one sees that $(\lambda + \lambda' + 1) + \kappa$ is an odd number and the Coulomb excitation functions appearing in eq. (12) are thus different from those used in first order perturbation theory. Therefore, they cannot be expressed by the orbital integrals $R_{\lambda\mu}$.

Similarly, one finds in coordinate system B

$$P^{B}_{(\lambda\lambda')k\kappa} = \sqrt{\frac{\pi}{2k+1}}\, \frac{(2k-1)!!}{(k-1)!} \sum_{\kappa'} D^{k}_{\kappa\kappa'}(0, -\tfrac{1}{2}\pi, 0)\, Y_{k\kappa'}(\tfrac{1}{2}\pi, 0)\, I_{\lambda+\lambda'+1,\kappa'}(\vartheta, \tilde{\xi}). \tag{13}$$

It is noted that for backwards scattering where the orbital integrals are independent of κ', the expression (13) reduces to

$$P^{B}_{(\lambda\lambda')k\kappa}(\pi, \tilde{\xi}) = \delta_{\kappa 0}\, \frac{(2k-1)!!}{2(k-1)!}\, I_{\lambda+\lambda'+1,0}(\pi, \tilde{\xi}). \tag{14}$$

If, furthermore, $\tilde{\xi} = 0$ one finds from (IV.2.10)

$$P^{B}_{(\lambda\lambda')k\kappa}(\pi, 0) = \delta_{\kappa 0}\, \frac{(2k-1)!!\,(\lambda+\lambda')!}{(k-1)!\,(2\lambda+2\lambda'+1)!!}. \tag{15}$$

§ 3. *Excitation probabilities and cross sections*

In this section we shall utilize the results of the previous sections to derive expressions for the excitation probabilities and cross sections to second order. A discussion of the application of these formulae will be given later in this chapter.

The excitation probability of the state $|f\rangle$ is found from eq. (III.1.2) by inserting the result (1.2), (IV.3.7) and (1.15). The result may be written

$$P_{f} = P^{(1)}_{f} + P^{(1,2)}_{f} + P^{(2)}_{f}, \tag{1}$$

where $P^{(1)}_{f}$ is the first-order excitation probability (IV.3.2), i.e.

$$P^{(1)}_{f} = \sum_{\lambda} |\chi^{(\lambda)}_{0\to f}|^{2} R^{2}_{\lambda}(\vartheta, \xi_{f0}). \tag{2}$$

The second term in eq. (1) is the interference between the first- and second-order amplitudes, i.e.

$$P_{\mathrm{f}}^{(1,2)} = \sum_{\lambda\lambda'\lambda''I_z} \sqrt{(2I_z+1)(2\lambda+1)(2\lambda'+1)(2\lambda''+1)}(-1)^{I_0+I_{\mathrm{f}}}$$

$$\times \begin{Bmatrix} \lambda & \lambda' & \lambda'' \\ I_z & I_{\mathrm{f}} & I_0 \end{Bmatrix} \chi_{0\to\mathrm{f}}^{(\lambda)} \chi_{0\to z}^{(\lambda')} \chi_{z\to\mathrm{f}}^{(\lambda'')} \sum_{\mu} R_{\lambda\mu}^{*}(\vartheta, \xi_{\mathrm{f}0})\, G_{(\lambda'\lambda'')\lambda\mu}(\vartheta, \xi_{z0}, \xi_{\mathrm{f}z}), \quad (3)$$

while the third term is the square of the second-order amplitude

$$P_{\mathrm{f}}^{(2)} = \frac{1}{4} \sum_{\substack{\lambda_1\lambda_1'\lambda_2\lambda_2' \\ I_z I_{z'} k}} \sqrt{(2I_z+1)(2I_{z'}+1)(2\lambda_1+1)(2\lambda_1'+1)(2\lambda_2+1)(2\lambda_2'+1)}$$

$$\times (2k+1) \begin{Bmatrix} \lambda_1 & \lambda_2 & k \\ I_{\mathrm{f}} & I_0 & I_z \end{Bmatrix} \begin{Bmatrix} \lambda_1' & \lambda_2' & k \\ I_{\mathrm{f}} & I_0 & I_{z'} \end{Bmatrix} \chi_{0\to z}^{(\lambda_1)} \chi_{z\to\mathrm{f}}^{(\lambda_2)}$$

$$\times \chi_{0\to z'}^{(\lambda_1')} \chi_{z'\to\mathrm{f}}^{(\lambda_2')} \sum_{\kappa} [R_{(\lambda_1\lambda_2)k\kappa}^{*} R_{(\lambda_1'\lambda_2')k\kappa} + G_{(\lambda_1\lambda_2)k\kappa}^{*} G_{(\lambda_1'\lambda_2')k\kappa}]. \quad (4)$$

The expressions (3) and (4) are most easily obtained in the coordinate systems A and B, where the orbital integrals are real. Since the combinations of these integrals appearing in (3) and (4) are scalars under rotation, the result is valid in any coordinate system. We have utilized that $\lambda + \lambda' + \lambda''$ or that $\lambda_1 + \lambda_1' + \lambda_2 + \lambda_2'$ are even numbers.

As was noted earlier, terms of a similar order of magnitude as $P^{(2)}$ may arise from interference between the first-order and the third-order excitation amplitudes. Only if the first-order amplitude is very small, as can happen e.g. if the spins of the involved states are such that the direct excitation must take place through a high multipole order, the second-order term $P^{(2)}$ is significant. The study of the interference of first and second order includes the so-called reorientation effect, while the study of the excitations which cannot take place through a first-order process is called double excitation. A more detailed discussion of the information which can be extracted from these two types of experiments will be given in § 5 and § 6 below.

A common problem for all studies of second- or higher-order processes is the summation over the intermediate states which occurs in eqs. (3) and (4). In principle, this summation should be carried out over all states of the nucleus including the very high excited states as well as the initial and final states. The contribution arising from a given intermediate state will depend on the product $\chi_{0\to z}\chi_{z\to\mathrm{f}}$ as well as on the values of the parameters ξ_{z0} and $\xi_{\mathrm{f}z}$. The dependence of the functions R and G on the ξ's which was studied in § 2 shows a significant difference for large values of the excitation energy of the intermediate state. Thus the function R vanishes exponentially for high values of the energy E_z of the intermediate state, while the function G decreases only as $(E_z - E_0)^{-1}$. From this argument one may thus not exclude the possibility that highly excited states could give a significant contribution.

In general, however, the product $\chi_{0\to z}\chi_{z\to f}$ becomes small for highly excited states, since the oscillator strength for low multipole orders except for $\lambda = 1$ is concentrated at low excitation energies. The conclusion is thus that the summations over the intermediate states should include the states in the low-lying spectrum which are strongly coupled to the ground state and that besides these states the high-lying states of the giant dipole resonance may give a contribution.

In the sudden approximation the contribution from the terms containing G must vanish, if the summation is carried out over all intermediate states. If the summations in eqs. (3) and (4) are carried out over a finite number of states, the quantities $G_{(\lambda\lambda')k\kappa}$ are, in general, non-vanishing. It is noticed, however, that the interference term $P^{(1,2)}$ in eq. (3) nevertheless vanishes identically as is seen in coordinate system B where the selection rules for $\xi = 0$ demand μ even for the orbital integrals $R_{\lambda\mu}$ and μ odd for the double integrals $G_{(\lambda\lambda')k\mu}$ (c.f. table 1).

§ 4. *Angular distribution of gamma rays*

The angular distribution of the de-excitation γ-rays can be evaluated to second order by inserting eqs. (1.2), (IV.3.7) and (1.15) into the general expressions of chapter III. In order to compare the result with the expressions obtained in first-order perturbation theory it is convenient to express the statistical tensor in a form similar to eq. (IV.4.14). The general structure of this expression relies on the form (IV.3.7) of the excitation amplitude, where the dependence on the spin quantum numbers is contained in a 3-j symbol. It is always possible to write an excitation amplitude in such a form, since we may generally expand $a_{I_f M_f, I_0 M_0}$ into tensor components according to the expression [ALD 62]

$$a_{I_f M_f, I_0 M_0} = \sum_l (-1)^{I_0 - M_0}\sqrt{2I_0 + 1}\begin{pmatrix} I_0 & l & I_f \\ -M_0 & \nu & M_f \end{pmatrix}[A_{l\nu} + iB_{l\nu}]. \qquad (1)$$

We have here introduced the two quantities $A_{l\nu}$ and $B_{l\nu}$ which transform as tensors of order l, ν under rotations and which are real in the coordinate systems A and B. Comparing expression (1) with eq. (IV.3.7) one observes that in first-order perturbation theory

$$A_{l\nu}^{(1)} = 0$$

and (2)

$$B_{l\nu}^{(1)} = -\sqrt{2l+1}\,\chi_{0\to f}^{(l)} R_{l\nu}(\vartheta, \xi).$$

The corresponding quantities to second order can be seen by inspection from eq. (1.15) to be given by

$$\begin{aligned} A_{l\nu} &= A_{l\nu}^{(1)} + A_{l\nu}^{(2)}, \\ B_{l\nu} &= B_{l\nu}^{(1)} + B_{l\nu}^{(2)}, \end{aligned} \tag{3}$$

where

$$A_{l\nu}^{(2)} = -\tfrac{1}{2} \sum_{\lambda\lambda' z} (2l+1)\sqrt{(2I_z+1)(2\lambda+1)(2\lambda'+1)}(-1)^{I_0+I_f+l+\lambda+\lambda'} \times \begin{Bmatrix} I_0 & I_f & l \\ \lambda' & \lambda & I_z \end{Bmatrix} \chi_{0\to z}^{(\lambda)} \chi_{z\to f}^{(\lambda')} R_{(\lambda\lambda')l\nu}(\vartheta, \xi_{z0}, \xi_{fz}) \tag{4}$$

and

$$B_{l\nu}^{(2)} = -\tfrac{1}{2} \sum_{\lambda\lambda' z} (2l+1)\sqrt{(2I_z+1)(2\lambda+1)(2\lambda'+1)}(-1)^{I_0+I_f+l+\lambda+\lambda'} \times \begin{Bmatrix} I_0 & I_f & l \\ \lambda' & \lambda & I_z \end{Bmatrix} \chi_{0\to z}^{(\lambda)} \chi_{z\to f}^{(\lambda')} G_{(\lambda\lambda')l\nu}(\vartheta, \xi_{z0}, \xi_{fz}). \tag{5}$$

Using the form (1) for the excitation amplitudes one finds that the statistical tensor (III.2.11) can be written in the

$$\rho_{k\kappa}(I_f, I_f) = \sqrt{2I_f+1} \sum_{ll'\nu\nu'} (-1)^{I_0+I_f+\nu} \begin{Bmatrix} l & l' & k \\ I_f & I_f & I_0 \end{Bmatrix} \times \begin{pmatrix} l & l' & k \\ \nu & -\nu' & -\kappa \end{pmatrix} [A_{l\nu} + iB_{l\nu}][A_{l'\nu'}^{*} - iB_{l'\nu'}^{*}]. \tag{6}$$

The integers l and l' appear in this formula as effective multipole orders for the excitation process. For the excitation of a state of spin I in an even-even nucleus one thus finds the unique values of l and l'

$$l = l' = I \tag{7}$$

irrespective of the excitation mechanism of this state.

In the following discussion we assume that the excitation amplitudes $A_{l\nu}$ and $B_{l\nu}$ have been evaluated in coordinate system B where especially simple symmetry relations hold for the statistical tensor (6). Thus, it follows from the symmetry relations of table III.1 that for even $k+\kappa$, only the real part of the product $(A_{l\nu} + iB_{l\nu})(A_{l'\nu'} - iB_{l'\nu'})$ contributes, while for odd $k+\kappa$, only the purely imaginary part of this product contributes. It is noted that in contrast to the first-order theory both types of terms appear in second order.

The explicit expressions for the γ-angular distribution under various geometries can be taken over directly from the results quoted in § IV.4. For the γ-angular distribution in the coordinate system C, where the z-axis is chosen along the direction of the incoming projectile one finds in analogy to eq. (IV.4.22)

$$W^{\mathrm{C}} = \sum_{k\kappa} \left\{ \sum_{ll'} F_k(ll'I_0I_f) b^{ll'}_{k\kappa} \right\} \left\{ \sum_{LL'} F_k(LL'I_{ff}I_f) \delta_L \delta^*_{L'} \right\} Y_{k\kappa}(\vartheta^{\mathrm{C}}_{\gamma}, \varphi^{\mathrm{C}}_{\gamma}), \tag{8}$$

where k is even, while κ may be both even and odd.

In analogy to eq. (IV.4.23) the particle parameters $b^{ll'}_{k\kappa}$ are given by

$$\begin{aligned} b^{ll'}_{k\kappa} = & -\{(2k+1)(2l+1)(2l'+1)\}^{-1/2} \begin{pmatrix} l & l' & k \\ 1 & -1 & 0 \end{pmatrix}^{-1} \\ & \times \sum_{\nu\nu'\kappa'} (-1)^{\nu'} \begin{pmatrix} l & l' & k \\ \nu & -\nu' & \kappa' \end{pmatrix} [A_{l\nu} + \mathrm{i}B_{l\nu}][A_{l'\nu'} - \mathrm{i}B_{l'\nu'}] \\ & \times D^{k*}_{\kappa'\kappa}(\tfrac{1}{2}\pi, \tfrac{1}{2}(\pi + \vartheta), \pi), \end{aligned} \tag{9}$$

where the amplitudes $A_{l\nu}$ and $B_{l\nu}$ are evaluated in the coordinate system B. With the help of the expressions (3), (8) and (9) it is possible to discuss in a rather simple way the second-order corrections to the first-order angular distributions given in § IV.4. One notices that there are two types of corrections.

Firstly, in the terms where κ is even and where the real part of eq. (9) contributes, the coefficients are changed in comparison to the first-order results (IV.4.23). The corrections are partly linear in χ due to interference between the terms $B^{(1)}$ and $B^{(2)}$, partly quadratic in χ due to terms of the type $A^{(2)}A^{(2)}$ or $B^{(2)}B^{(2)}$. As was noted earlier, quadratic terms in χ of similar order of magnitude may arise from the interference between first- and third-order amplitudes. Only if the first-order amplitude is very small the quadratic corrections in χ are significant.

Secondly, there will appear new terms with odd κ. These terms are partly linear in χ, arising from the interference of $A^{(2)}$ with $B^{(1)}$ and partly quadratic in χ (terms of the type $A^{(2)}B^{(2)}$). It is thus in principle possible to distinguish experimentally between second-order contributions of type $A_{l\nu}$ and $B_{l\nu}$. Here again, it should be emphasized that the angular distribution of γ-quanta (8) is not very sensitive to second-order effects. As was discussed in chapter III one finds especially for scattering in the backwards hemisphere that the angular distribution is essentially determined purely geometrically, since the excitation mainly proceeds under conservation of the magnetic quantum number in the coordinate system B.

§5. *Double excitations*

In the previous sections we have derived the general expressions for Coulomb excitation cross sections and angular distributions to second-order perturbation theory. A major problem in the evaluation of these expressions is the summation over the intermediate states $|z\rangle$. In order to obtain a survey over the effects of various intermediate states, we shall in this and the following sections consider the idealized case where only one state $|z\rangle$ is of importance. Double excitation occurs when this intermediate state is a low-lying energy level. Static moment effects (§ 6) occur when the intermediate state coincides with the initial or the final state and the polarization effect (§ 7) is the contribution of high-lying intermediate states.

In this section we shall thus consider the situation in which a state with spin I_f can be excited directly or through a low-lying state of spin I_z (see fig. 1).

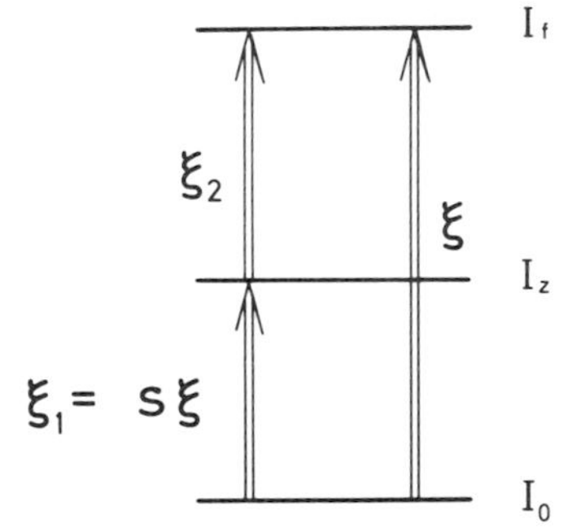

Fig. 1. Schematic picture of double excitation. The figure illustrates the excitation of the state of spin I_f through a second-order process via a state of spin I_z. It is assumed that the direct excitation $I_0 \rightarrow I_f$ is forbidden or small. The adiabaticity parameters ξ are indicated.

A pure double excitation is realized in practice if the direct excitation $I_0 \rightarrow I_f$ is small or vanishing, i.e. if

$$\chi_{0\rightarrow f} \ll \chi_{0\rightarrow z}\chi_{z\rightarrow f}. \tag{1}$$

Under these circumstances the second-order excitation probability $P^{(2)}$ in eq. (3.4) is larger than the interference term $P^{(1,2)}$ in eq. (3.3), and the terms

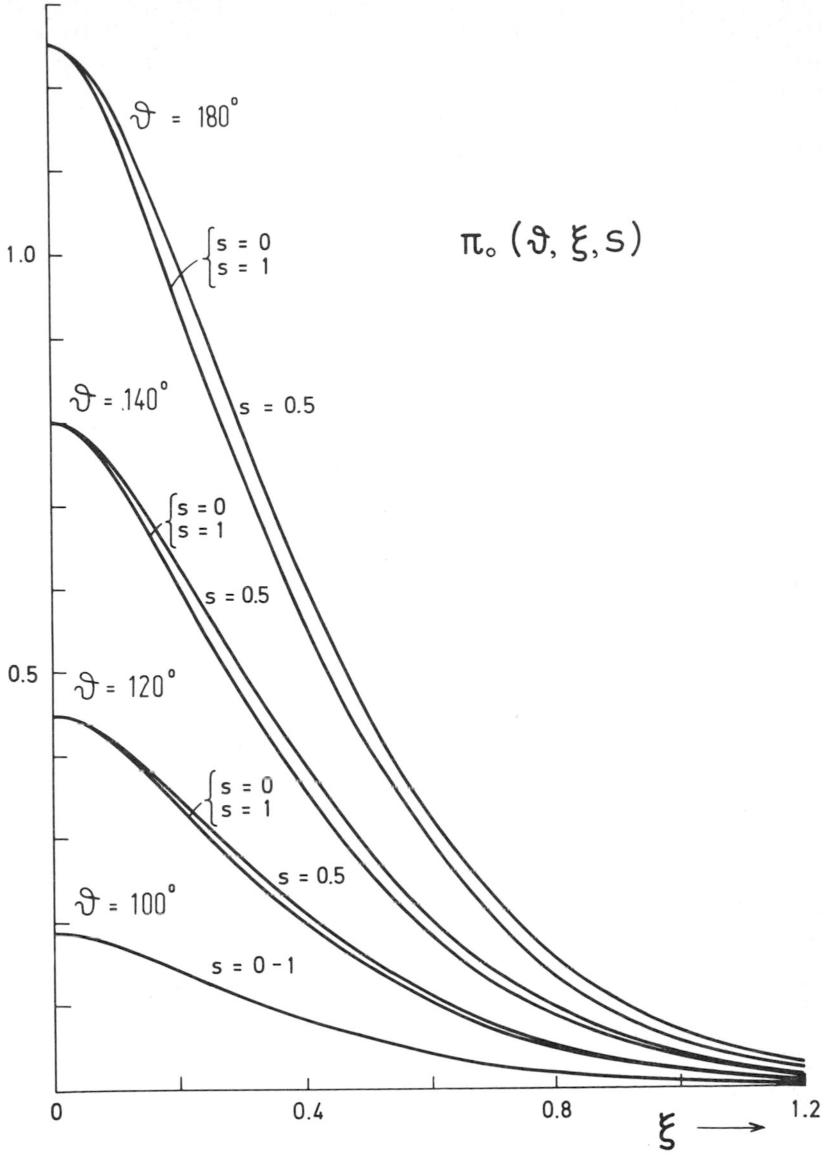

Fig. 2. Double excitation of a state with spin 0 via a 2^+ state. The relative excitation probability $\pi_0(\vartheta, s, \xi)$ is plotted as a function of the total ξ, corresponding to a transition 0 to 0 for $\vartheta = 180°$, 140°, 120° and 100°. The ξ-value of the intermediate state is defined as $s\xi$. For each angle, the curves for different values of $0 < s < 1$ are indicated.

arising from the static moments as well as the terms from third-order perturbation can be neglected. While the general formula for $P^{(2)}$ is given in § 3 we shall here discuss only two special cases which are of practical interest and which serve as illustrations of double excitation.

Firstly, we consider a pure double quadrupole excitation in an even-even nucleus of a state with spin 0^+ through an intermediate state of spin 2^+. A direct transition $0^+ \to 0^+$ is not possible and we therefore obtain to second order the following excitation probability

$$P_0^{(2)} = |\chi_{0\to 2}^{(2)}|^2 |\chi_{2\to 0}^{(2)}|^2 \pi_0(\vartheta, s, \xi) \tag{2}$$

with

$$\pi_0(\vartheta, s, \xi) = \tfrac{25}{4}\{|R_{(22)00}|^2 + |G_{(22)00}|^2\}. \tag{3}$$

Through the measurements of the excitation cross section one may thus determine the product $|\chi_{0\to 2}^{(2)}\chi_{2\to 0}^{(2)}|^2$ and since $\chi_{0\to 2}^{(2)}$ can be measured separately by a first-order excitation of the 2^+ state one may therefore determine the magnitude of the quadrupole matrix element between the excited states of spin 2 and 0. The maximum value of $P_0^{(2)}$ is obtained for $\vartheta = \pi$ and $\xi_1 = \xi_2 = 0$, where

$$P_0^{(2)}(\pi, 0, 0) = \tfrac{5}{4}|\chi_{0\to 2}^{(2)}|^2 |\chi_{2\to 0}^{(2)}|^2. \tag{4}$$

In fig. 2 we have plotted the relative probability $\pi_0(\vartheta, s, \xi)$ as a function of the ξ value which corresponds to a direct transition to the 0^+ state, i.e.

$$\xi = \xi_1 + \xi_2. \tag{5}$$

As a second parameter we choose the ratio

$$s = \frac{\xi_1}{\xi_1 + \xi_2} = \frac{E_z - E_0}{E_f - E_0} \tag{6}$$

which is always a positive number. When s varies from zero to one, the intermediate state energy changes from the ground state to the excited 0^+ state. From fig. 2 it is seen that the excitation probability does not depend very strongly on the value of s. The curves which are obtained are similar to the first-order excitation probabilities. Comparing the results with the relative excitation probabilities in table IV.3 it is seen, however, that the double excitation depends less strongly on ξ than in first-order quadrupole excitation. The ξ-dependence is actually rather similar to the ξ-dependence of an octupole excitation.

As an example we illustrate in figs. 3a–c the case of ^{72}Ge where the lowest excited state at 691 keV has spin 0^+. It has been excited by a double E2 process via the 834 keV 2^+ state and was detected by observing the conversion electrons of the 691 keV and 834 keV transitions.

A case very similar to the one discussed above is that of a double quadru-

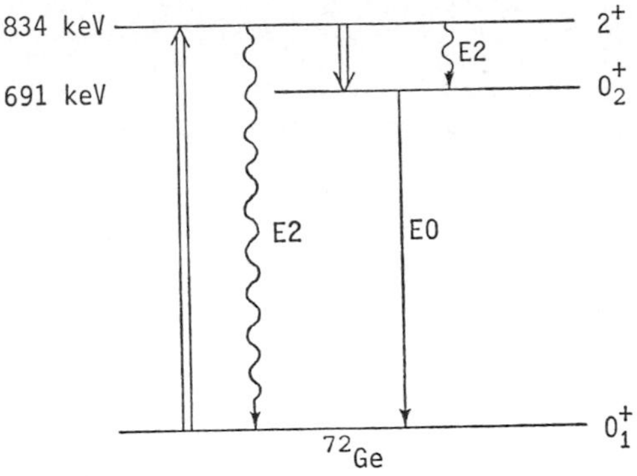

Fig. 3a. Level scheme of ^{72}Gc. The parameter s, defined in eq. (6), is equal to 1.21. The 691 keV state decays by E0 conversion. The conversion coefficient of the 834 keV transition is 5.43×10^{-3}.

pole excitation leading to a state of spin 4^+ through an intermediate state of spin 2^+. The direct excitation can only take place through an E4 interaction which is usually quite weak.

Inserting the appropriate spins in eq. (3.4) one obtains the following expression for the double excitation probability

$$P_4^{(2)} = |\chi_{0\to 2}^{(2)}|^2 |\chi_{2\to 4}^{(2)}|^2 \pi_4(\vartheta, s, \xi) \tag{7}$$

with

$$\pi_4(\vartheta, s, \xi) = \tfrac{25}{4} \sum_\kappa \{|R_{(22)4\kappa}|^2 + |G_{(22)4\kappa}|^2\}. \tag{8}$$

Similarly one may here determine the quantity $|\chi_{2\to 4}|^2$ from a measurement of the excitation probability $P_4^{(2)}$. In this case the maximum value of the probability is given by

$$P_4^{(2)}(\pi, 0, 0) = \tfrac{5}{14} |\chi_{0\to 2}^{(2)}|^2 |\chi_{2\to 4}^{(2)}|^2 . \tag{9}$$

In fig. 4 we have plotted the relative excitation probability $\pi_4(\vartheta, s, \xi)$ as a function of ξ for various values of ϑ and s (see eqs. (5) and (6)). The results are very similar to the results displayed in fig. 2. Actually, one may show that, except for the normalization factor $\frac{2}{7}$ the relative excitation probabilities are identical for $\vartheta = 180°$ as well as for $\xi = 0$.

The validity of the expression (7) for the excitation probability of the 4^+ state depends on the smallness of the direct transition $0 \to 4$, i.e. more precisely on the condition

$$|x| = |\chi_{0\to 4}^{(4)}| / |\chi_{0\to 2}^{(2)} \chi_{2\to 4}^{(2)}| \ll 1. \tag{10}$$

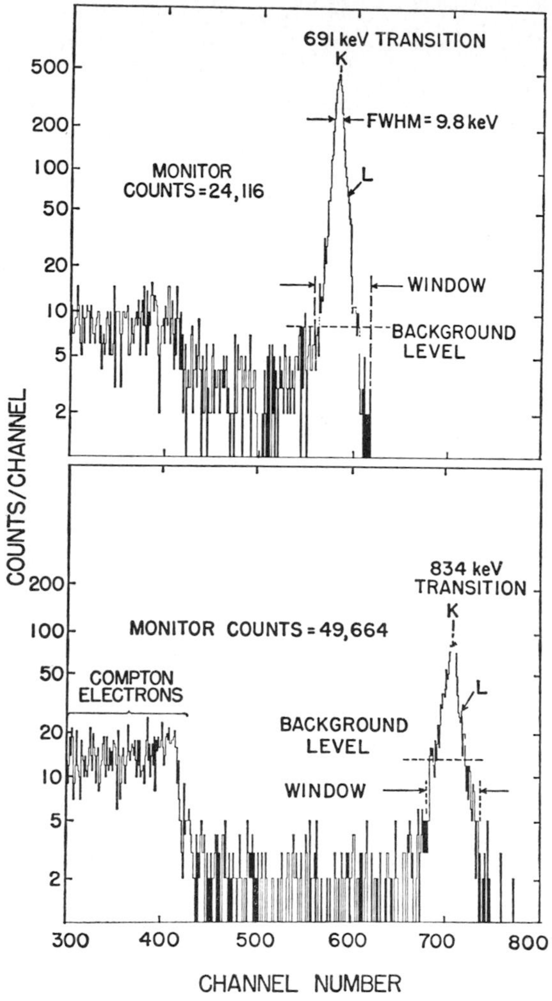

Fig. 3b. Conversion electron spectra during the bombardment of natural germanium with 36 MeV ^{16}O ions showing the two peaks of interest (from Haight, R. C., 1972, Phys. Rev. **C5**, 1984).

We may evaluate the quantity x explicitly from eq. (IV.3.6), in terms of the reduced matrix elements, i.e.

$$x = \sqrt{\frac{5}{4\pi}}\,\frac{5}{7}\,\frac{\hbar v}{Z_1 e}\,\frac{\langle 0\|\,\mathscr{M}(E4)\,\|4\rangle}{\langle 0\|\,\mathscr{M}(E2)\,\|2\rangle\,\langle 2\|\,\mathscr{M}(E2)\,\|4\rangle}. \tag{11}$$

In the cases where there is a sizeable double excitation this quantity is usually small, but still it may be possible to observe the interference between the

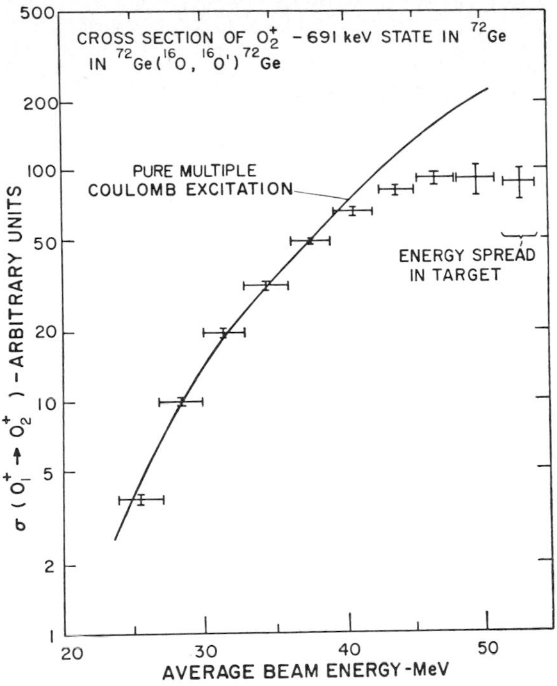

Fig. 3c. Excitation function of the 691-keV (0_2^+) state in ^{72}Ge (from Haight, R. C., 1972, Phys. Rev. **C5**, 1984).

direct transition and the double excitation even when the first-order excitation, being quadratic in $\chi^{(4)}_{0\to 4}$, can not be detected. Correcting expression (7) for the interference, one obtains, according to eq. (3.3)

$$P_4 = |\chi^{(2)}_{0\to 2}|^2|\chi^{(2)}_{2\to 4}|^2\pi_4(\vartheta, s, \xi)[1 + d(\vartheta, s, \xi)x + a(\vartheta, s, \xi)x^2], \qquad (12)$$

where

$$d(\vartheta, s, \xi) = \frac{5}{\pi_4}\sum_\mu R_{4\mu}(\vartheta, \xi_{0\to 4})G_{(22)4\mu}(\vartheta, \xi_{0\to 2}, \xi_{2\to 4}) \qquad (13)$$

and

$$a(\vartheta, s, \xi) = R_4^2(\vartheta, \xi)/\pi_4(\vartheta, s, \xi). \qquad (14)$$

In the excitation probability (12) we have also included the direct first-order E4 transition which is quadratic in x. According to the observation made above that the double excitation probabilities behave very much like a first-order process as a function of ϑ and ξ independent of s, the coefficient a

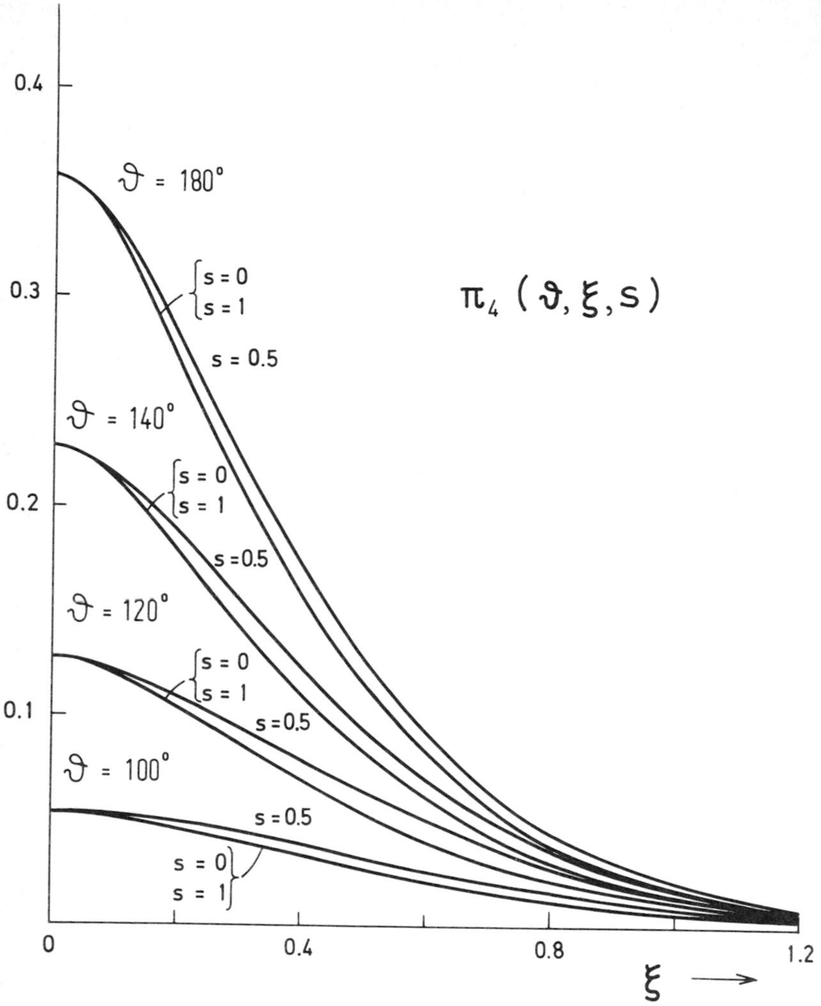

Fig. 4. Double excitation of a state with spin 4 via a 2^+ state. The relative excitation probability $\pi_4(\vartheta, s, \xi)$ is plotted as a function of the total ξ corresponding to a transition 0 to 4 for $\vartheta = 180°$, 140°, 120° and 100°. The ξ-value of the intermediate state is defined as $s\xi$. For each angle, the curves for different values of $0 < s < 1$ are indicated.

does not depend very strongly on these parameters. The coefficient a may therefore be estimated by the value at $\vartheta = \pi$ and $\xi = 0$, i.e.

$$a(\pi, s, 0) = \tfrac{14}{5} \tag{15}$$

being slightly larger for other values of ϑ and ξ. The coefficient $d(\vartheta, s, \xi)$ is

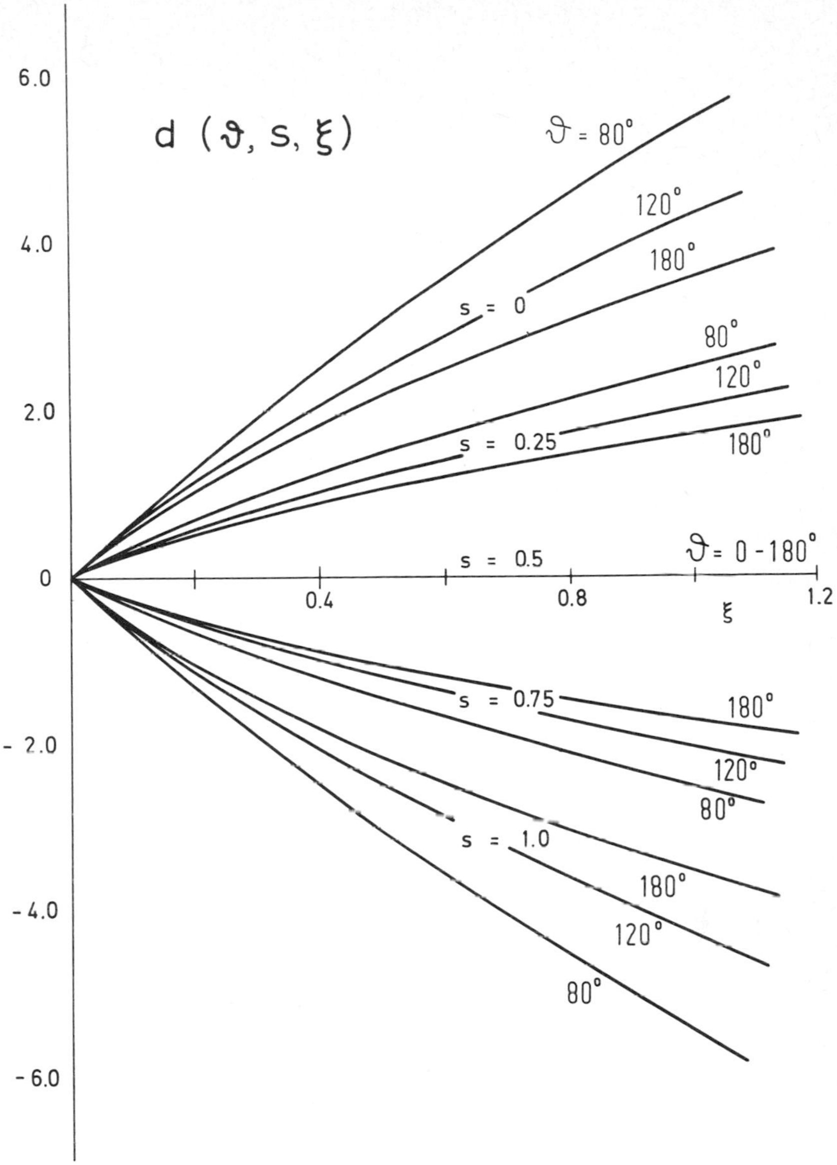

Fig. 5. The interference between double excitation of a 4^+ state and the direct excitation. The coefficient $d(\vartheta, s, \xi)$ is plotted as a function of ξ for various values of s and ϑ (see eq. (5.13)).

illustrated in fig. 5 as a function of ξ for various values of s and ϑ. It is seen that d is roughly proportional to ξ and to the deviation of s from the value $\frac{1}{2}$. Hence, it vanishes for $s = \frac{1}{2}$, which is the case where the energy of the intermediate state is just the half of the energy of the 4^+ state. An example for the double E2 excitation of a 4^+ rotational state is shown in figs. 6a–c.

In most cases the direct excitation competes strongly with the double excitation. We shall discuss in the remainder of this section the important example which applies to an even-even nucleus where a state of spin 2^+ may be excited directly as well as through another state of spin 2. In such a case of strong interference one should neglect the contribution $P^{(2)}$ and the excitation probability of the state 2′ is thus given by

$$P_{2'} = |\chi^{(2)}_{0\to 2'}|^2 \sum_{\mu} |R_{2\mu}(\vartheta, \xi_{0\to 2'})|^2 + 5\chi^{(2)}_{0\to 2'}\chi^{(2)}_{0\to 2}\chi^{(2)}_{2\to 2'} \sum_{\mu} R_{2\mu}(\vartheta, \xi_{0\to 2'}) G_{(22)2\mu}(\vartheta, \xi_{0\to 2}, \xi_{2\to 2'}). \tag{16}$$

Through a measurement of the excitation probability of the state 2′ as a function of projectile charge or energy or as a function of the scattering angle one may utilize that the two terms in (16) have a different behavior as a function of these quantities and one may thus determine $|\chi^{(2)}_{0\to 2'}|^2$ and $\chi^{(2)}_{0\to 2'}\chi^{(2)}_{0\to 2}\chi^{(2)}_{2\to 2'}$ separately. The feasibility of such experiments can be discussed easily if one writes the excitation probability (16) in the form

$$P_{2'} = |\chi_{0\to 2'}|^2 R_2^2(\vartheta, \xi)[1 + yc(\vartheta, s, \xi)], \tag{17}$$

where

$$y = \frac{\chi_{0\to 2}\chi_{2\to 2'}}{\chi_{0\to 2'}} = \chi_{0\to 2}\frac{1}{\sqrt{5}}\frac{\langle 2\| \mathscr{M}(\mathrm{E2})\| 2'\rangle}{\langle 0\| \mathscr{M}(\mathrm{E2})\| 2'\rangle} \tag{18}$$

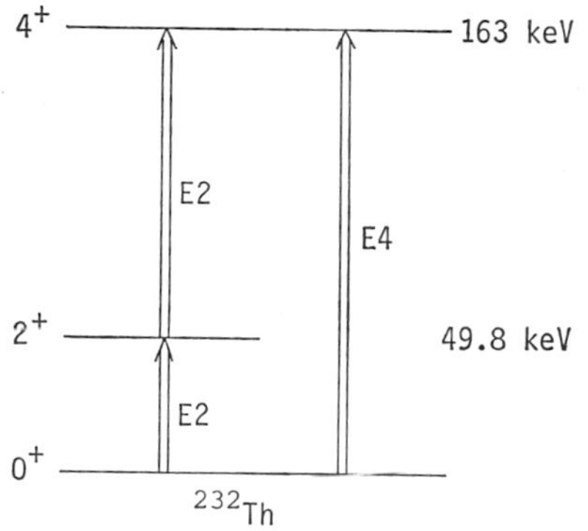

Fig. 6a. Level diagram of ^{232}Th. The parameter s, defined in eq. (6), is equal to 0.304.

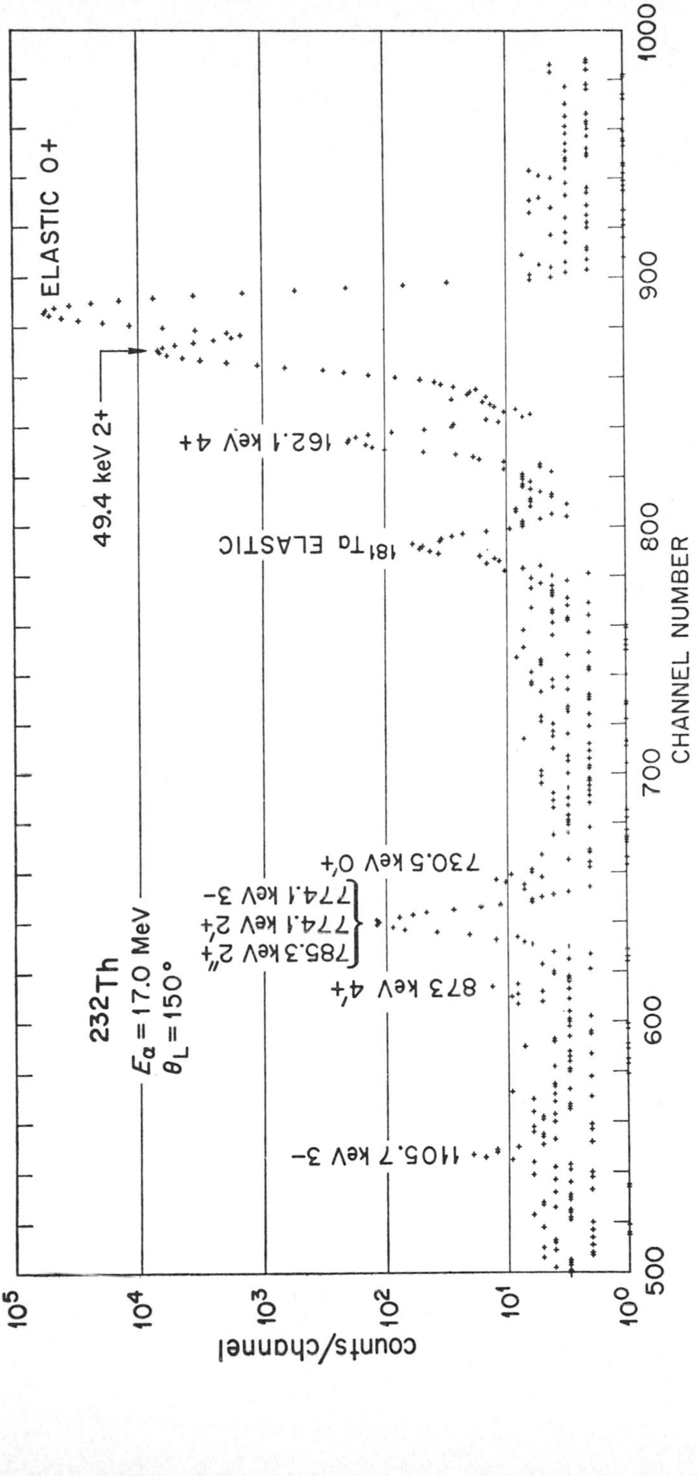

Fig. 6b. Elastically and inelastically scattered 17 MeV ^{4}He ions from ^{232}Th at a laboratory angle of 150° using a carbon foil as the target backing (from Bemis, C. E. et al., 1973, Phys. Rev., **C8**, 1466).

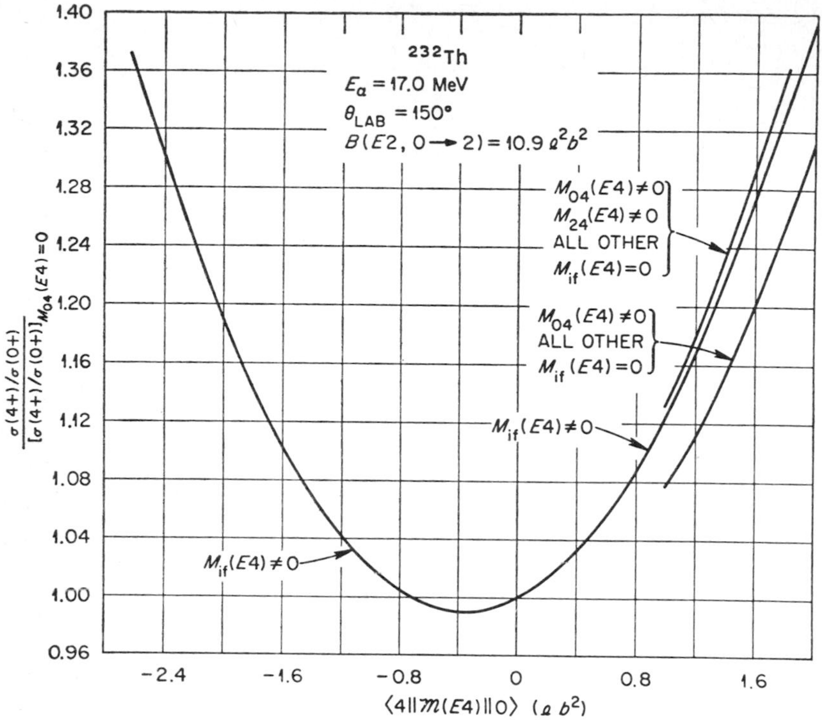

Fig. 6c. Influence of the E4 matrix element on the cross section of the 4^+ state. In the actual evaluation of the experiments, multiple excitation was taken into account (see Bemis, C. E. et al., 1973, Phys. Rev. **C8,** 1466).

and

$$c(\vartheta, s, \xi) = 5 \sum_{\mu} R_{2\mu}(\vartheta, \xi) G_{(22)2\mu}(\vartheta, \xi_{0\to2}, \xi_{2\to2'})/R_2^2(\vartheta, \xi). \tag{19}$$

In figs. 7a–c the coefficient $c(\vartheta, s, \xi)$ has been plotted as a function of ξ for various values of the deflection angle ϑ and the parameter s given in eq. (6).

It is noted that a measurement of the quantity y leads to a determination of the relative sign of the quadrupole matrix-elements between the states of spin 0, 2 and 2′, which is a quantity independent of the phase convention of the nuclear states. It should be noted that, in general, terms of equal magnitude may arise from the interference with the static quadrupole moment of the 2′ state discussed in the following section.

For the sake of completeness we illustrate in fig. 8 the relative excitation probability of a 2^+ state through pure double excitation with an intermediate

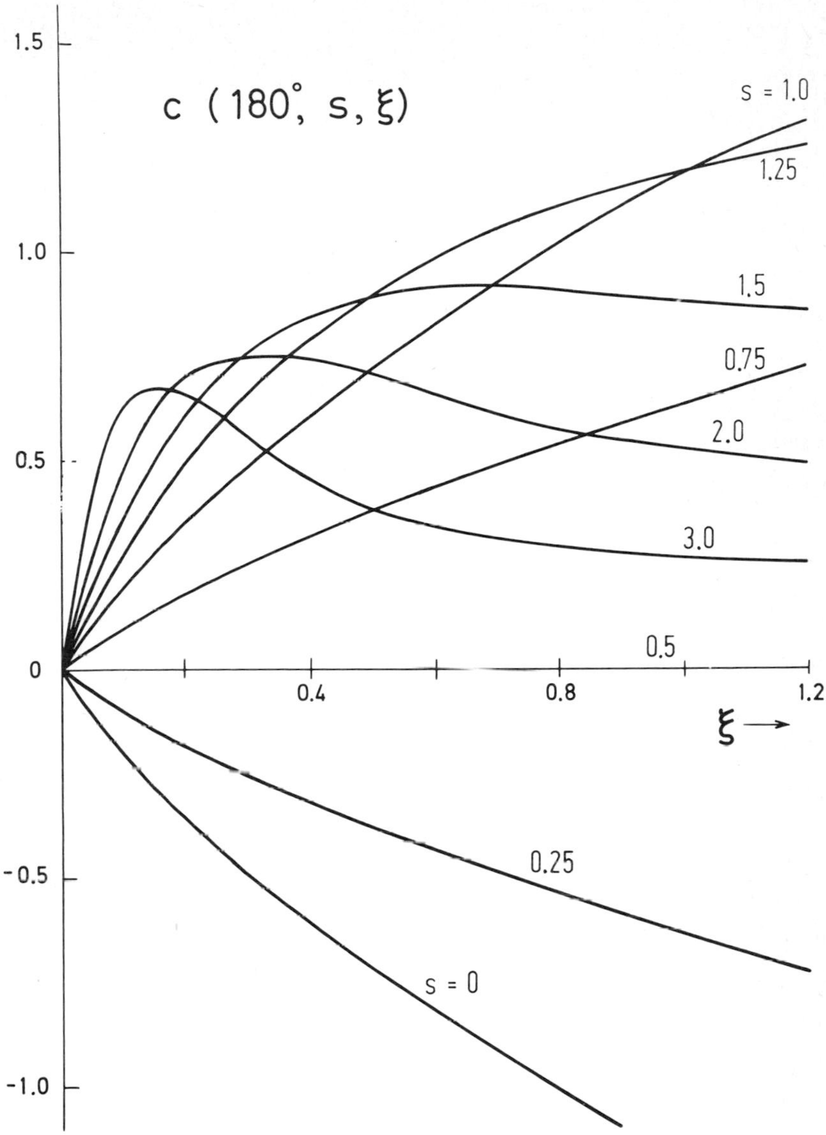

Fig. 7a.

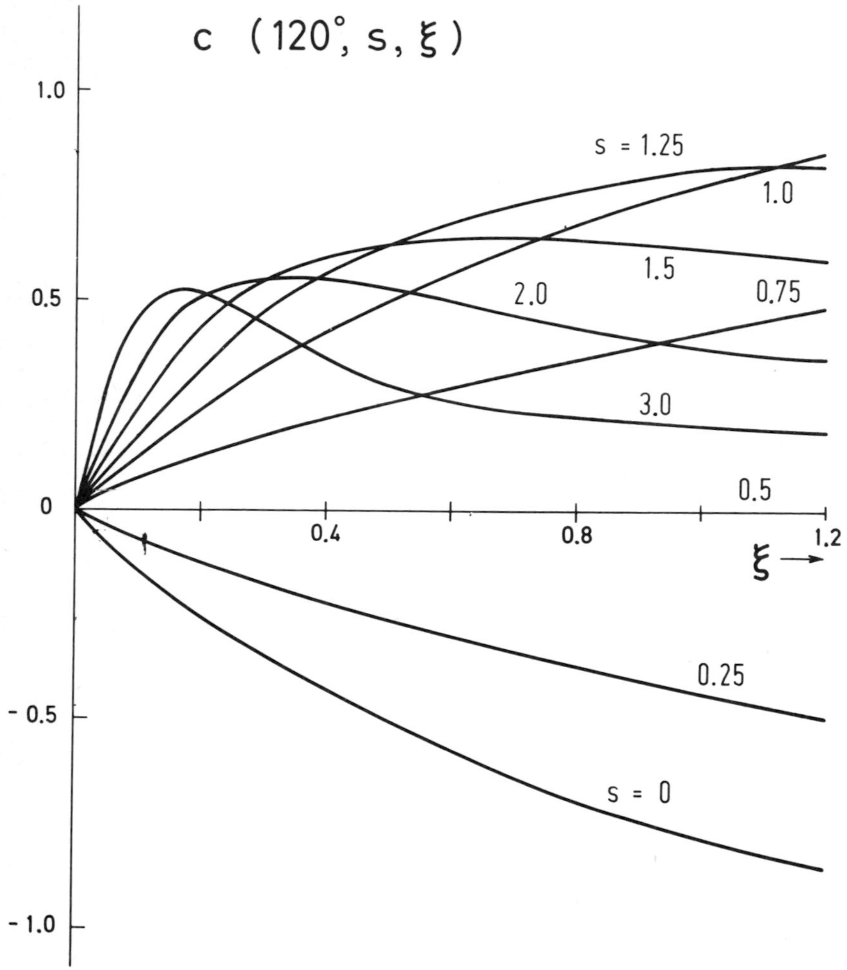

Fig. 7b.

state of spin 2^+. This function is defined in analogy to eqs. (2) and (7), i.e.

$$P_2^{(2)} = |\chi_{0\to 2}^{(2)}|^2 |\chi_{2\to 2'}^{(2)}|^2 \pi_2(\vartheta, s, \xi), \tag{20}$$

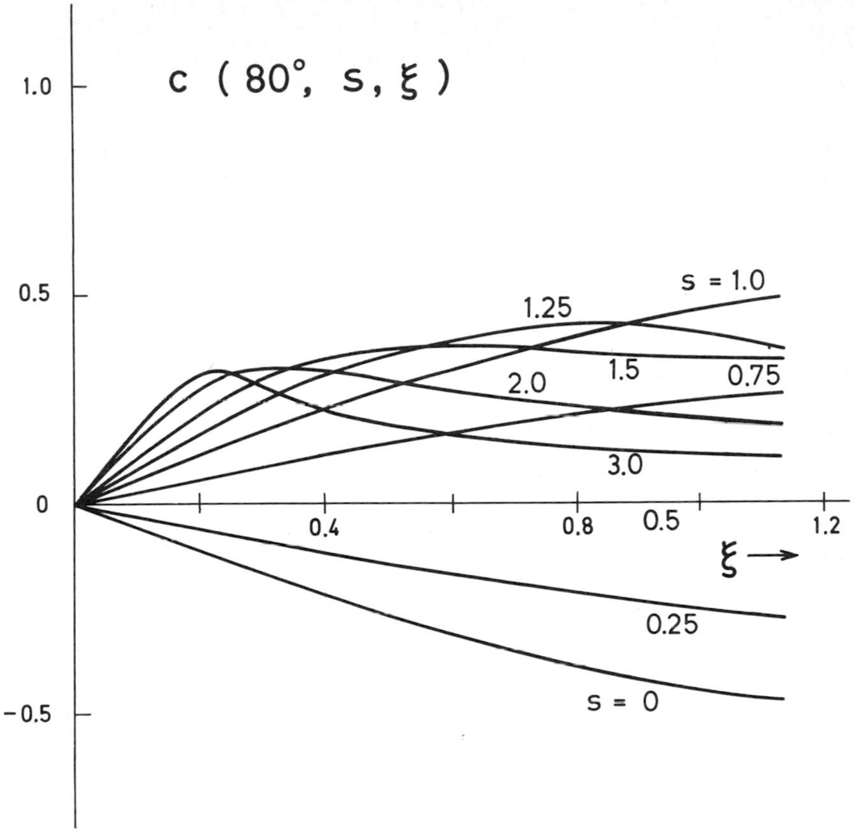

Fig. 7c.

Fig. 7. The interference between first-order Coulomb excitation and double excitation. The coefficient $c(\vartheta, s, \xi)$ defined in eq. (5.19) is plotted as a function of ξ for $\vartheta = 180°$, 120° and 80° for various values of the parameter s defined in eq. (5.6). The case $s > 1$ indicates a situation where the intermediate state lies above the final state.

where

$$\pi_2(\vartheta, s, \xi) = \tfrac{25}{4} \sum_{\kappa} \{|R_{(22)2\kappa}|^2 + |G_{(22)2\kappa}|^2\}. \tag{21}$$

As was noted above, terms of similar order may arise from the interference between first-order excitation and triple excitation.

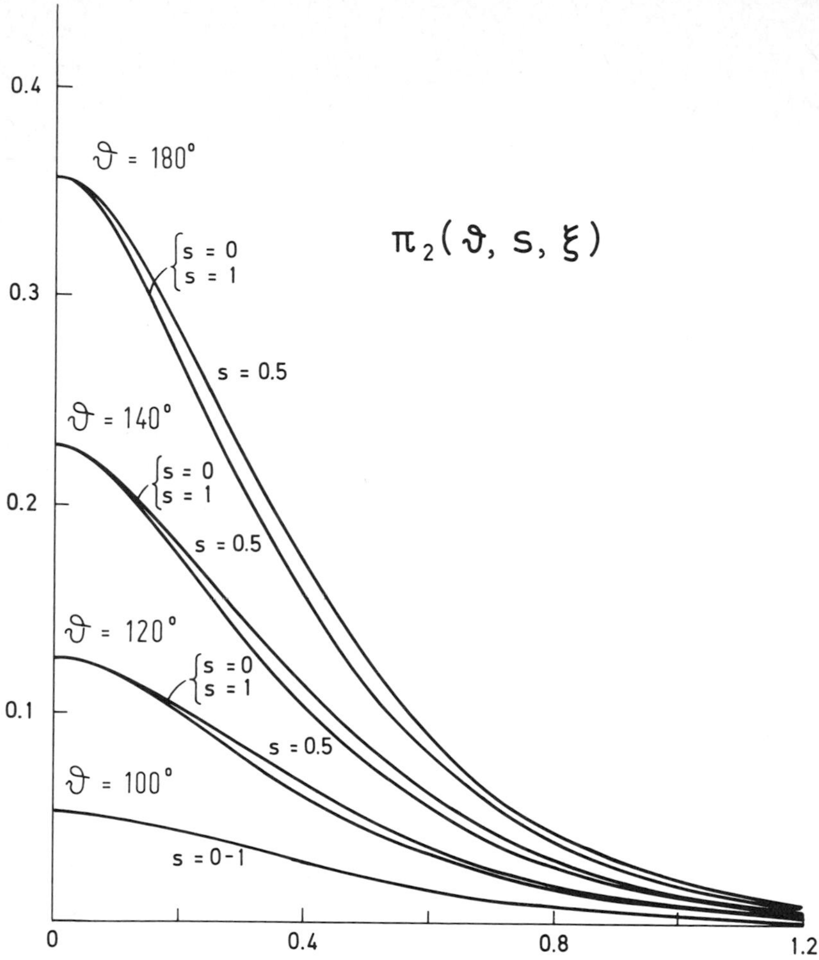

Fig. 8. Double excitation of a state with spin 2 via a 2^+ state. The relative excitation probability $\pi_2(\vartheta, s, \xi)$ is plotted as a function of the total ξ corresponding to a direct transition to the final state for $\vartheta = 180°$, 140°, 120° and 100°. The ξ-value of the intermediate state is defined as $s\xi$. For each angle the curves for different values of $0 < s < 1$ are indicated.

§ 6. *Effects of static moments*

As a second application of the expressions in § 3 we shall consider the case where the intermediate state is identical to the initial or the final state in a transition which may take place in first-order Coulomb excitation (see fig. 9).

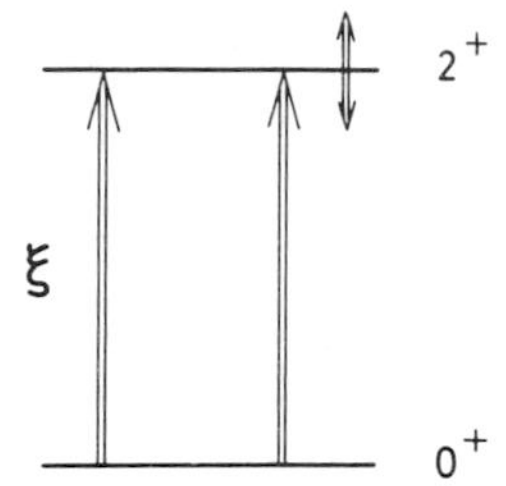

Fig. 9. Schematic picture of the effect of static moments. The figure illustrates the excitation of a 2^+ state in an even-even nucleus, where the direct first-order transition interferes with the second-order process, where the 2^+ state itself acts as intermediate state.

We consider first the excitation of a 2^+ state in an even-even nucleus, and we assume that the second-order contributions are small. For this situation we may use the expression (5.17) for the excitation probability with $2' = 2$, i.e.

$$P_2 = |\chi^{(2)}_{0\to 2}|^2 R_2^2(\vartheta, \xi)[1 + \chi^{(2)}_{2\to 2} c(\vartheta, s = 1, \xi)], \tag{1}$$

where $\xi = \xi_{0\to 2}$.

Through the measurement of P_2 one may determine both the magnitude and the sign of $\chi^{(2)}_{2\to 2}$. This quantity is proportional to the static quadrupole moment Q_2 of the 2^+ state, i.e.

$$\chi^{(2)}_{2\to 2} = \frac{4}{15}\sqrt{\frac{\pi}{5}}\frac{Z_1 e}{\hbar v}\frac{1}{a^2}\langle 2\| \mathscr{M}(\mathrm{E}2)\|2\rangle = \sqrt{\frac{7}{90}}\frac{Z_1 e}{\hbar v} Q_2$$

$$= 8.474 \frac{A_1^{1/2} E_{\mathrm{MeV}}^{3/2} Q_{\mathrm{barn}}}{Z_1 Z_2^2 (1 + A_1/A_2)^2}. \tag{2}$$

In this expression the bombarding energy E_{MeV} is measured in MeV and the quadrupole moment Q_{barn} in $e \times 10^{-24}\ \mathrm{cm}^2$. The function $c(\vartheta, s = 1, \xi)$ is given in fig. 7. It is seen from fig. 7 that c is almost proportional to ξ which is a consequence of the fact that $G_{(22)2\kappa}(\vartheta, \xi_1, \xi_2)$ is almost linear in $\xi_1 - \xi_2$ for small values of this quantity. One may utilize this fact to write (1) in the form [BOE 68]

$$P_2 = P^{(1)}[1 + qK(\vartheta, \xi)], \tag{3}$$

where $P_2^{(1)}$ is the first-order excitation probability while

$$q = \frac{A_1 \Delta E_{\text{MeV}} \langle 2 \| \mathscr{M}(\text{E2}) \| 2 \rangle}{Z_2(1 + A_1/A_2)}. \tag{4}$$

The excitation energy is here measured in MeV, while the reduced matrix element $\langle 2 \| \mathscr{M}(\text{E2}) \| 2 \rangle$ is measured in the unit $e \times 10^{-24}$ cm^2. The connection between this quantity and the static quadrupole moment Q_2 of the 2^+

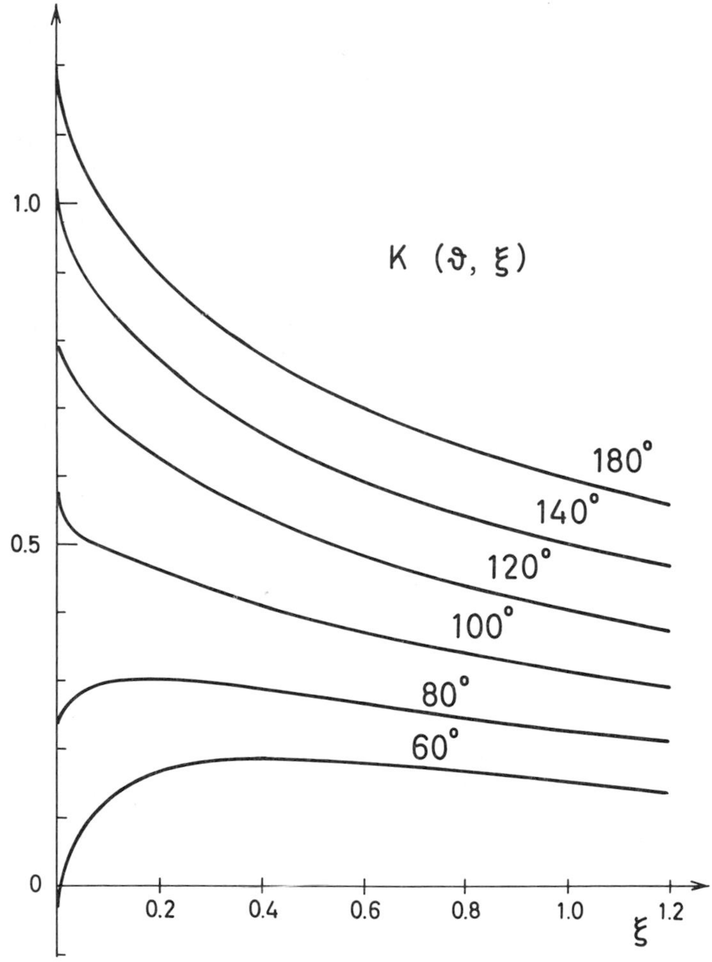

Fig. 10. The coefficient $K(\vartheta, \xi)$ for the effects of static moments which is defined in eq. (6.6) is plotted as a function of ξ for various values of the deflection angle. For small values of ξ the coefficient diverges logarithmically.

state is given by

$$\langle 2\| \mathscr{M}(\mathrm{E2}) \|2\rangle = \sqrt{\frac{7}{2\pi}\frac{5}{4}}\, Q_2 = \frac{1}{0.7579}\, Q_2. \tag{5}$$

The quantity

$$K(\vartheta, \xi) = \frac{0.5056}{\xi}\, c(\vartheta, s = 1, \xi) \tag{6}$$

which is illustrated in fig. 10 turns out to be rather independent of ξ and, for backwards scattering angles, is of the order of unity.

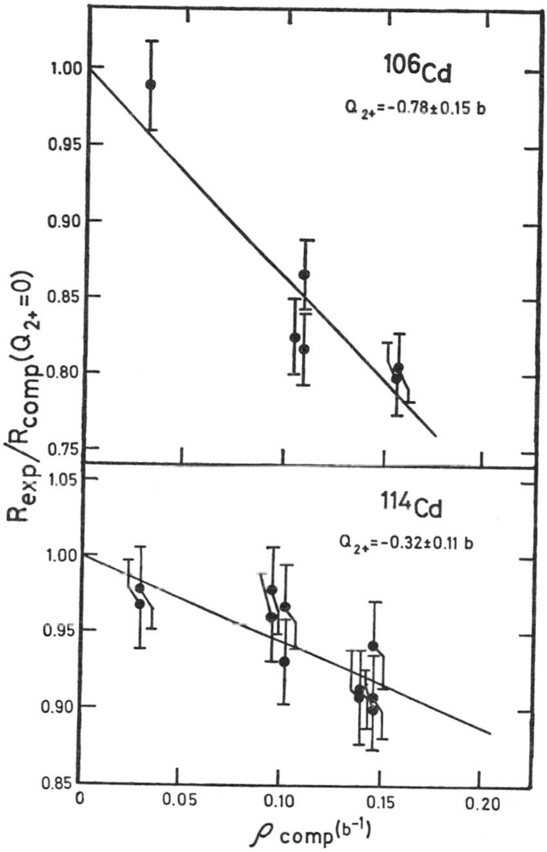

Fig. 11. Experimental excitation probabilities measured by recording the deexcitation gamma rays in coincidence with ^{4}He, ^{16}O and ^{32}S projectiles backscattered into a ring counter. The ratios of the measured probabilities to those calculated for $Q_2 = 0$ as a function of $\rho = qK/Q_2$ are shown. Q_2 is determined from the shape of the curve. The plots have been drawn for a positive sign of the interference terms involving higher 2^+ states. (From Kleinfeld, A. M. et al., 1970, Nucl. Phys. **A158**, 81.)

It should be noted that, in general, the excitation probability may receive contributions of a similar magnitude from the virtual excitation of higher (or lower) lying 2^+ states (see § 5).

As can be seen from eqs. (3) and (4) Q_2 can be determined by a measurement of the relative cross sections for different projectiles ($q \sim A_1$) or for different scattering angles (dependence of K on ϑ). An example of the first method is shown in fig. 11, while two different experimental techniques (particle spectroscopy and particle-gamma coincidences) for the second method are illustrated in figs. 12a–b and 13a–b.

If the groundstate spin is larger than or equal to one, there will appear two interference terms, one arising from the static moment of the excited state, another from the quadrupole moment of the ground state. For pure quadrupole excitation the excitation probability will be of the form (1) where $\chi^{(2)}_{2\to 2}$ should be replaced by

$$\chi^{(2)}_{2\to 2} \to \sqrt{5}(-1)^{I_0+I_f}\left\{\sqrt{2I_f+1}\begin{Bmatrix} 2 & 2 & 2 \\ I_f & I_f & I_0 \end{Bmatrix}\chi^{(2)}_{I_f\to I_f} - \sqrt{2I_0+1}\begin{Bmatrix} 2 & 2 & 2 \\ I_0 & I_0 & I_f \end{Bmatrix}\chi^{(2)}_{I_0\to I_0}\right\}. \quad (7)$$

We have here used the symmetry properties of the function $G_{(22)2\mu}$ (see table 1). Through a measurement of the excitation cross sections one may thus only determine the linear combination (7) of the quadrupole moments. However, by a measurement of the angular distribution of subsequent γ-rays,

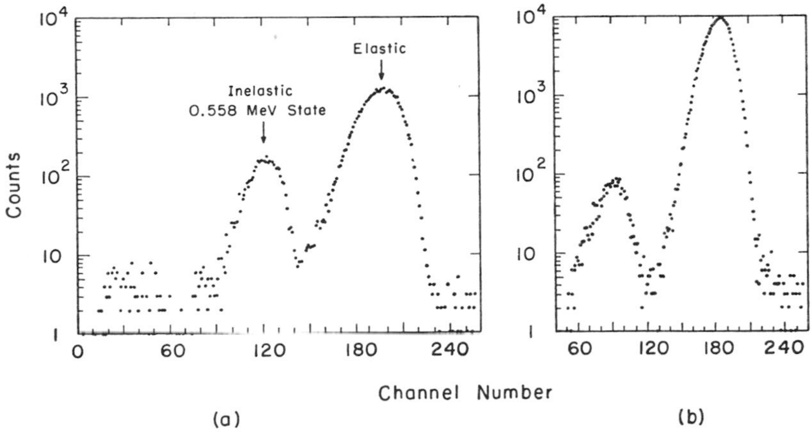

Fig. 12a. Typical spectra of 42 MeV ^{16}O ions scattered from ^{114}Cd at (a) $\theta_{\mathrm{lab}} = 143°$ and (b) $\theta_{\mathrm{lab}} = 48°$. (From Saladin, J. X. et al., 1969, Phys. Rev. **186**, 1241.)

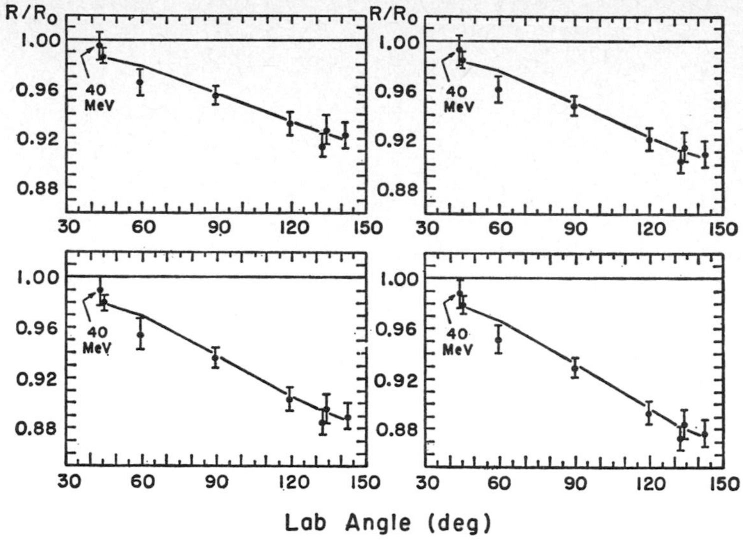

	S_1	S_2	B(E2,0→2) e^2b^2	Q_{2+} e·b
a	+	+	0.563	−0.45
b	+	−	0.562	−0.52
c	−	+	0.561	−0.64
d	−	−	0.561	−0.74

S_1 = Sign of $(M_{12} \cdot M_{24} \cdot M_{41} \cdot M_{22})$

S_2 = Sign of $(M_{12} \cdot M_{27} \cdot M_{71} \cdot M_{22})$

Fig. 12b. Results of the least-squares-fit analysis for ^{114}Cd. The points represent the experimental ratios R divided by those (R_0) calculated for $Q_2 = 0$. The solid lines connect values of R/R_0 obtained from the fit. The four fits are for the four possible combinations of the signs of the interference terms involving the two higher 2^+ states. It is seen that the experiments do not allow a distinction between the four possible choices. (From Saladin, J. X. et al., 1969, Phys. Rev. **186**, 1241.)

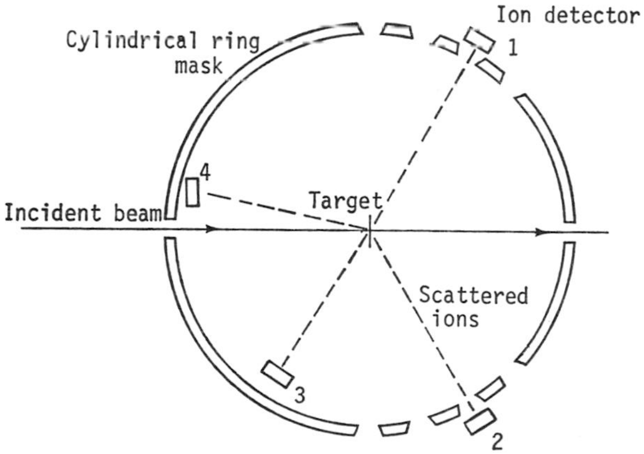

Fig. 13a. Illustration of the experimental setup where the inelastically scattered particles are recorded at several angles in coincidence with the deexcitation gamma rays detected by a large NaI counter placed with its symmetry axis perpendicular to the reaction plane. (From Thomson, J. A. et al., 1971, Phys. Rev. **C4**, 1699.)

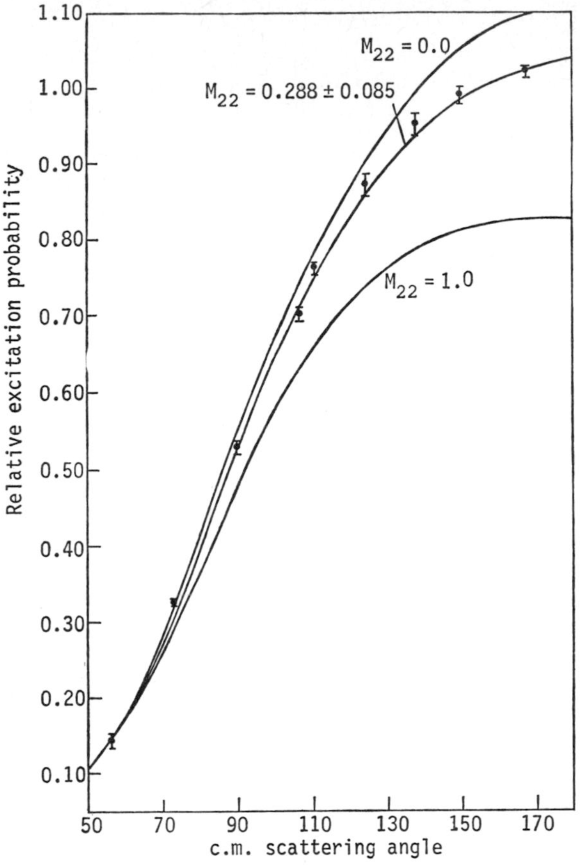

Fig. 13b. Relative excitation probability for the 0.847 MeV state of ^{56}Fe excited by 25 MeV ^{16}O ions. For these bombarding conditions the analysis is insensitive to transitions via higher lying states. (From Thomson, J. A. et al., 1971, Phys. Rev. **C4**, 1699.)

the quadrupole moments will appear in other combinations and both moments may thus be determined in this way, at least in principle.

The change in the angular distribution with respect to the angular distribution that was obtained in first-order theory, is caused by Coulomb-excitation-induced transitions between different magnetic substates of the excited state. This reorientation is also proportional to the static quadrupole moment [BRE 56]. The angular distribution is easily obtained from the general expression in § 4. For the case considered here of a $0^+ \to 2^+$ transition one finds

$$\begin{aligned} A_{2\nu} &= -\tfrac{5}{2}\sqrt{5}\chi^{(2)}_{0\to 2}\chi^{(2)}_{2\to 2}R_{(22)2\nu}, \\ B_{2\nu} &= -5\chi^{(2)}_{0\to 2}R_{2\nu} - \tfrac{5}{2}\sqrt{5}\chi^{(2)}_{0\to 2}\chi^{(2)}_{2\to 2}G_{(22)2\nu}. \end{aligned} \tag{8}$$

Especially we shall consider the angular distribution where the particle has been observed in a ring counter corresponding to a scattering angle ϑ. In this case the result (IV.4.24)–(IV.4.25) applies when the amplitudes (8) are inserted instead of the first-order amplitude.

To lowest order in χ_{22} one finds the following result for the particle parameter

$$a_{k0}^{\mathrm{E2}}(\vartheta, \xi) = (a_{k0}^{\mathrm{E2}}(\vartheta, \xi))_{1.\mathrm{order}} + \chi_{22} D_k(\vartheta, \xi). \tag{9}$$

The first term in eq. (9) is the normalized particle parameter (IV.4.25) in first order. The coefficients D_2 and D_4 are illustrated in figs. 14a-b as functions

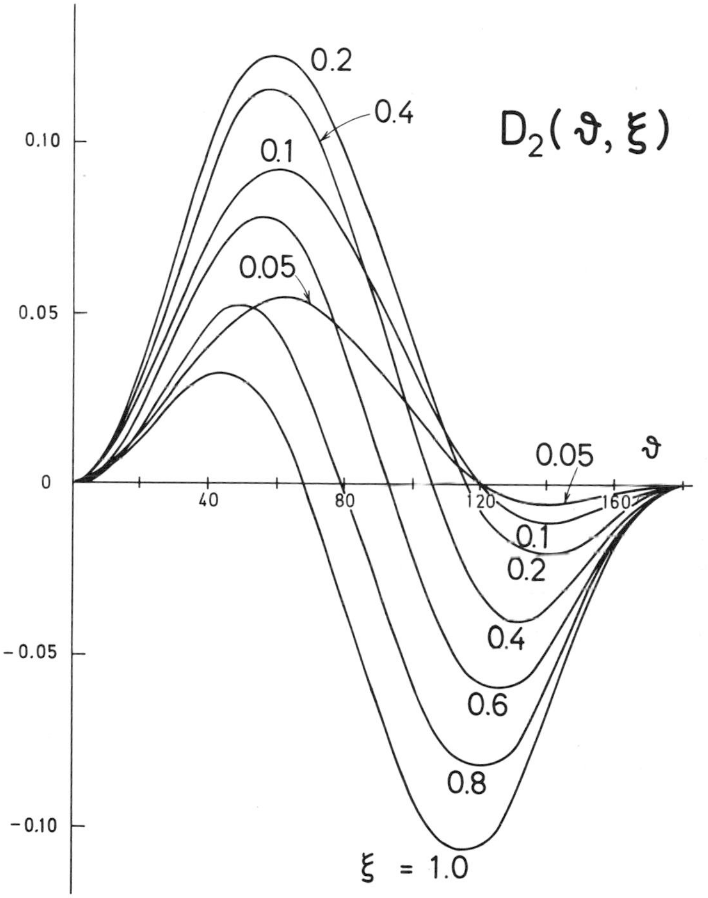

Fig. 14a. The coefficient D_2 describing the reorientation effect for the angular distribution of γ-quanta in the first excited state in even-even nuclei. The coefficient which is defined in eq. (6.9) is plotted as a function of ϑ for various values of ξ.

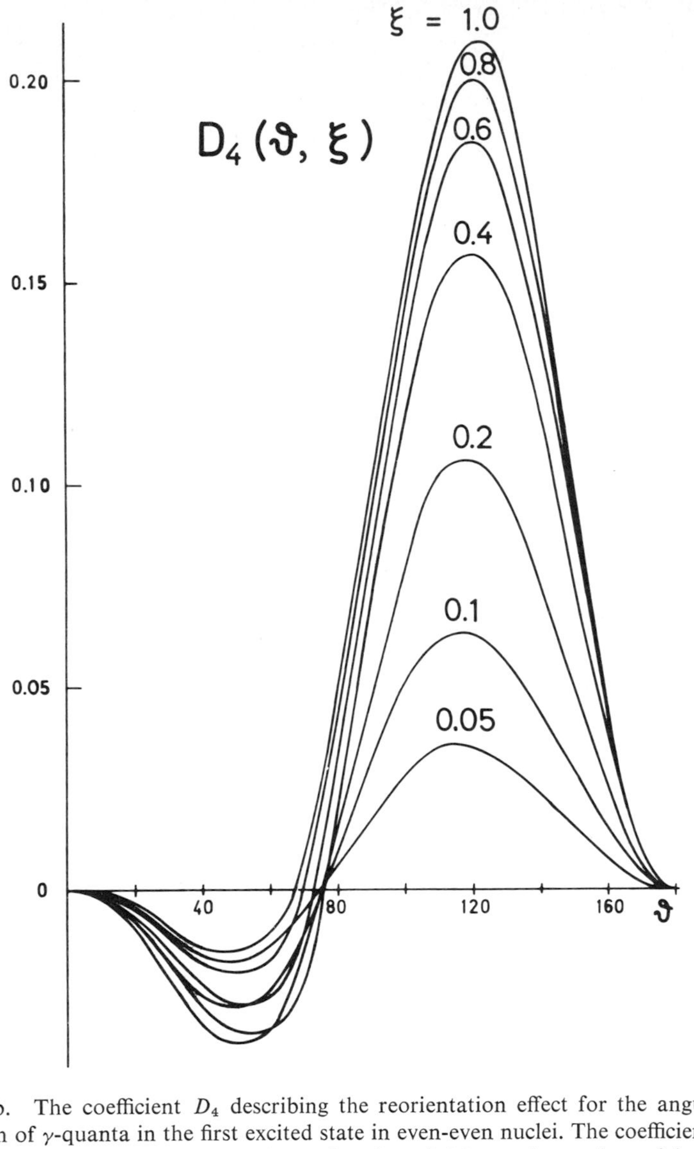

Fig. 14b. The coefficient D_4 describing the reorientation effect for the angular distribution of γ-quanta in the first excited state in even-even nuclei. The coefficient which is defined in eq. (6.9) is plotted as a function of ϑ for various values of ξ.

of ϑ for various values of ξ, while the particle parameters a_{k0}^{E2} in first order are illustrated in fig. IV.17.

Some properties of these curves can be understood from the discussion in chapter III. It is thus seen that

$$D_k(\pi, \xi) = 0 \tag{10}$$

which follows from the selection rule $M_i = M_f$. Furthermore, one finds for $\xi = 0$,

$$D_k(\vartheta, 0) = 0 \tag{11}$$

according to the discussion in § 4.

A measurement of the static moments by reorientation experiments should therefore be performed for not too small values of ξ and for scattering angles about 120 degrees, and it is seen from fig. 14 that the particle parameter with $k = 4$ is more sensitive to the static quadrupole moment than the particle parameter for $k = 2$. The rather small effect of Q_2 on the gamma ray angular-distribution is illustrated in figs. 15a-b.

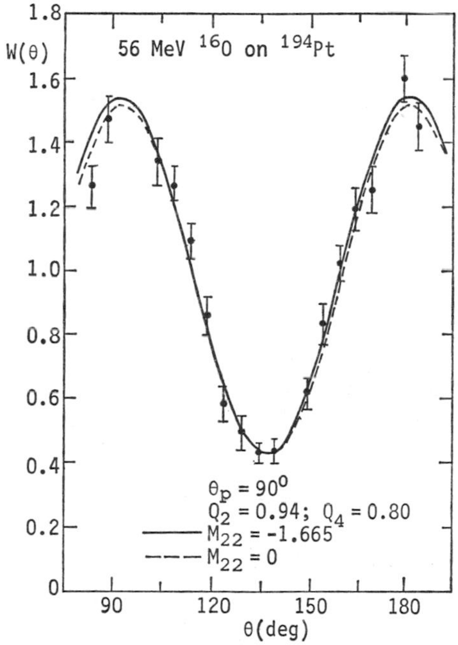

Fig. 15a. The observed and expected angular distributions of de-excitation gamma rays from the first 2^+ state of ^{194}Pt in coincidence with oxygen projectiles scattered through a laboratory angle of 90°. The gamma-ray angle, θ, is measured with respect to the beam direction in the scattering plane. The curves are theoretical calculations. (From Grodzins, L. et al., 1973, J. Phys. Soc. Japan **34**, 187.)

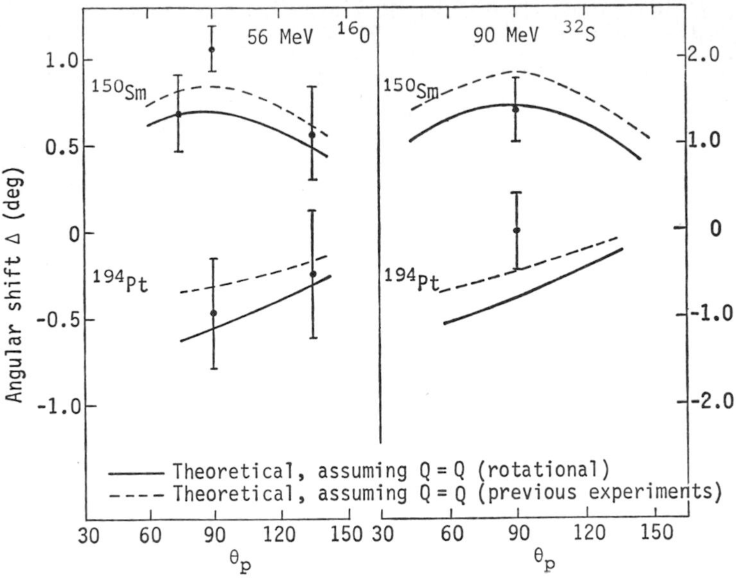

Fig. 15b. Measured distribution of γ-ray precession angles Δ for 56 MeV ^{16}O and 90 MeV ^{32}S beams exciting the first 2^+ states of ^{150}Sm and ^{194}Pt. The dashed curves are theoretically expected assuming that the Q_2 values are those measured by other methods. The solid curves use Q values derived from the B(E2) values of a pure rotational model. The angular shifts Δ are measured with respect to the recoil axis and are opposite for the two states investigated. (From Grodzins, L. et al., 1973, Phys. Rev. Lett. **30**, 453.)

§ 7. *Polarization effects*

The effect of the high-lying states in second-order Coulomb excitation is related to the polarization effects discussed in chapter II. In the limiting case where the excitation of low-lying states takes place through a virtual transition via high-lying states, the formulae given in § 2 and § 3 are considerably simplified and we shall in this section discuss in more detail the effects associated with such virtual excitations.

We shall consider the effect of an intermediate state $|z\rangle$ whose energy is so high that it will not be excited. The ξ-value ξ_1 for the transition $I_0 \to I_z$ and the ξ-value ξ_2 for the transition $I_z \to I_f$ are then large compared to one, while

$$\xi = \xi_1 + \xi_2 \tag{1}$$

is of the order of unity or smaller (see fig. 16).

In this situation the real part of the excitation amplitude (1.15) vanishes and the imaginary part can be written in the form (2.10). The resulting excita-

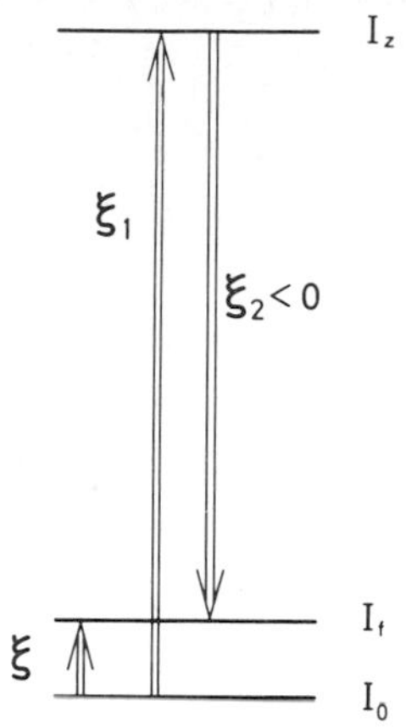

Fig. 16. Schematic picture of the polarizing effect. The figure illustrates the excitation of a state of spin I_f, partly directly, and partly via an intermediate state of high energy (spin I_z). The ξ-values for the various transitions are indicated.

tion amplitude to second order is thus given by

$$a_{I_f M_f, I_0 M_0} = -\mathrm{i}(-1)^{I_0-M_0}\sqrt{2I_0+1} \sum_{\lambda\lambda'\lambda'' I_z} \sqrt{2\lambda+1} \begin{pmatrix} I_0 & \lambda & I_f \\ -M_0 & \mu & M_f \end{pmatrix}$$
$$\times \Big[\chi^{(\lambda)}_{0\to f} R_{\lambda\mu}(\vartheta, \xi) - \sqrt{(2\lambda+1)(2\lambda'+1)(2\lambda''+1)(2I_z+1)}(-1)^{I_0+I_f}$$
$$\times \begin{Bmatrix} \lambda'' & \lambda' & \lambda \\ I_0 & I_f & I_z \end{Bmatrix} \chi^{(\lambda')}_{0\to z} \chi^{(\lambda'')}_{z\to f} \begin{pmatrix} \lambda' & \lambda'' & \lambda \\ 0 & 0 & 0 \end{pmatrix} \frac{(2\lambda'-1)!!\,(2\lambda''-1)!!\,(\lambda-1)!}{(\lambda'-1)!\,(\lambda''-1)!\,(2\lambda-1)!!}$$
$$\times \frac{1}{2\xi} P_{(\lambda'\lambda'')\lambda\mu}(\vartheta, \xi) \Big]. \tag{2}$$

This formula could also be obtained from the polarization potential introduced in § II.6. In fact, the result (2) is identical to the first-order solution of the coupled equations (II.6.7), where one has corrected the multipole interaction by a polarization potential $V_{\mathrm{pol}}(t)$. The appearance of the factor $1/\bar{\xi}$ in eq. (2) corresponds to the energy denominator in the polarization potential. The result (2.12) that the function $P_{(\lambda'\lambda'')\lambda\mu}$ can be expressed in terms of an orbital integral is connected with the fact that the polarization effects can be expressed in terms of a polarization potential which is only a function of the position of the projectile.

The excitation probability can directly be obtained from eq. (2). To second order one finds

$$P_f = \sum_{\lambda} |\chi^{(\lambda)}_{0\to f}|^2 R^2_{\lambda}(\vartheta, \xi) \Big[1 - \sum_{\lambda'\lambda''} z(\lambda'\lambda''\lambda) E_{(\lambda'\lambda'')\lambda}(\vartheta, \xi) \Big] \tag{3}$$

where

$$E_{(\lambda'\lambda'')\lambda}(\vartheta, \xi) = \frac{2}{R_\lambda^2(\vartheta, \xi)} \sum_\mu R_{\lambda\mu}^*(\vartheta, \xi) P_{(\lambda'\lambda'')\lambda\mu}(\vartheta, \xi). \tag{4}$$

The nuclear structure coefficient $z_{0\to f}(\lambda'\lambda''\lambda)$ is accordingly defined by

$$\begin{aligned} z_{0\to f}(\lambda'\lambda''\lambda) &= \sum_z \frac{\chi_{0\to z}^{(\lambda')}\chi_{z\to f}^{(\lambda'')}}{\chi_{0\to f}^{(\lambda)} 2\xi} \sqrt{2I_z+1} \begin{Bmatrix} \lambda & \lambda' & \lambda'' \\ I_z & I_f & I_0 \end{Bmatrix} (-1)^{I_0+I_f} \\ &\quad \times \sqrt{(2\lambda+1)(2\lambda'+1)(2\lambda''+1)} \begin{pmatrix} \lambda' & \lambda'' & \lambda \\ 0 & 0 & 0 \end{pmatrix} \frac{(2\lambda'-1)!!\,(2\lambda''-1)!!\,(\lambda-1)!}{(\lambda'-1)!\,(\lambda''-1)!\,(2\lambda-1)!!} \\ &\approx \frac{4\sqrt{\pi}E}{Z_2} a^{\lambda-\lambda'-\lambda''} \begin{pmatrix} \lambda' & \lambda'' & \lambda \\ 0 & 0 & 0 \end{pmatrix} \frac{(2\lambda+1)^{3/2}}{[(2\lambda'+1)(2\lambda''+1)]^{1/2}} \\ &\quad \times \sum_z (-1)^{I_0+I_f} \begin{Bmatrix} \lambda & \lambda' & \lambda'' \\ I_z & I_f & I_0 \end{Bmatrix} \frac{\langle I_0 \| \mathscr{M}(E\lambda') \| I_z \rangle \langle I_z \| \mathscr{M}(E\lambda'') \| I_f \rangle}{e(E_z - E_0) \langle I_0 \| \mathscr{M}(E\lambda) \| I_f \rangle}, \end{aligned} \tag{5}$$

where E is the centre-of-mass energy, while E_z and E_0 are the energies of the nuclear states $|z\rangle$ and $|0\rangle$, respectively. The summation in (5) should be performed over all the highly excited states $|z\rangle$ which may contribute to given multipolarities λ' and λ''.

The main polarization effects are expected to arise from the giant dipole resonance where the multipolarities λ' and λ'' are both equal to one. The fact that the energy of the giant dipole resonance is very high is not too serious since the coefficients $z_{0\to f}(\lambda'\lambda''\lambda)$ defined in eq. (5) are inversely proportional to the energy of the intermediate state. For this case the magnitude of the structure coefficient is estimated in appendix J for final states $|f\rangle$ which may be described as surface vibrational or rotational states. For the special case $\lambda' = \lambda'' = 1$ and $\lambda = 2$, the dipole polarization function $E_{(11)2}(\vartheta, \xi)$ is given by

$$E_{(11)2}(\vartheta, \xi) = \frac{2I_{20}I_{30} + 3I_{22}I_{32} + 3I_{2,-2}I_{3,-2}}{|I_{20}|^2 + \frac{3}{2}|I_{22}|^2 + \frac{3}{2}|I_{2,-2}|^2}, \tag{6}$$

where $I_{\lambda\mu}(\vartheta, \xi)$ is defined by eq. (IV.2.7).

The function $E_{(11)2}$ is illustrated in fig. 17 as a function of ξ for different scattering angles. If one uses the estimate for z given in appendix J

$$z = 0.5 \times 10^{-2} \frac{E_{\text{MeV}} A_2}{Z_2^2(1 + A_1/A_2)} \tag{7}$$

one sees that the polarization effect may amount to about 10% for excitation produced by the heaviest projectiles, while for excitation produced by oxygen ions it is only of the order of 1 to 2%. An experimental indication of a larger polarization effect is shown in fig. 18.

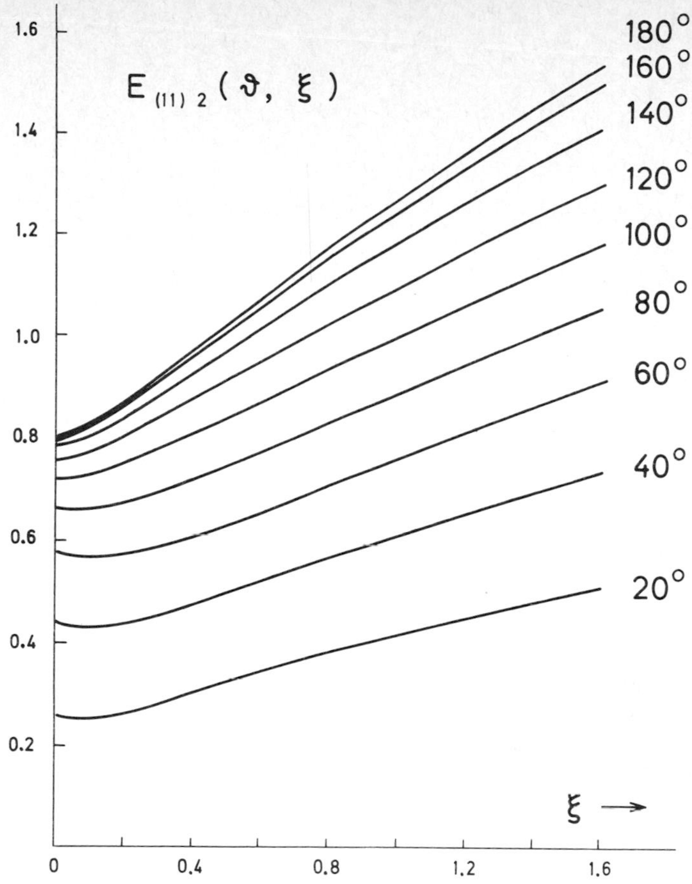

Fig. 17. The function $E_{(11)2}(\vartheta, \xi)$ describing the dipole polarization effect in quadrupole excitation. The function which is defined in eq. (7.6) is plotted as a function of ξ for various values of the scattering angle ϑ.

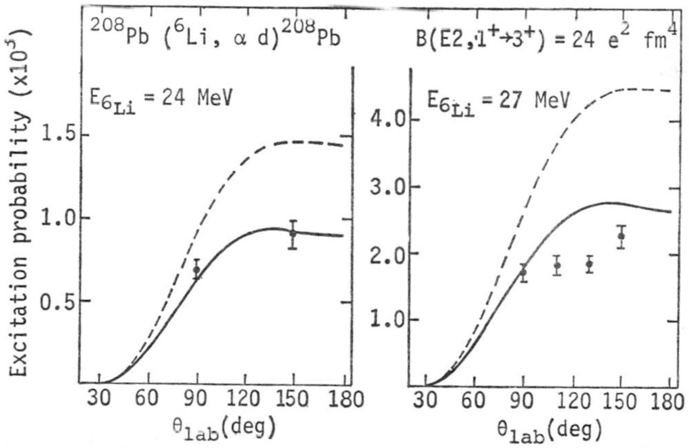

Fig. 18. Comparison of the experimental excitation probabilities for projectile excitation of ^{6}Li with theoretical shapes evaluated using $B(\mathrm{E2}; 1^+ \rightarrow 3^+) = 24e^2\mathrm{fm}^4$. Dashed line: without taking into account the projectile polarization; solid line: this effect included. (From Disdier, D. L. et al., 1971, Phys. Rev. Lett. **27**, 1391.)

§ 8. *Higher-order perturbation theory*

Formally the evaluation of higher-order perturbation theory offers no difficulties in principle [MAS 65]. However, the resulting formulae become so elaborate that they are of little practical use. Furthermore, as we discussed in § 1, the perturbation theory converges rather slowly, so that in cases where one has to include third-order effects, also effects of still higher order may often be important.

In this section we shall thus only consider the structure of the third-order theory in the general case. In the special case of the sudden approximation the perturbation theory has a very simple structure which makes it possible to carry out the perturbation expansion to any order. This expansion which may be of practical interest especially in connection with computers will be discussed in chapter VI.

The third-order contribution to the excitation amplitude can be evaluated from eq. (II.3.8). The amplitude may thus be written

$$a_n = a_n^{(1)} + a_n^{(2)} + a_n^{(3)}, \tag{1}$$

where $a_n^{(1)}$ and $a_n^{(2)}$ are the first- and second-order amplitudes discussed in § 1, while the third-order amplitudes are given by

$$\begin{aligned} a_n^{(3)} = \sum_{zz'} \left(\frac{-\mathrm{i}}{\hbar}\right)^3 \int_{-\infty}^{+\infty} & \mathrm{d}t \, \langle n| \, V_{\mathrm{E}}(t) \, |z'\rangle \exp(\mathrm{i}\omega_{nz'}t) \\ \times \int_{-\infty}^{t} & \mathrm{d}t' \, \langle z'| \, V_{\mathrm{E}}(t') \, |z\rangle \exp(\mathrm{i}\omega_{z'z}t') \\ \times \int_{-\infty}^{t'} & \mathrm{d}t'' \, \langle z| \, V_{\mathrm{E}}(t'') \, |0\rangle \exp(\mathrm{i}\omega_{z0}t''). \end{aligned} \tag{2}$$

The decomposition of this complex amplitude in real and imaginary parts (in coordinate systems A and B) follows directly from eq. (II.3.36), i.e.

$$a^{(3)}_{I_{\mathrm{f}}M_{\mathrm{f}},I_0M_0} = \langle I_{\mathrm{f}}M_{\mathrm{f}}| - \tfrac{1}{2}(RG + GR) + \mathrm{i}(H - \tfrac{1}{4}R^3) \, |I_0M_0\rangle . \tag{3}$$

The amplitude can thus be expressed in terms of the orbital integrals $R_{\lambda\mu}$ and $G^{\lambda\lambda'}_{\mu\mu'}$ and a new integral defined by

$$\begin{aligned} H^{\lambda_1\lambda_2\lambda_3}_{\mu_1\mu_2\mu_3}(\vartheta, \xi_1, \xi_2, \xi_3) &= -\frac{1}{\pi^2} \mathscr{P} \int_{-\infty}^{+\infty} \frac{\mathrm{d}q}{q} \mathscr{P} \int_{-\infty}^{+\infty} \frac{\mathrm{d}r}{r} R_{\lambda_1\mu_1}(\vartheta, \xi_1 + q) \\ &\quad \times R_{\lambda_2\mu_2}(\vartheta, \xi_2 - q + r) R_{\lambda_3\mu_3}(\vartheta, \xi_3 - r) \\ &= -\frac{1}{\pi} \mathscr{P} \int_{-\infty}^{+\infty} \frac{\mathrm{d}q}{q} R_{\lambda_1\mu_1}(\vartheta, \xi_1 + q) G^{\lambda_2\lambda_3}_{\mu_2\mu_3}(\vartheta, \xi_2 - q, \xi_3), \end{aligned} \tag{4}$$

which is a real quantity in coordinate systems A and B.

By means of the amplitudes (3) one may evaluate the excitation probability to fourth order. The fourth-order correction arises from the interference between $a^{(1)}$ and the last term in $a^{(3)}$ and an evaluation of the corrections to the excitation probabilities in § 3 thus involves the computations of the orbital integrals (4).

A simple result is obtained in the limit of the sudden approximation where according to eq. (II.3.34) the orbital integral H can be evaluated to give

$$H^{\lambda_1\lambda_2\lambda_3}_{\mu_1\mu_2\mu_3}(\vartheta, 0, 0, 0) = -\tfrac{1}{3}R_{\lambda_1\mu_1}(\vartheta, 0)R_{\lambda_2\mu_2}(\vartheta, 0)R_{\lambda_3\mu_3}(\vartheta, 0). \tag{5}$$

In the sudden approximation the perturbation expansion can be carried out to any order. According to eq. (II.3.35) one finds

$$\begin{aligned} a_n &= \langle n|\exp\left\{-\frac{\mathrm{i}}{\hbar}\int_{-\infty}^{+\infty} V_{\mathrm{E}}(t)\,\mathrm{d}t\right\}|0\rangle = \langle n|\,\mathrm{e}^{\mathrm{i}R}\,|0\rangle \\ &= \langle n|\,1 + \mathrm{i}R + \tfrac{1}{2}(\mathrm{i}R)^2 + \tfrac{1}{6}(\mathrm{i}R)^3 + \cdots + |0\rangle. \end{aligned} \tag{6}$$

If one takes into account only a finite number of intermediate states, the operator R is represented by a finite matrix and (6) can be evaluated by repeated matrix multiplication. The matrix elements of $\mathrm{i}R$ are the excitation amplitudes in first order (eq. (IV.3.7)) between two arbitrary states with $\xi = 0$, i.e.

$$\langle I_n M_n|\,R\,|I_m M_m\rangle = -\sum_{\lambda}\zeta^{\lambda\mu}_{nm}R_{\lambda\mu}(\vartheta, 0), \tag{7}$$

where

$$\zeta^{\lambda\mu}_{nm} = (-1)^{I_m - M_m}\sqrt{(2I_m + 1)(2\lambda + 1)}\begin{pmatrix} I_m & \lambda & I_n \\ -M_m & \mu & M_n\end{pmatrix}\chi^{(\lambda)}_{m\to n}. \tag{8}$$

We note finally that an improvement of the sudden approximation is obtained by a series expansion of the S-matrix where one includes besides the term R for finite values of ξ also the first-order correction G (c.f. eq. (II.3.30)). The excitation amplitude is then given by

$$a_n = \langle n|\,1 + \mathrm{i}(R + G) + \tfrac{1}{2}[\mathrm{i}(R + G)]^2 + \cdots + |0\rangle. \tag{9}$$

For a finite number of states the operator $R + G$ can again be represented by a finite matrix where the matrix elements of R and G may be written (see (1.11))

$$\langle I_n M_n|\,R\,|I_m M_m\rangle = -\sum_{\lambda}\zeta^{\lambda\mu}_{nm}R_{\lambda\mu}(\vartheta, \xi_{nm}) \tag{10}$$

and

$$\langle I_n M_n|\,G\,|I_m M_m\rangle = -\tfrac{1}{2}\sum_{\substack{\lambda\lambda'\mu\mu' \\ z}}\zeta^{\lambda'\mu'}_{nz}\zeta^{\lambda\mu}_{zm}G^{\lambda\lambda'}_{\mu\mu'}(\vartheta, \xi_{zm}, \xi_{nz}). \tag{11}$$

In the summation over intermediate states $|z\rangle$ in (11) one may include states which are otherwise neglected in the finite matrix as e.g. high-lying states which only contribute through the polarization effect (see § II.6).

CHAPTER VI

Multiple Excitations

In the two previous chapters we have discussed the perturbation treatment of the Coulomb excitation process. As was noted in chapter V, the perturbation expansion must be carried out to high order to give a sufficient accuracy in cases where the parameter $\chi(\vartheta)$ is of the order of unity or larger. Already the second-order treatment is of considerable complexity and higher-order treatments become impracticable. Still, it should be remembered that even in cases where the parameters χ are large, $\chi(\vartheta)$ will become small for forward scattering angles and therefore a perturbation treatment can be applied for $\vartheta < \vartheta_0$ where $\chi(\vartheta_0) \ll 1$ for all transitions from the ground state.

For the treatment of cases where $\chi(\vartheta) \gtrsim 1$ one must look for methods that avoid the perturbation expansion. A number of such methods have been explored. They may be classified into two types, one in which one considers only a finite number of nuclear states and one in which all nuclear states of a definite degree of freedom are taken into account in a nuclear model.

The study of the Coulomb excitation for nuclear models is of special interest because multiple Coulomb excitation takes place especially among the nuclear states which are described quite well by the excitation of collective (vibrational and rotational) degrees of freedom. An experimental example of the multiple excitation of a large number of levels belonging to rotational bands in ^{235}U is given in figs. 1a–c.

For a hypothetical nucleus possessing only pure harmonic vibrational states the excitation amplitude can be evaluated exactly within the semiclassical approximation. The simple results which are found for this case cannot be applied directly to actual nuclei, but they offer, through their simplicity, a useful survey of the effects which come into play in multiple Coulomb excitation. This case is discussed in chapter VII.

The more realistic case of the excitation of rotational states in deformed

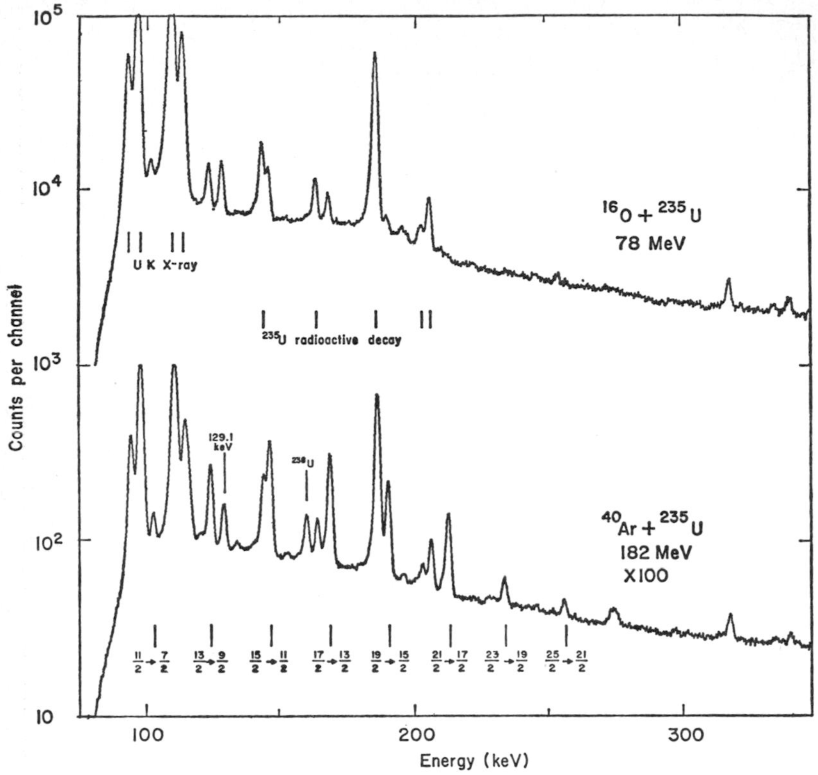

Fig. 1a. Gamma-ray spectra of ^{235}U bombarded with ^{16}O and ^{40}Ar projectiles in the region 100–350 keV. This is the energy region where the rotational crossover transitions in the ground state band occur and these are indicated on the figure, as are a few other lines of interest. (From Stephens, F. S. et al., 1968, Nucl. Phys. **A115**, 129.)

nuclei can only be evaluated explicitly in the sudden approximation where one neglects the finite excitation energies (see chapter VIII).

In the case where one limits oneself to a finite number of nuclear states one may, in the sudden approximation, obtain quite simple results by means of a diagonalization procedure which can be generalized to include deviations from the sudden approximation using the expansion (II.3.30). With the use of computers it has also become possible to solve directly the set of coupled differential equations (II.4.5) which determine the amplitudes on a finite number of nuclear states during the collision.

In this chapter we shall discuss the evaluation of the excitation amplitude by the various methods described above. For the computation of cross sections and angular distributions of γ-rays we refer to chapter III.

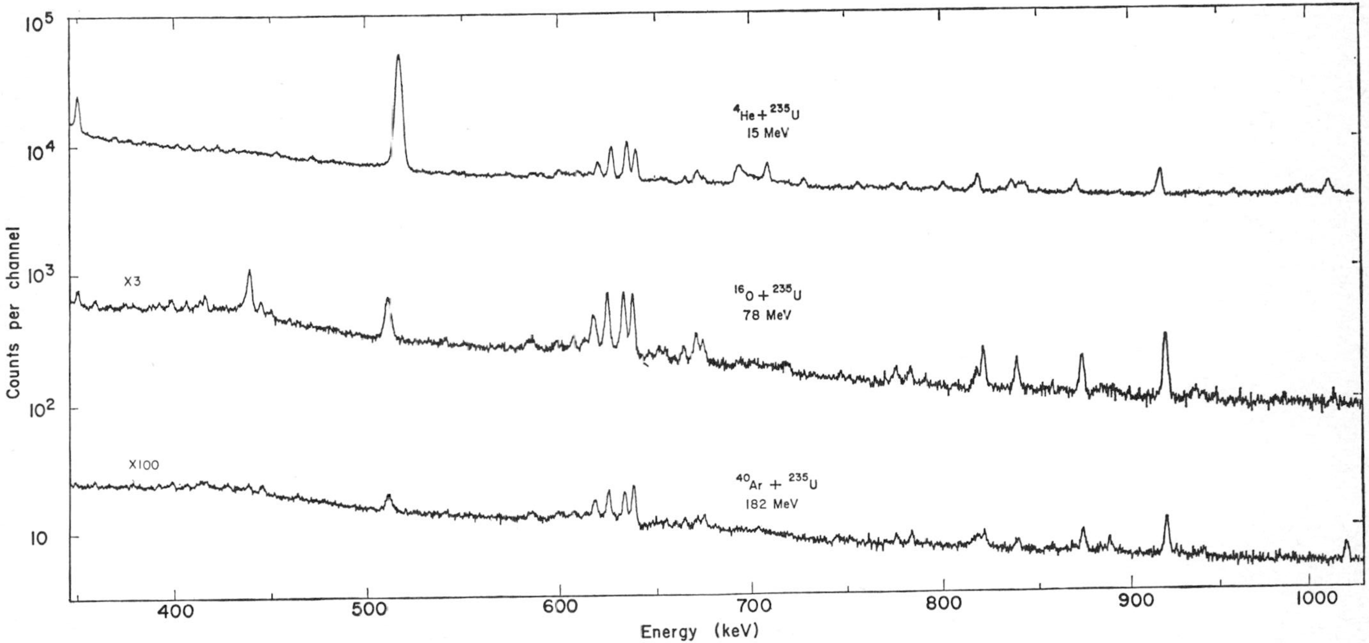

Fig. 1b. Gamma-ray spectra of ^{235}U in the region 400–1000 keV. This is the region of the gamma transitions between different rotational bands. (From Stephens, F. S. et al., 1968, Nucl. Phys. **A115**, 129.)

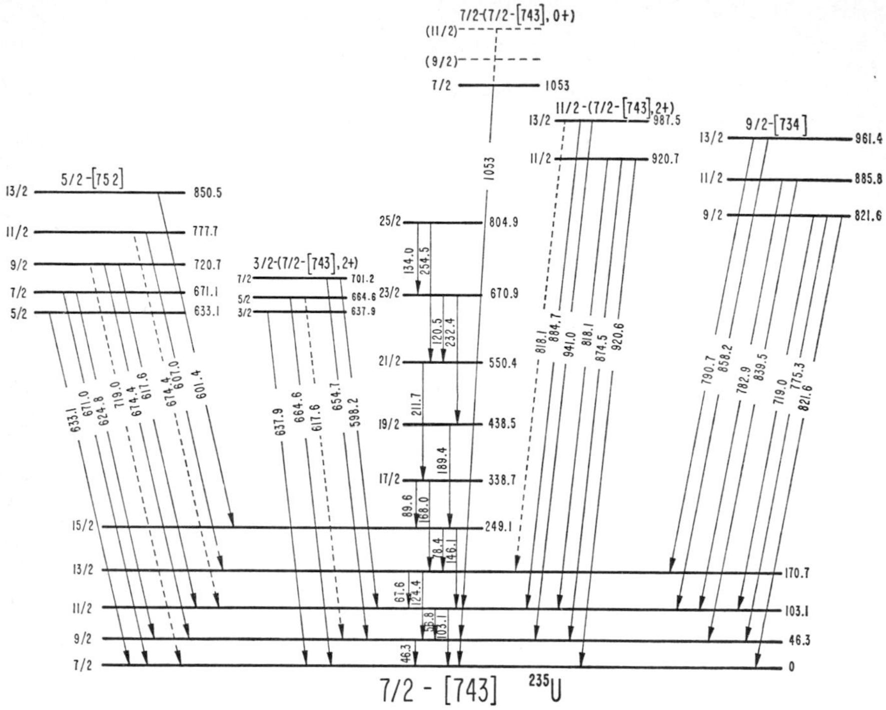

Fig. 1c. Levels Coulomb excited in ^{235}U. (From Stephens, F. S. et al., 1968, Nucl. Phys. **A115**, 129.)

§ 1. *The coupled differential equations*

In the semiclassical approximation the amplitudes $a_n(t)$, of the nuclear state vector at time t, on the eigenstates $|n\rangle$ of the undisturbed nucleus, satisfy the coupled differential equations (II.4.5)

$$\mathrm{i}\hbar\dot{a}_n = \sum_m \langle n|\, V_{\mathrm{E}}(t)\, |m\rangle \exp\left\{\frac{\mathrm{i}}{\hbar}(E_n - E_m)t\right\} a_m(t). \tag{1}$$

We shall specify the nuclear states by their spin quantum numbers I and M and introduce the multipole expansion (II.1.12) of the interaction $V_{\mathrm{E}}(t)$. Introducing furthermore the parametrization in § II.9 of the hyperbolic orbit, we may by means of eqs. (IV.1.8) and (IV.3.6) write eq. (1) in the form

$$\frac{\mathrm{d}}{\mathrm{d}w}[a_{I_nM_n}(w)] = -\mathrm{i} \sum_{\lambda\mu I_m M_m} Q_{\lambda\mu}(\varepsilon, w)\sqrt{2I_m+1}\, \chi^{(\lambda)}_{m\to n}$$

$$\times \exp\{\mathrm{i}\xi_{nm}[\varepsilon \sinh w + w]\}\sqrt{2\lambda+1}(-1)^{I_m - M_m}\begin{pmatrix} I_m & \lambda & I_n \\ -M_m & \mu & M_n \end{pmatrix} a_{I_mM_m}(w), \tag{2}$$

where (see eq. (IV.2.5))*

$$\xi_{nm} = \frac{a}{v}\frac{E_n - E_m}{\hbar}. \tag{3}$$

In order to find the excitation amplitudes after the collision

$$a_{I_nM_n}(w = +\infty) = a_{I_nM_n, I_0M_0}, \tag{4}$$

one should solve the set of coupled differential equations (2) with the initial condition

$$a_{I_nM_n}(w = -\infty) = \delta_{I_nI_0}\delta_{M_nM_0}. \tag{5}$$

In principle, the summation in eq. (2) should be extended over all nuclear states. In practice, as was discussed in § IV.1 one may in first approximation neglect the states outside the χ-submatrix that contain all the states strongly coupled to the ground state. Once the amplitudes for the states within the ground state submatrix have been determined as functions of w, the excitation of other states can be obtained by a perturbation treatment. The classification of the χ-matrix into submatrices is illustrated in fig. 2. The group of states which contain the ground state is denoted by G, other groups are X and Y. The matrices are defined in such a way that all elements $\chi_{nm}(\mathrm{X}, \mathrm{Y})$ between two different groups X and Y are small.

The procedure for solving (2) may thus be the following. First one solves

	G	X	Y
G	χ_{nm} (G,G)	χ_{nm} (G,X)	χ_{nm} (G,Y)
X	χ_{nm} (X,G)	χ_{nm} (X,X)	χ_{nm} (X,Y)
Y	χ_{nm} (Y,G)	χ_{nm} (Y,X)	χ_{nm} (Y,Y)

Fig. 2. Schematic picture of the decomposition of the χ-matrix into submatrices. The submatrix containing the ground state is indicated by G, while X and Y denote other groups of strongly coupled states.

* It should be noted that for the quantities v and a, appearing in eq. (3), one should use symmetrized expressions as discussed in § IV.7 and § 8 below.

the equations

$$\frac{\mathrm{d}}{\mathrm{d}w}[a_n(w)] = -\mathrm{i}\sum_{\substack{\lambda\mu \\ m\in \mathrm{G}}} Q_{\lambda\mu}(\varepsilon, w)\exp\{\mathrm{i}\xi_{nm}(\varepsilon \sinh w + w)\}\zeta^{\lambda\mu}_{nm}(\mathrm{G},\mathrm{G})a_m(w) \quad (6)$$

for the states belonging to the group G.

In (6) we have introduced the abbreviations n for I_n, M_n and m for I_m, M_m. Furthermore

$$\zeta^{\lambda\mu}_{nm}(\mathrm{X},\mathrm{Y}) = \sqrt{2I_m+1}\,\chi^{(\lambda)}_{m\to n}\sqrt{2\lambda+1}$$
$$\times(-1)^{I_m-M_m}\begin{pmatrix} I_m & \lambda & I_n \\ -M_m & \mu & M_n\end{pmatrix}, \qquad \begin{matrix} n\in \mathrm{X} \\ m\in\mathrm{Y}.\end{matrix} \quad (7)$$

The matrix $\zeta^{\lambda\mu}_{nm}(\mathrm{X},\mathrm{Y})$, where n belongs to X and m belongs to Y, has the symmetry property

$$\zeta^{\lambda\mu}_{nm}(\mathrm{X},\mathrm{Y}) = (-1)^{\mu}\zeta^{\lambda-\mu}_{mn}(\mathrm{Y},\mathrm{X}). \quad (8)$$

Once the amplitudes $a_n(w)$ within the group G have been determined the amplitudes on the states within a group X can be obtained from the following set of coupled inhomogeneous equations

$$\frac{\mathrm{d}}{\mathrm{d}w}[a_k(w)] + \mathrm{i}\sum_{\substack{\lambda\mu \\ p\in \mathrm{X}}} Q_{\lambda\mu}(\varepsilon, w)\exp\{\mathrm{i}\xi_{kp}(\varepsilon \sinh w + w)\}\zeta^{\lambda\mu}_{kp}(\mathrm{X},\mathrm{X})a_p(w)$$
$$= -\mathrm{i}\sum_{\substack{\lambda\mu \\ m\in \mathrm{G}}} Q_{\lambda\mu}(\varepsilon, w)\exp\{\mathrm{i}\xi_{km}(\varepsilon \sinh w + w)\}\zeta^{\lambda\mu}_{km}(\mathrm{X},\mathrm{G})a_m(w), \quad (9)$$

where $a_m(w)$ are known from the solution of eq. (6). This set of equations should be solved with the initial condition

$$a_k(w=-\infty) = 0, \quad (10)$$

where k belongs to X.

From the criteria given above the number of states which should be included within the various groups can still be very large. One may limit this number, however, by two further considerations. Firstly, the excitation energy cannot be too large. The amplitude on such high-lying states will be small due to the rapid oscillation of the exponential function. Still, these states may contribute through a virtual excitation which amounts to a polarization effect as was described in § II.6. As was shown in this section the effect of polarization caused by these states can be included in the coupled equations for the low-lying states by a polarization potential which should be added to the usual multipole interaction. The explicit form of the polarization potential depends on the multipole orders of the virtual transitions involved. The matrix elements of V_{pol} between two given low-lying states will depend on the product of the matrix elements connecting these two states with the high excited intermediate states. A convenient form for the matrix elements of the polarization potential is given in § V.7. The substitution of $V_E(t)$ with $V_E(t) + V_{pol}(t)$ is brought about by replacing the collision function $Q_{\lambda\mu}(\varepsilon, w)$ in eqs. (6)–(9) with the collision function $Q^{eff}_{\lambda\mu}(\varepsilon, w)$ given by

$$\chi^{(\lambda)}_{r\to s} Q^{eff}_{\lambda\mu}(\varepsilon, w) = \chi^{(\lambda)}_{r\to s} Q_{\lambda\mu}(\varepsilon, w)\left[1 - \sum_{\lambda'\lambda''} z_{r\to s}(\lambda'\lambda''\lambda)(\varepsilon \cosh w + 1)^{\lambda-\lambda'-\lambda''-1}\right], \tag{11}$$

where $z_{r\to s}(\lambda'\lambda''\lambda)$ is given by eq. (V.7.5). For the contribution from the giant resonance this quantity is estimated in appendix J. A second consideration has to do with the fact that even if the excitation energy is neglected, there exists an approximate upper limit for the number of transitions which can occur in a multiple excitation. It thus follows from the differential equations (1) or (2) that a state which can be reached directly from the ground state will be populated within a time $\tau/\chi(\vartheta)$ where τ is the collision time. A state, which is connected to this state, but not to the ground state can only start to be populated after this time has elapsed and it will therefore be populated within the time $2\tau/\chi(\vartheta)$ etc. where we have assumed that all χ's are of the same order of magnitude. The number of steps, n, within the collision time is thus of the order $\chi(\vartheta)$. This can also be seen by means of the perturbation theory where the lowest-order contribution to the excitation probability in the sudden approximation is of the magnitude (see eq. (V.8.6))

$$P_n \approx [\chi(\vartheta)^n/n!]^2. \tag{12}$$

The rule that the number of steps is of the order of magnitude $\chi(\vartheta)$ is connected with the classical result that there is a maximum value of the angular momentum which can be transferred to the nucleus in a collision. Thus, classically, the angular momentum transfer along the z-axis is in the sudden approximation given by

$$\Delta l_z = \int_{-\infty}^{+\infty} (\mathbf{r} \times \boldsymbol{\nabla})_z V_E(t)\, dt = \int_{-\infty}^{+\infty} iL_z V_E(t)\, dt. \tag{13}$$

For a given multipolarity λ the maximum value of this integral is given by

$$\Delta l \lesssim \hbar\lambda\chi^{(\lambda)}(\vartheta) \tag{14}$$

in agreement with the above considerations. The magnitude of Δl for a given value of $\chi^{(\lambda)}$ can be estimated from fig. 3, where the ratio $\chi^{(\lambda)}(\vartheta)/\chi^{(\lambda)}$ is illustrated as a function of ϑ for $\lambda = 1, 2, 3$. In chapters VII and VIII we shall see the effect of the limitation (14) on the multiple excitation of rotational and vibrational states.

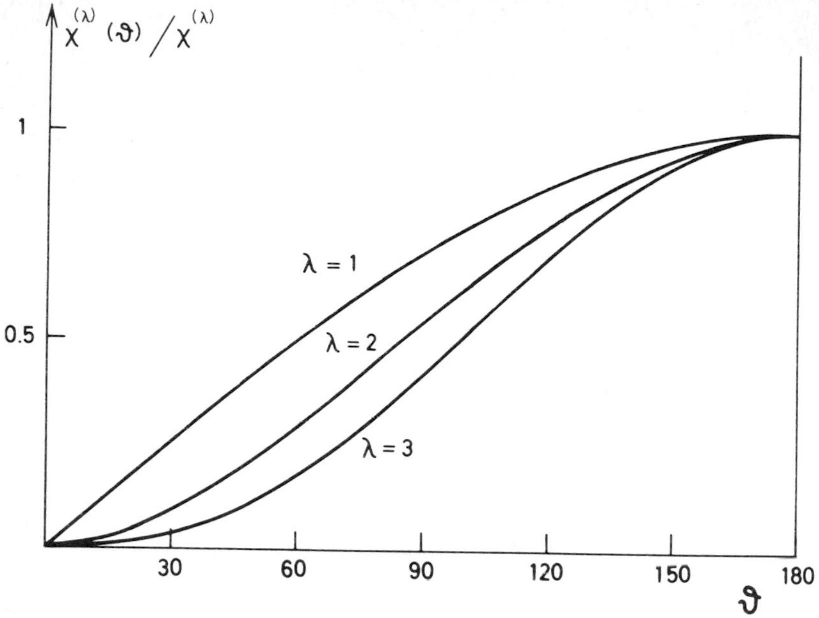

Fig. 3. The ratio $\chi^{(\lambda)}(\vartheta)/\chi^{(\lambda)}$ as a function of ϑ for $\lambda = 1$, 2 and 3. The quantity $\chi^{(\lambda)}(\vartheta)$ is defined in eq. (IV.3.8).

In the differential equations (2), (6) or (9) magnetic substates of all the levels involved must be treated separately and the number of equations in (6) is therefore the total number N of magnetic quantum numbers in the group G

$$N = \sum_{n \in G} (2I_n + 1). \tag{15}$$

In order to evaluate the Coulomb excitation of unpolarized targets these equations should be solved $(2I_0 + 1)$ times corresponding to the initial conditions (5).

The number of equations may be reduced by means of the symmetry relations discussed in § II.8. In *coordinate system* B it thus follows from the expression (II.8.19) i.e.

$$a_{I_n M_n, I_0 M_0} = (-1)^{I_n - I_0 + \pi_n - \pi_0} a_{I_n - M_n, I_0 - M_0} \tag{16}$$

that one need only evaluate the excitation amplitudes for initial magnetic quantum numbers $M_0 \geqslant 0$, i.e. the differential equations need only be solved $I_0 + 1$ or $I_0 + \frac{1}{2}$ times.

An important simplification arises for the case of backward scattering, where the magnetic quantum number (see (II.8.24)) is conserved in coordinate system B during the collision. This can also be seen directly from the differential equations, since for $\vartheta = \pi$

$$Q^{\mathrm{B}}_{\lambda\mu}(w) = 0, \qquad \text{for } \mu \neq 0. \tag{17}$$

This means that for each initial M_0 one need only consider states which have the magnetic quantum number $M_n = M_0$.

For deflection angles different from π, the excitation of the states with magnetic quantum number M_0 is still the most important. Especially in the neighbourhood of $\vartheta = \pi$ the collision functions $Q^{\mathrm{B}}_{\lambda\mu}(w)$ are small for $\mu \neq 0$ and one may in first order neglect the amplitudes with $M_n \neq M_0$. We may estimate the excitation amplitude of such states by means of the argument given above. Thus, to populate a state with magnetic quantum number M_n by an interaction with multipolarity λ, μ one must at least go through $|M_n - M_0|/\mu$ steps. Since the number of steps is limited by

$$\chi^{(\lambda)} \int_{-\infty}^{+\infty} |Q^{\mathrm{B}}_{\lambda\mu}(\vartheta, w)| \, \mathrm{d}w, \tag{18}$$

the maximum change in magnetic quantum number can thus be estimated by

$$|M_n - M_0| \leqslant \mu\chi^{(\lambda)} \int_{-\infty}^{+\infty} |Q^{\mathrm{B}}_{\lambda\mu}(\vartheta, w)| \, \mathrm{d}w. \tag{19}$$

The integrals in (19) are illustrated in fig. 4 for $\lambda = 1$, 2 and 3. It is noted that for even μ the integral is equal to $R^{\mathrm{B}}_{\lambda\mu}(\vartheta, 0)$. It is seen from these curves that for a given value of $\chi^{(\lambda)}$ the maximum change in the magnetic quantum number occurs for intermediate angles where it is of the order of magnitude

$$|M_n - M_0| \lessapprox \tfrac{1}{5}\chi^{(\lambda)}. \tag{20}$$

In a first approximation one may leave out the states for which the magnetic quantum numbers exceed the estimate (19). The population of these states can be taken into account in a perturbation treatment of the same type as the one discussed above (see eq. (9)).

In *coordinate system* A a reduction in the number of coupled equations follows from the symmetry relation (II.8.17), i.e.

$$a_{I_n M_n, I_0 M_0} = (-1)^{M_n - M_0 + \pi_n - \pi_0} a_{I_n M_n, I_0 M_0} \tag{21}$$

which also holds as a function of time. This equation shows that for a given initial magnetic quantum number M_0 one need in the coupled equations only consider those states for which

$$(-1)^{M_n - M_0 + \pi_n - \pi_0} = 1. \tag{22}$$

This reduces the number of coupled equations by about a factor of two. In contrast, however, to the situation in coordinate system B the equations should be solved for all initial values of M_0.

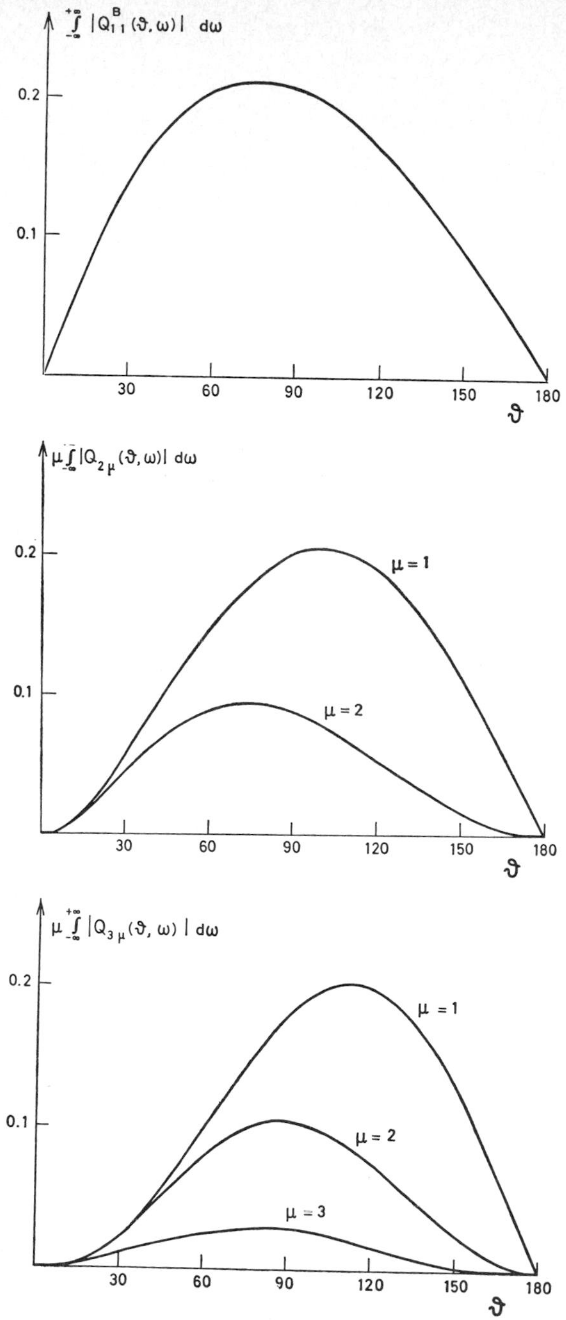

Fig. 4. The orbital integrals which describe the maximum angular momentum transfer along the z-axis in coordinate system B. The integrals which are defined in eq. (1.18) are given for $\lambda = 1, 2$ and 3 as functions of the scattering angle for the possible values of μ.

§ 2. *The diagonalization procedure*

As was discussed in the preceding section one may often justify that only a finite number of nuclear states take an active part in the multiple Coulomb excitation process. While the method in the preceding section was based directly on the differential equations for the excitation amplitude one may instead start from the formal solution (II.3.13) as it is given in terms of the S-matrix (II.3.19). If the series (II.3.30) converges so rapidly that one may limit oneself to the first few terms the excitation amplitude can be evaluated explicitly by a diagonalization method, where the polarization effect due to high-lying states may even be included.

According to eqs. (II.3.13) and (II.3.19) the excitation amplitude is given by

$$a_n = \langle n | \exp(2iQ) | 0 \rangle, \tag{1}$$

where the Hermitian phase shift operator Q is given by eq. (II.3.30), i.e.

$$2Q = R + G + (H - \tfrac{1}{12}R^3) + \cdots. \tag{2}$$

In order to evaluate (1) we introduce a unitary matrix U which diagonalizes Q, i.e.

$$\langle n | U^\dagger Q U | m \rangle = \lambda_n \delta_{nm}. \tag{3}$$

Since Q is Hermitian, the eigenvalues λ_n are real numbers. They are the classical eigenphase shifts for Coulomb excitation. The unitary transformation U will also diagonalize the operator $\exp(2iQ)$ and we thus find the following explicit expression for the excitation amplitude

$$\begin{aligned} a_n &= \langle n | UU^\dagger \exp(2iQ) UU^\dagger | 0 \rangle = \langle n | U \exp(2iU^\dagger QU) U^\dagger | 0 \rangle \\ &= \sum_m \langle n | U | m \rangle \langle m | U^\dagger | 0 \rangle \exp(2i\lambda_m) \\ &= \sum_m \langle 0 | U | m \rangle^* \langle n | U | m \rangle \exp(2i\lambda_m). \end{aligned} \tag{4}$$

The problem of determining the Coulomb excitation amplitudes is thus reduced to the problem of solving the eigenvalue equation (3).

We shall here discuss the case where only the two first terms in eq. (II.3.30) are taken into account, i.e.

$$2Q = R + G. \tag{5}$$

The matrix elements of R and G can be expressed in terms of the orbital integrals introduced in chapters IV and V and one finds according to eqs. (V.8.10) and (V.8.11)

$$\langle I_p M_p | \, 2Q \, | I_q M_q \rangle = -\sum_{\lambda\mu} \zeta_{pq}^{\lambda\mu} R_{\lambda\mu}(\vartheta, \xi_{pq}) + \tfrac{1}{2} \sum_{\substack{\lambda\lambda'\mu\mu' \\ z}} \zeta_{zq}^{\lambda\mu} \zeta_{pz}^{\lambda'\mu'} G_{\mu'\mu}^{\lambda'\lambda}(\vartheta, \xi_{pz}, \xi_{zq}). \quad (6)$$

With our conventions for the phases of the nuclear wave functions which lead to real ζ matrices and choosing either the coordinate system A or B leading to real functions R and G, the Q-matrix elements are seen to be real and symmetric under exchange of initial and final states. Thus it can be diagonalized by a real orthogonal matrix U.

In the summation over intermediate states $|z\rangle$ one may include states which do not belong to the finite number of states considered in the diagonalization (see § II.6, § V.7 and eq. (1.11)).

It is noted that, although the amplitudes (4) have to be evaluated for the $(2I_0 + 1)$ substates belonging to the state $|0\rangle$, the diagonalization (3) has to be performed only once.

While the rank N of the matrix to be diagonalized is given by

$$N = \sum_n (2I_n + 1), \quad (7)$$

an important simplification arises in coordinate system B for backward scattering, where the magnetic quantum number is conserved (see § 1) and where the rank is equal to the number of nuclear states considered. In chapter VIII we shall give some explicit applications of this method.

The usefulness of the diagonalization method relies on the fact that the expansion (2) can be terminated after the first or second term. The convergence of this series is mainly related to the magnitude of the parameters ξ and χ. Thus for $\xi = 0$ one knows (see § II.3) that the series (2) breaks off after the first term. The magnitude for $\xi \neq 0$ of the second term as compared to the first term in (2) can be estimated from the fact that usually the contribution from the orbital integral G is proportional to ξ for small values of ξ (see table V.1). Since the ζ-matrices are of the order of χ, the ratio between the two first terms is then of the order $\chi\xi$. One may thus expect that the method is applicable as long as the product $\chi\xi$ remains small.

As an example of the application of the diagonalization procedure we consider the excitation of a rotational band in an even-even nucleus. The nuclear matrix elements have been evaluated from the pure rotational model (see chapter VIII) and the five lowest nuclear states, i.e. the states of spin $I = 0, 2, 4, 6$ and 8, have been taken into account in the evaluation of the amplitudes according to eqs. (1)–(6). The excitation probabilities have been evaluated for $\vartheta = 180°$, as functions of $\chi_{0\to 2}$ for different values of $\xi_{0\to 2}$. The results are illustrated in figs. 5a-c where they are compared also with the result of a computer calculation of the coupled differential equation for the five states according to § 1. It is seen that the two calculations are in good agreement for small values of the product $\chi\xi$.

In two important cases the expansion of Q breaks off after a finite number of terms. Thus, in the sudden approximation it breaks off after the first term, while for excitation of pure vibrational states it breaks off after the second term. The first case is studied in the following section, while the second case is treated in chapter VII.

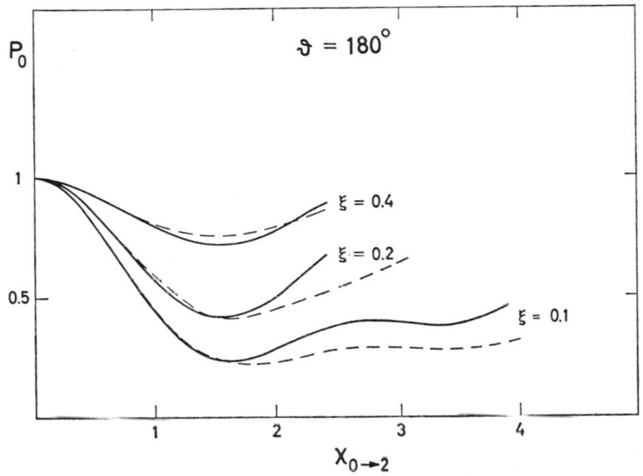

Fig. 5a.

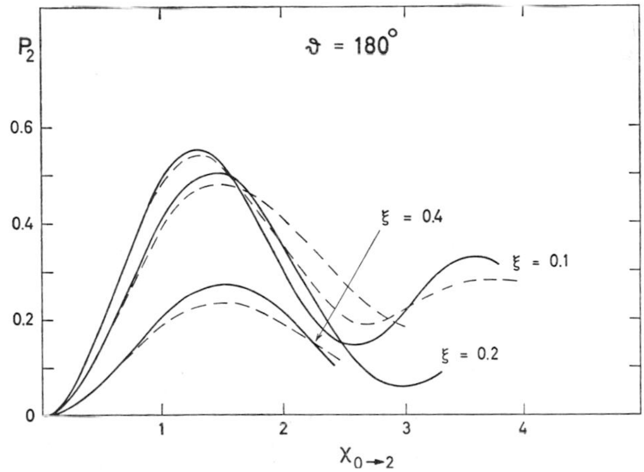

Fig. 5b.

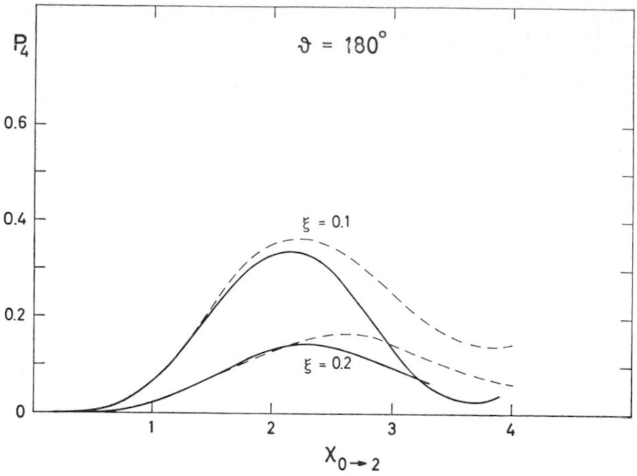

Fig. 5c.

Fig. 5. The excitation probabilities P_0, P_2 and P_4 of the three lowest states in a rotational band of an even-even nucleus. The probabilities are given for backwards scattering as a function of χ for $\xi = 0.1$, 0.2 and 0.4. Both parameters correspond to the $0^+ \to 2^+$ transition. The solid curve indicates the result of the diagonalization procedure described in § VI.2 which takes into account the five lowest states, while the dashed curves indicate the solution of the coupled differential equations including the same states. For $\xi = 0$ the two methods give identical results and are therefore not given. They can be found in fig. 6.

§ 3. *The sudden approximation*

For the different procedures described in § 1 and § 2 the equations which determine the multiple Coulomb excitation amplitudes must be solved anew if one changes the bombarding conditions, i.e. if one changes the deflection angle or bombarding energy. This is connected with the fact that the parameters χ and ξ are different functions of the projectile energy. In the case, however, where only one multipole order is important and where one may neglect the energy differences and set all ξ equal zero, the dependence on the bombarding energy can be extracted. It is also possible in this approximation to extract, in an approximate way, the dependence on the deflection angle.

In the sudden approximation, according to eqs. (2.6) and (V.8.8), the matrix elements of the operator Q is given by

$$\begin{aligned} \langle I_p M_p | \, 2Q \, | I_q M_q \rangle &= -\zeta_{pq}^{\lambda\mu} R_{\lambda\mu}(\vartheta, 0) \\ &= \chi_{q\to p}^{(\lambda)} (-1)^{I_q - M_q + 1} \sqrt{(2\lambda+1)(2I_q+1)} \\ &\quad \times \begin{pmatrix} I_q & \lambda & I_p \\ -M_q & \mu & M_p \end{pmatrix} R_{\lambda\mu}(\vartheta, 0). \end{aligned} \tag{1}$$

The energy dependence of the matrix element (1) is contained in the factor $\chi^{(\lambda)}_{q\to p}$ and the dependence is uniquely determined by λ. This common energy dependence may thus be extracted e.g. by writing

$$\langle I_p M_p|\, 2Q\, |I_q M_q\rangle = \chi^{(\lambda)}_{0\to 1}\rho^{\lambda\mu}_{pq} R_{\lambda\mu}(\vartheta, 0), \tag{2}$$

where $\chi^{(\lambda)}_{0\to 1}$ is the parameter describing the transition between the states 0 and 1 which could be e.g. the ground state and the first excited state. The matrix ρ is given by

$$\rho^{\lambda\mu}_{pq} = (-1)^{I_q - M_q + 1}\sqrt{(2\lambda+1)(2I_0+1)}\begin{pmatrix} I_q & \lambda & I_p \\ -M_q & \mu & M_p \end{pmatrix}\frac{\langle I_q\|\,\mathscr{M}(\mathrm{E}\lambda)\,\|I_p\rangle}{\langle I_0\|\,\mathscr{M}(\mathrm{E}\lambda)\,\|I_1\rangle} \tag{3}$$

and it is noted that ρ is defined to be independent of bombarding energy and deflection angle.

The unitary transformation U which diagonalizes $\rho R_{\lambda\mu}$ will also diagonalize the Q matrix, independent of the value of $\chi^{(\lambda)}_{0\to 1}$. If we solve the eigenvalue equation (2.3)

$$\sum_{pq} U^{\dagger}_{np}\rho^{\lambda\mu}_{pq} R_{\lambda\mu}(\vartheta, 0) U_{qm} = 2\nu_m \delta_{nm} \tag{4}$$

we thus find the following result for the excitation amplitudes

$$a_n = \sum_m \langle 0|\, U\, |m\rangle^* \langle n|\, U\, |m\rangle \exp(2\mathrm{i}\chi^{(\lambda)}_{0\to 1}\nu_m). \tag{5}$$

It is now seen explicitly that the energy dependence only enters through the parameters $\chi^{(\lambda)}_{0\to 1}$ in the exponent of eq. (5). As an illustration the excitation probabilities have been evaluated for the rotational states in an even-even nucleus for $\vartheta = 90°$. Five states with spin $I = 0, 2, 4, 6$ and 8 have been taken into account and the matrix elements are taken from the pure rotational model. The results are given in fig. 6. We note that in the sudden approximation the orbital integrals in coordinate system B, $R^{\mathrm{B}}_{\lambda\mu}$, vanish for odd values of μ (cf. table IV.1) and that, therefore, the only magnetic substates which should be considered are those which differ by an even integer from the ground state quantum number M_0. This reduces the rank of the matrix to be diagonalized by about a factor of two as compared to the general case.

A further simplification occurs for backward scattering where in this coordinate system only magnetic substates with $M = M_0$ need to be considered. As was discussed in § 1 and § 2 one may in first order neglect the excitation of states with $M \neq M_0$ also for $\vartheta \neq \pi$. In the sudden approximation where the orbital integrals for odd μ vanish, this approximation holds to better accuracy than in the general case where the terms with $\mu = 1$

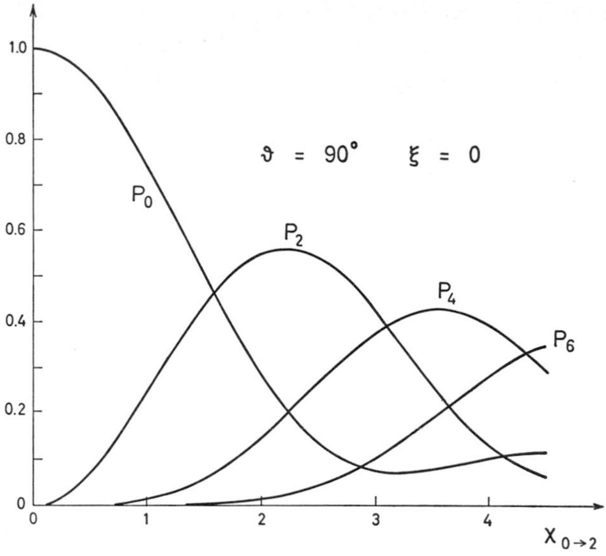

Fig. 6. The excitation probabilities P_I for the four lowest states in the ground state rotational band of an even-even nucleus. The results have been obtained in the sudden approximation by the diagonalization procedure described in § 3. The results are given for $\vartheta = 90°$ as a function of the χ-parameter for the transition $0 \to 2$.

give the most important corrections. It is a further simplification in the sudden approximation that by keeping only the $\mu = 0$ terms, the angular dependence of the excitation amplitudes can be extracted together with the energy dependence. In this approximation the Q-matrix (2) takes the form

$$\langle I_p M_p | \, 2Q \, | I_q M_q \rangle = [\chi^{(\lambda)}_{0\to 1} R^{\mathrm{B}}_{\lambda 0}(\vartheta, 0)] \rho^{\lambda 0}_{pq} \delta_{M_p M_q} . \tag{6}$$

In this case the ρ-matrix depends only on the common magnetic quantum number $M_p = M_q = M_0$ and on the spins. The unitary transformation U which diagonalizes ρ, i.e.

$$\sum_{pq} U^{\dagger}_{np} \rho^{\lambda 0}_{pq} U_{qm} = 2\mu_m \delta_{nm} \tag{7}$$

determines the excitation amplitudes through the expression

$$a_n = \sum_m \langle 0 | \, U \, | m \rangle^* \langle n | \, U \, | m \rangle \exp\{2\mathrm{i}[\chi^{(\lambda)}_{0\to 1} R^{\mathrm{B}}_{\lambda 0}(\vartheta, 0) \mu_m]\} . \tag{8}$$

It should be noted that although the rank of the matrix to be diagonalized is greatly reduced, the diagonalization for an unpolarized target has to be performed $I_0 + 1$ or $I_0 + \frac{1}{2}$ times.

It is seen from eq. (8) that in this approximation the excitation for arbitrary scattering angles can be reduced to the case of backward scattering. In the latter case the orbital integral $R^{\mathrm{B}}_{\lambda 0}(\pi, 0)$ is unity and the results for a scattering angle ϑ are thus obtained from those for backward scattering by the substitution

$$\chi^{(\lambda)}_{0\to 1} \to \chi^{(\lambda)}_{0\to 1} R^{\mathrm{B}}_{\lambda 0}(\vartheta, 0). \tag{9}$$

The product χR thus enters as an effective χ parameter, which we shall denote by χ_{eff}, i.e.

$$\chi_{\mathrm{eff}} = \chi^{(\lambda)}_{0\to 1} R^{\mathrm{B}}_{\lambda 0}(\vartheta, 0). \tag{10}$$

It should be noted that the parameter χ_{eff} which we have introduced deviates only slightly from the parameter $\chi^{(\lambda)}_{0\to 1}(\vartheta)$ that was introduced in eq. (IV.3.8). The ratio between the two quantities is given in table 1. As can be seen from

TABLE 1

The ratio of the parameters $\chi^{(\lambda)}(\vartheta)$ and $\chi^{(\lambda)}_{\mathrm{eff}}(\vartheta)$ for $\lambda = 1, 2$ and 3 is given as a function of the scattering angle ϑ. For $\lambda = 1$ the two quantities are identical.

	$\chi(\vartheta)/\chi_{\mathrm{eff}}(\vartheta)$		
ϑ	$\lambda = 1$	$\lambda = 2$	$\lambda = 3$
180	1.0000	1.0000	1.0000
160	1.0000	1.0000	1.0001
140	1.0000	1.0002	1.0006
120	1.0000	1.0012	1.0031
100	1.0000	1.0041	1.0101
80	1.0000	1.0106	1.0256
60	1.0000	1.0239	1.0550
40	1.0000	1.0482	1.1036
20	1.0000	1.0896	1.1745

this table, the quantities $\chi(\vartheta)$ and χ_{eff} differ mostly at forward scattering angles. For these angles the excitation process can essentially be treated by the first-order perturbation theory where the excitation probability of the states which can be reached from the ground state is $|\chi_{0\to p}(\vartheta)|^2$. If we thus substitute $\chi(\vartheta)$ for χ_{eff} in the approximations (6)–(10) we have made a change only of the order $|R_{\lambda 2}(\vartheta)/R_{\lambda 0}(\vartheta)|^2$ but on the other hand obtained an expression which leads to the correct result for the excitation probability at the forward angles. The validity of this approximation which we call the $\chi(\vartheta)$-approximation will be discussed in detail in the following section, where the contribution from the terms with $\mu \neq 0$ will be estimated. As an illustration of the accuracy of the $\chi(\vartheta)$-approximation the results of the diagonalization procedure for $\vartheta = 180°$ in the sudden approximation are given in fig. 7 as a function of $\chi_{0\to2}$. In the same figure the corresponding results for $\vartheta = 90°$ are indicated as functions of $\chi_{0\to2}(90°)$. It is seen that the difference between the two results is quite small.

In the $\chi(\vartheta)$-approximation one may thus for $\xi = 0$ obtain comprehensive expressions for the excitation amplitudes both as a function of energy (i.e. χ)

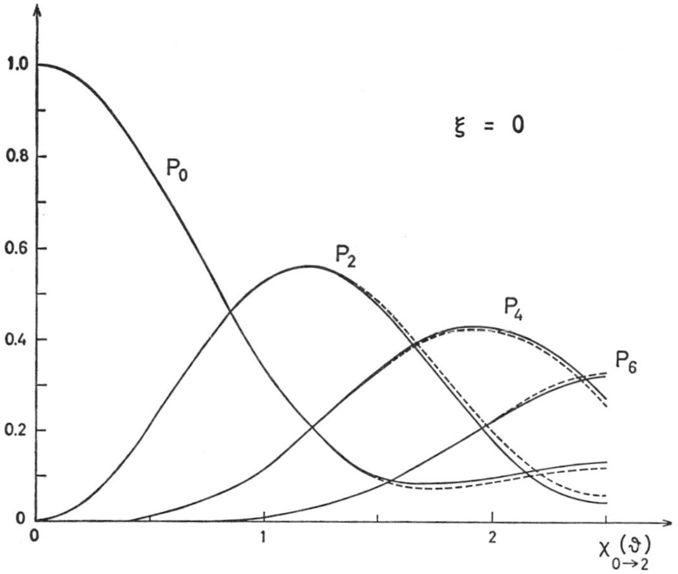

Fig. 7. The excitation probabilities of the four lowest states in a rotational band of an even-even nucleus. The results are evaluated by a five state diagonalization procedure in the sudden approximation. The solid curves indicate the results for $\vartheta = 180°$, while the dashed curves indicate the result of the calculation given in fig. 5 when it is plotted as a function of $\chi(90°)$.

and deflection angle by a single diagonalization. As an example, we consider again the case of a pure rotational model with five nuclear states. The resulting excitation amplitude may be written [ALD 60]

$$\begin{Bmatrix} a_{0,0} \\ a_{2,0} \\ a_{4,0} \\ a_{6,0} \\ a_{8,0} \end{Bmatrix} = \begin{Bmatrix} 0.296 & 0.269 & 0.219 & 0.150 & 0.066 \\ -0.308 & -0.131 & 0.094 & 0.208 & 0.138 \\ 0.260 & -0.141 & -0.277 & 0.008 & 0.151 \\ -0.188 & 0.312 & -0.046 & -0.203 & 0.125 \\ 0.098 & -0.250 & 0.288 & -0.207 & 0.070 \end{Bmatrix} \begin{Bmatrix} e^{i1.044\chi(\vartheta)} \\ e^{i0.488\chi(\vartheta)} \\ e^{-i0.430\chi(\vartheta)} \\ e^{-i1.392\chi(\vartheta)} \\ e^{-i2.063\chi(\vartheta)} \end{Bmatrix}. \quad (11)$$

For small values of $\chi(\vartheta)$ the series expansion of (11) must coincide with the perturbation expansion given in § V.8. In fig. 8 the excitation probabilities are plotted as functions of $\chi(\vartheta)$ together with the third-order perturbation calculation.

As we have seen, it is possible with the help of the sudden approximation to obtain an overview over the multiple excitation process. The results can be improved by taking deviations into account in a perturbation expansion as will be discussed in the next section.

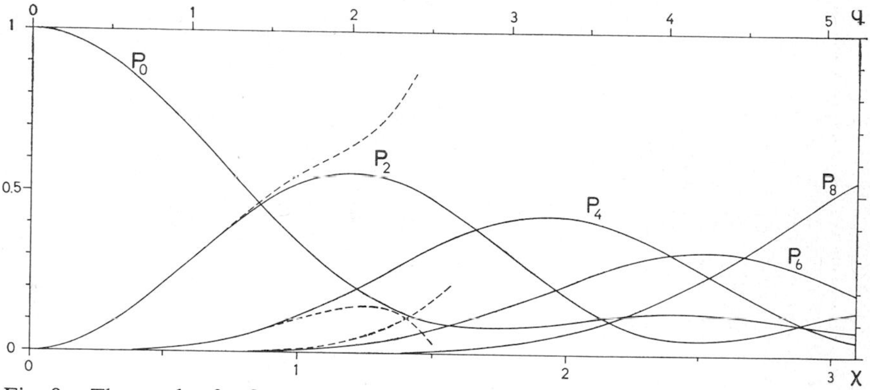

Fig. 8. The result of a five state diagonalization for the ground state rotational band in an even-even nucleus. The excitation probabilities for the five states are given as functions of $\chi_{0\to2}$. The alternative variable q defined in (VIII.2.10) is also indicated. The dashed curves show the results of the third-order perturbation calculation.

The usefulness of the sudden approximation is not limited to the diagonalization procedure. In cases where the nuclear wave functions are known, the excitation amplitude in the sudden approximation can be evaluated exactly, i.e. including infinitely many states. It thus follows from eq. (II.3.35) that the excitation amplitude can be written

$$a_n = \langle n| \exp\left\{-\frac{\mathrm{i}}{\hbar}\int_{-\infty}^{+\infty} V_{\mathrm{E}}(t)\,\mathrm{d}t\right\}|0\rangle. \tag{12}$$

In this expression we may introduce directly the multipole expansion (II.1.12) which after integration may be expressed by

$$a_{I_{\mathrm{f}}M_{\mathrm{f}},I_0M_0} = \langle I_{\mathrm{f}}M_{\mathrm{f}}| \exp\left\{-\mathrm{i}\sum_{\lambda\mu}\chi_{0\to1}^{(\lambda)}\frac{(-1)^{\mu}\sqrt{(2\lambda+1)(2I_0+1)}}{\langle 0\|\mathscr{M}(\mathrm{E}\lambda)\|1\rangle}\right.$$
$$\left.\times\,\mathscr{M}(\mathrm{E}\lambda,-\mu)R_{\lambda\mu}(\vartheta,0)\right\}|I_0M_0\rangle. \tag{13}$$

If explicit expressions are known for the nuclear wave functions (and multipole moments) the computation of the excitation amplitudes is reduced to the evaluation of definite integrals. In chapter VIII we shall carry through the explicit evaluation of this expression for the rotational model.

If only one multipole order is responsible for the excitation, it is seen that the energy dependence of the excitation amplitude only enters through the parameter $\chi_{0\to1}^{(\lambda)}$. Similar to the above discussion the main contribution will arise from the $\mu = 0$ term and one may thus use the $\chi(\vartheta)$-approximation neglecting the terms $\mu \neq 0$.

§ 4. *Deviations from the sudden approximation*

The sudden approximation offers through its simplicity a useful illustration of the multiple Coulomb excitation process, especially if one uses at the same time the $\chi(\vartheta)$-approximation. Although powerful methods of evaluating Coulomb excitation without these approximations are available, it may nevertheless be instructive to discuss how the first-order deviation from the sudden approximation and the $\chi(\vartheta)$-approximation can be calculated. In this and the following section we shall outline how these calculations can be done without entering into a detailed numerical evaluation of the corrections.

An estimate of the deviation from the sudden approximation can either be based on the expression (II.3.13) or on the expansion (II.3.30).

Starting from eq. (II.3.13) which reads

$$a_n = \langle n| \mathscr{T} \exp\left\{ -\frac{\mathrm{i}}{\hbar} \int_{-\infty}^{+\infty} \exp\left(\frac{\mathrm{i}}{\hbar} H_0 t \right) V_{\mathrm{E}}(t) \exp\left(-\frac{\mathrm{i}}{\hbar} H_0 t \right) \mathrm{d}t \right\} |0\rangle, \quad (1)$$

one may expand the exponent by means of the formula

$$\exp\left(\frac{\mathrm{i}}{\hbar} H_0 t \right) V_{\mathrm{E}}(t) \exp\left(-\frac{\mathrm{i}}{\hbar} H_0 t \right) = V_{\mathrm{E}}(t) + \frac{\mathrm{i}}{\hbar} t [H_0, V_{\mathrm{E}}(t)] + \left(\frac{\mathrm{i}}{\hbar} t \right)^2 [H_0, [H_0, V_{\mathrm{E}}(t)]] + \cdots. \quad (2)$$

The first term in this expansion leads exactly to the sudden approximation and the matrix elements of the second term are proportional to ξ. Neglecting higher terms, one obtains

$$a_n = a_n^{(0)} + \Delta a_n, \quad (3)$$

where $a_n^{(0)}$ is given by eq. (3.12) while the first correction term, according to eq. (II.3.41), is given by

$$\Delta a_n = \frac{1}{\hbar^2}\langle n|\int_{-\infty}^{+\infty} \mathrm{d}t\, t \exp\left\{-\frac{\mathrm{i}}{\hbar}\int_t^{\infty} V_{\mathrm{E}}(t')\,\mathrm{d}t'\right\}$$

$$\times [H_0, V_{\mathrm{E}}(t)] \exp\left\{-\frac{\mathrm{i}}{\hbar}\int_{-\infty}^{t} V_{\mathrm{E}}(t')\,\mathrm{d}t'\right\}|0\rangle. \tag{4}$$

An application to the case of a rotational model will be given in chapter VIII. In the diagonalization procedure, expression (4) can be evaluated rather easily in the $\chi(\vartheta)$-approximation in so far as only one multipole order contributes. In this case the time dependence of $V_{\mathrm{E}}(t)$ can be extracted and the unitary transformation U which diagonalizes $\int_{-\infty}^{+\infty} V_{\mathrm{E}}(t)\,\mathrm{d}t$, will also diagonalize the indefinite integral $\int_{-\infty}^{t} V_{\mathrm{E}}(t)\,\mathrm{d}t$, as well as $V_{\mathrm{E}}(t)$.

The explicit evaluation of (4) is given in [ALD 60]. In principle it is not difficult to generalize the present expansion to include higher powers in ξ. In this connection, however, one meets the problem that the time integrations become divergent. This is due to the fact that the function $V_{\mathrm{E}}(t)$ is proportional to $t^{-\lambda-1}$ for large values of t and that the nth term in the expansion gives rise to a factor t^n. It is thus noted that already the term (4) is divergent for dipole excitation. This difficulty is connected with the fact that the expansion of the orbital integrals $R_{\lambda\mu}(\vartheta, \xi)$ for small values of ξ contain a term $\xi^{\lambda}\log\xi$.

A more elegant expansion method is obtained from the expressions (II.3.30) or (2.5). The first term is directly related to the sudden approximation as was discussed in § 3. If the parameters ξ are small numbers the operator G may be considered as a perturbation. Expanding the exponential function in eq. (2.1) according to eq. (II.3.38) one may write the excitation amplitude in the form (3) with

$$a_n^{(0)} = \langle n|\exp\left\{-\frac{\mathrm{i}}{\hbar}\int_{-\infty}^{+\infty} \tilde{V}(t')\,\mathrm{d}t'\right\}|0\rangle \tag{5}$$

and

$$\Delta a_n = \mathrm{i}\int_0^1 \mathrm{d}x\,\langle n|\,\mathrm{e}^{\mathrm{i}xR} G\, \mathrm{e}^{\mathrm{i}(1-x)R}\,|0\rangle$$

$$= \mathrm{i}\int_0^1 \mathrm{d}x\,\langle n|\exp\left\{-\frac{\mathrm{i}}{\hbar}x\int_{-\infty}^{+\infty} \tilde{V}(t')\,\mathrm{d}t'\right\}$$

$$\times G \exp\left\{-\frac{\mathrm{i}}{\hbar}(1-x)\int_{-\infty}^{+\infty} \tilde{V}(t')\,\mathrm{d}t'\right\}|0\rangle. \tag{6}$$

We shall here only discuss the case where we set $\tilde{V}(t) = V_{\rm E}(t)$ and where the first term $a_n^{(0)}$ leads to the sudden approximation. The correction term Δa_n can then be evaluated if the nuclear wave functions are known, and an example will be discussed in chapter VIII. It may also be evaluated more generally by means of the diagonalization procedure. Thus we introduce the unitary transformation U which diagonalizes $\int_{-\infty}^{+\infty} V_{\rm E}(t')\,{\rm d}t'$ (see eq. (3.4)), i.e.

$$\langle n|\, U^{\dagger}\left\{-\frac{\rm i}{\hbar}\int_{-\infty}^{+\infty} V_{\rm E}(t')\,{\rm d}t'\right\}U\,|m\rangle = 2{\rm i}\nu_m\chi_{0\to 1}^{(\lambda)}\delta_{nm}. \tag{7}$$

We have here limited ourselves to the case where only one multipole order is important. Inserting (7) into (6) and performing the integration over x one obtains

$$\Delta a_n = \sum_{mpqs} \langle m|\, U\,|q\rangle^* \langle n|\, U\,|q\rangle \langle 0|\, U\,|s\rangle^* \langle p|\, U\,|s\rangle$$
$$\times \frac{\exp(2{\rm i}\nu_q\chi_{0\to 1}^{(\lambda)}) - \exp(2{\rm i}\nu_s\chi_{0\to 1}^{(\lambda)})}{2\chi_{0\to 1}^{(\lambda)}[\nu_q - \nu_s]}\langle m|\, G\,|p\rangle, \tag{8}$$

where according to eq. (2.6) the matrix element of G is given by

$$\langle m|\, G\,|p\rangle = \tfrac{1}{2}\sum_{\mu\mu' z} \zeta_{zp}^{\lambda\mu}\zeta_{mz}^{\lambda\mu'} G_{\mu'\mu}^{\lambda\lambda}(\vartheta, \xi_{mz}, \xi_{zp}). \tag{9}$$

In the evaluation of (9) one may include any intermediate state $|z\rangle$ which could be of importance (see § II.6, § V.7 and eq. (1.11)).

§ 5. *Deviations from the $\chi(\vartheta)$-approximation*

The $\chi(\vartheta)$-approximation which was discussed in § 3 leads to an important simplification of the diagonalization procedure, since the rank of the matrix which has to be diagonalized is reduced to the number of nuclear states included. The accuracy of the $\chi(\vartheta)$-approximation was discussed in § 1 where it was shown that the change in magnetic quantum number could be estimated by (1.19). The magnetic substates which are deleted in the solution of the coupled equations can be included by a perturbation theory.

We shall first base our discussion on the general expressions (3.12)–(3.13) which we shall write in the form

$$a_n = \langle n|\, \exp\left\{2{\rm i}\sum_{\mu} Q_\mu\right\}|0\rangle, \tag{1}$$

where the μ denotes the magnetic quantum number in the multipole expansion and

$$2Q_\mu = -\sum_{\lambda}\chi_{0\to 1}^{(\lambda)}\frac{(-1)^\mu\sqrt{(2\lambda+1)(2I_0+1)}}{\langle 0\|\, \mathcal{M}({\rm E}\lambda)\,\|1\rangle}\mathcal{M}({\rm E}\lambda, -\mu)\,R_{\lambda\mu}(\vartheta, 0). \tag{2}$$

In the $\chi(\vartheta)$-approximation one neglects all terms with $\mu \neq 0$, i.e. the excitation amplitudes are given by

$$a_n^{(0)} = \langle n| \exp\{2iQ_0\} |0\rangle . \tag{3}$$

We note that the operators Q commute. Expanding the exponential function we may therefore write the amplitude (1) in the form

$$a_n = a_n^{(0)} + \Delta^{(1)}a_n + \Delta^{(2)}a_n , \tag{4}$$

where the first-order correction $\Delta^{(1)}a_n$ to the $\chi(\vartheta)$-approximation is given by

$$\begin{aligned}\Delta^{(1)}a_n &= i\sum_z \langle n|\, 2 \sum_{\mu\neq 0} Q_\mu \,|z\rangle \langle z| \exp(2iQ_0)\, |0\rangle \\ &= i\sum_z \langle n|\, 2 \sum_{\mu\neq 0} Q_\mu \,|z\rangle\, a_z^{(0)} .\end{aligned} \tag{5}$$

It is seen from this expression that the first-order corrections can be reduced to a simple combination of the already known amplitudes $a^{(0)}$. The first-order corrections give rise to excitation of states where $M_0 - M = \mu$ and consequently no corrections to the states with $M = M_0$ occur in first order. If one is interested in the corrections of the excitation probabilities one has to compute the second-order corrections to the states with $M = M_0$. The interference between these and the zero-order term is of the same magnitude as the square of terms with $M \neq M_0$. For these second-order corrections one finds

$$\Delta^{(2)}a_n = -\tfrac{1}{2}\sum_z \langle n| \Big(\sum_{\mu\neq 0} 2Q_\mu\Big)^2 |z\rangle\, a_z^{(0)} . \tag{6}$$

An illustration of the application of this method is given in [ALD 60].

§ 6. *Excitation of two weakly coupled groups of states*

We discussed in § 1 how the nuclear states can be classified into groups of states where multiple excitation will occur within a group, if any state in the group is populated. The χ-parameters connecting states in two different groups are small and the excitation of the groups which do not contain the ground state is thus small. If the multiple Coulomb excitation amplitudes are computed directly from the coupled differential equations, it was shown in § 1 that the excitation of these groups could be computed by the solution of the inhomogeneous differential equations (1.9). These equations result from a perturbation expansion in the small χ's connecting the different groups.

In a diagonalization procedure or in the sudden approximation a similar expansion can be performed. The expansion may again be based on either expression (II.3.13) or on (II.3.30) for the excitation amplitude in terms of the S-matrix. In both cases one decomposes the Coulomb interaction Hamiltonian in the parts

$$\tilde{V}(t) = \tilde{V}_{XX}(t) + \tilde{V}_{YY}(t) + \tilde{V}_{XY}(t) + \tilde{V}_{YX}(t) + \cdots, \tag{1}$$

where e.g.

$$\tilde{V}_{XY}(t) = \sum_{\substack{n\in X\\ m\in Y}} |n\rangle \langle n|\, \tilde{V}(t)\, |m\rangle \langle m| . \tag{2}$$

The symbols X and Y denote the various groups, $\tilde{V}_{XX}(t)$ being the part of the interaction which couples the group X to itself while $\tilde{V}_{XY}(t)$ is responsible for the weak coupling between the groups X and Y. It is noted that the operators $\tilde{V}_{XX}$ and $\tilde{V}_{YY}$ commute, while $\tilde{V}_{XY}$ in general commutes neither with $\tilde{V}_{XX}$ nor with $\tilde{V}_{YY}$. We now consider the excitation of a state n_X belonging to the group X which is different from the ground state group G. The excitation amplitude is then given by

$$a_{n_X} = \langle n_X|\, \mathscr{T} \exp\left\{ -\frac{i}{\hbar} \int_{-\infty}^{+\infty} [\tilde{V}_{GG}(t') + \tilde{V}_{XX}(t') + \tilde{V}_{XG}(t')]\, dt' \right\} |0\rangle \tag{3}$$

where we have neglected couplings to groups different from X and G. Using eq. (II.3.41) we may expand the expression (3) to first order in V_{XG}, i.e.

$$a_{n_X} = \frac{1}{i\hbar} \langle n_X| \int_{-\infty}^{+\infty} dt \left(\mathscr{T} \exp\left\{ -\frac{i}{\hbar} \int_{t}^{+\infty} \tilde{V}_{XX}(t')\, dt' \right\} \right)$$

$$\times\ \tilde{V}_{XG}(t) \left(\mathscr{T} \exp\left\{ -\frac{i}{\hbar} \int_{-\infty}^{t} \tilde{V}_{GG}(t')\, dt' \right\} \right) |0\rangle . \tag{4}$$

This expression is especially useful if one may apply the sudden approximation within the groups X and G. In this case the time ordering of the two exponential functions can be left out. It is noted that a finite energy difference may be kept in the matrix elements of $\tilde{V}_{XG}$.

The expression (4) is most easily evaluated by means of the diagonalization procedure if besides the sudden approximation the $\chi(\vartheta)$-approximation is applied and only one multipolarity contributes. In this case the unitary transformation which diagonalizes the exponential functions is time independent (see [ALD 60]).

The expression (4) can also be evaluated in the sudden approximation for various nuclear models. In chapter VIII we shall give an application of (4) to the case of the excitation of coupled rotational bands.

§ 7. *Simultaneous and mutual excitation*

In the collision between two heavy ions both target and projectile nuclei can usually be excited. An example of this is shown in fig. 9. In this connection we distinguish between simultaneous excitation of target and projectile and mutual excitation of target and projectile. This distinction is connected with the classification of the Coulomb interaction into monopole-multipole and multipole-monopole interaction on the one hand and the multipole-multipole interaction on the other hand (cf. § II.1). Since the monopole moments, i.e. the electric charges, are undisturbed by excitations, the excitations caused by the monopole-multipole interactions and vice versa proceed independently. As was seen in § II.2 this is reflected by the fact that the wave function of the two nuclei at any time can be written as a product wave function and, that the excitation amplitudes and the excitation probabilities are simply a product of the amplitudes or probabilities for target and projectile excitation.* This simultaneous excitation can thus be calculated by the methods described above. Since, however, the multipole moments depend on the nuclear states, the multipole-multipole interaction cannot be treated in a similar simple way. As we have seen in chapter IV.6 the effect of this interaction is small and one thus expects that mutual excitation can be handled in a perturbation treatment.

Since the states which may be populated by mutual excitation are already expected to be excited strongly by the monopole-multipole fields, a perturbation treatment must be based on the result given by eq. (II.4.36). Thus one solves first the coupled differential equations for the monopole-multipole interaction $V_{\mathrm{E}}(t)$. The excitation amplitude in zero order is

$$a_{0\to n}^{(\mathrm{sim})}(t) = a_{I_0{}^1M_0{}^1\to I_n{}^1M_n{}^1}(t)\cdot a_{I_0{}^2M_0{}^2\to I_n{}^2M_n{}^2}(t), \tag{1}$$

where the first factor is the excitation amplitude of the projectile under the influence of the monopole moment of the target, while the second factor is the excitation amplitude of the target under the influence of the projectile. Next one solves the same equation backwards in time from the initial condition that the projectile is in the state $|I_{\mathrm{f}}^1M_{\mathrm{f}}^1\rangle$ and the target in the state $|I_{\mathrm{f}}^2M_{\mathrm{f}}^2\rangle$. The solution is of the form

$$\bar{a}_{\mathrm{f}\to m}^{(\mathrm{sim})}(t) = \bar{a}_{I_{\mathrm{f}}{}^1M_{\mathrm{f}}{}^1\to I_m{}^1M_m{}^1}(t)\;\bar{a}_{I_{\mathrm{f}}{}^2M_{\mathrm{f}}{}^2\to I_m{}^2M_m{}^2}(t), \tag{2}$$

where the first factor is the backwards solution for the excitation amplitude of the projectile under the influence of the monopole moment of the target, which at time $t = +\infty$ satisfies the equation

$$\bar{a}_{I_{\mathrm{f}}{}^1M_{\mathrm{f}}{}^1\to I_m{}^1M_m{}^1}(+\infty) = \delta_{I_{\mathrm{f}}{}^1I_m{}^1}\delta_{M_{\mathrm{f}}{}^1M_m{}^1} \tag{3}$$

* This is only correct when the symmetrization discussed in § 8 below is neglected.

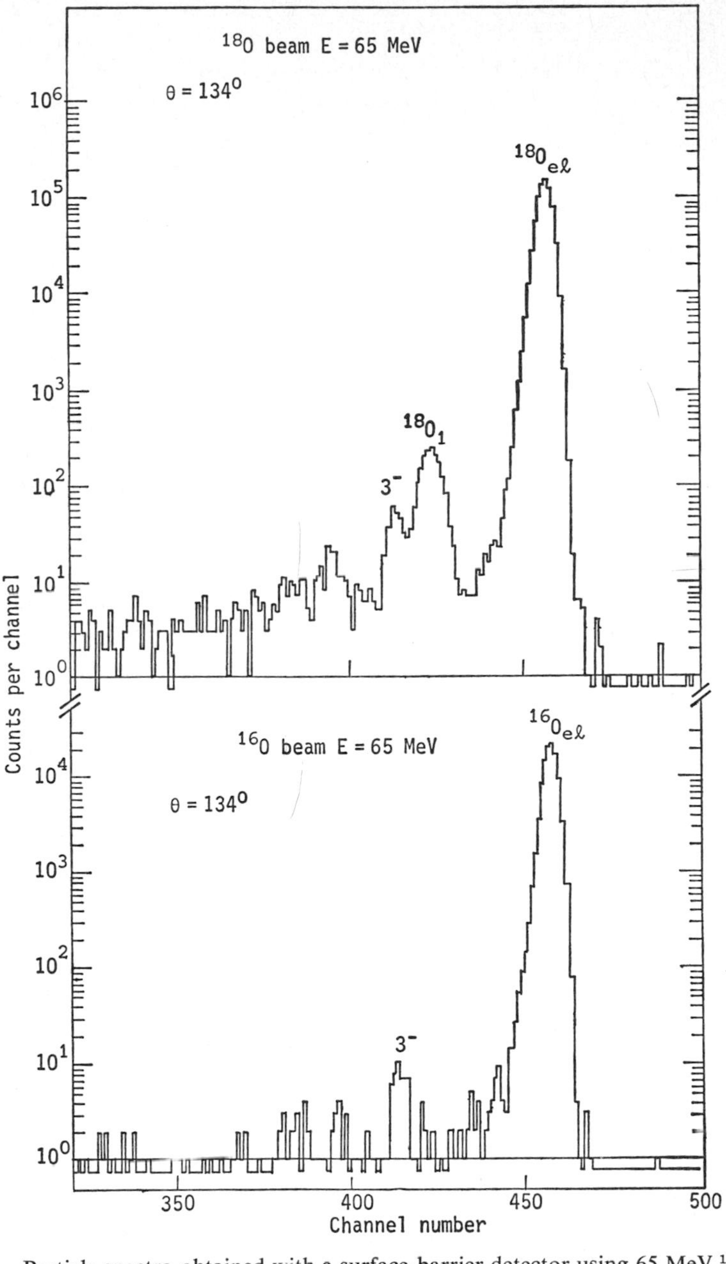

Fig. 9. Particle spectra obtained with a surface-barrier detector using 65 MeV ^{18}O and ^{16}O beams on a thin ^{208}Pb target. The 2.61 MeV 3^- level in ^{208}Pb and the 1.98 MeV 2^+ level in ^{18}O have been excited. (From Häusser, O., in: *Nuclear Spectroscopy*, ed. J. Cerny, Chapter VIIB.)

while the second factor is the corresponding factor for target excitation. Inserting these unperturbed solutions in eq. (II.4.36) one finds

$$a_{0\to f}^{(\mathrm{mut})}(\infty) = \frac{1}{i\hbar}\sum_{nm}\int_{-\infty}^{+\infty} \bar{a}_{I_f{}^1M_f{}^1\to I_m{}^1M_m{}^1}(t)\bar{a}_{I_f{}^2M_f{}^2\to I_m{}^2M_m{}^2}(t)$$

$$\times \langle I_m^1M_m^1I_m^2M_m^2|\ W(1,2;t)\ |I_n^1M_n^1I_n^2M_n^2\rangle$$

$$\times\ a_{I_0{}^1M_0{}^1\to I_n{}^1M_n{}^1}(t)\ a_{I_0{}^2M_0{}^2\to I_n{}^2M_n{}^2}(t)\ \exp\left\{\frac{i}{\hbar}t[E_m^1+E_m^2-E_n^1-E_n^2]\right\}\mathrm{d}t. \tag{4}$$

Explicit expressions for the multipole-multipole matrix elements in (4) are given in § IV.6. The total excitation amplitude is the sum of the excitation amplitudes (1) and (4), i.e.

$$a_{0\to f} = a_{0\to f}^{(\mathrm{sim})} + a_{0\to f}^{(\mathrm{mut})}. \tag{5}$$

§ 8. *Symmetrization of the parameters*

In the semiclassical description of Coulomb excitation that has been used above, the energy loss of the projectile in the collision is neglected and only one asymptotic velocity v is used. The change in the results which would be obtained by inserting for v instead of the bombarding velocity the relative velocity after the excitation, gives rise to an uncertainty in the results which in principle lies outside the scope of the semiclassical approximation. The uncertainty may be especially important for high excited states since the cross sections then depend rather sensitively on ξ.

This problem was discussed already in § IV.7 in first-order perturbation theory and a more systematic discussion of the improvement of the semiclassical theory will be given in chapter IX. In this section we shall nevertheless give an elementary argument for a symmetrization procedure which in most cases will take into account some major correction to the pure semiclassical picture.

As a starting point we recall the results of the first-order perturbation theory where the parameter ξ should be replaced by the expression

$$\xi_{f0} = \frac{Z_1Z_2e^2}{\hbar}\left(\frac{1}{v_f}-\frac{1}{v_0}\right), \tag{1}$$

instead of (1.3) where v_0 and v_f are the asymptotic relative velocities in initial and final state, respectively. One finds furthermore that one should use an orbit with the following symmetrized expression

$$a_{0f} = Z_1 Z_2 e^2 / m_0 v_0 v_f \tag{2}$$

for half the distance of closest approach. From the excitation probabilities P computed with the help of these parameters, one obtains the differential cross section in the center-of-mass system by the expression

$$d\sigma(0 \to f) = \frac{v_f}{v_0} P(0 \to f;\, \xi,\, \vartheta) \tfrac{1}{4} a_{0f}^2 \frac{1}{\sin^4(\frac{1}{2}\vartheta)}\, d\Omega. \tag{3}$$

It was found that this procedure leads to results which are in very close agreement with the results of a quantum mechanical description of the first-order Coulomb excitation.

In multiple Coulomb excitation, where several nuclear states are excited, a substitution analogous to eq. (2) would introduce a different value of a_{0f} for each final velocity v_f. For each level, the excitation would then be calculated for a different orbit of the projectile, violating hereby the condition

$$\sum_f P(0 \to f) = 1. \tag{4}$$

The fundamental rules, which have to be observed in the construction of a symmetrization are, besides (4), the relation

$$(2I_0 + 1) v_0^2\, d\sigma(0 \to f) = (2I_f + 1) v_f^2\, d\sigma(f \to 0) \tag{5}$$

between the cross sections for the two processes $0 \to f$ and $f \to 0$. The rule (5) must hold for any inelastic scattering according to time reversal invariance. Eq. (5) is fulfilled for expression (3) since in perturbation theory

$$P(0 \to f)(2I_0 + 1) = P(f \to 0)(2I_f + 1). \tag{6}$$

In order to study the possible symmetrization in multiple Coulomb excitation it is noted that the condition (4) is fulfilled if the matrix elements of the interaction $\tilde{V}(t)$ satisfy the relation

$$\langle n|\, \tilde{V}(\boldsymbol{r}(t))\, |m\rangle^* = \langle m|\, \tilde{V}(\boldsymbol{r}(t))\, |n\rangle. \tag{7}$$

It is thus seen that one may substitute for the classical trajectory $\boldsymbol{r}(t)$ in (7) a different expression $\boldsymbol{r}_{nm}(t)$ of each pair of nuclear states $|n\rangle$ and $|m\rangle$, provided only that $\boldsymbol{r}_{nm} = \boldsymbol{r}_{mn}$.

A natural choice for $\boldsymbol{r}_{nm}$ would be to use the hyperbolic orbits of § II.9 with a symmetrized expression for the distances of closest approach, i.e.

$$a = a_{nm} = Z_1 Z_2 e^2 / m_0 v_n v_m \tag{8}$$

and an average velocity

$$v = \tfrac{1}{2}(v_n + v_m), \tag{9}$$

where the velocities v_n and v_m are the asymptotic relative velocities corresponding to the nucleus being excited to the states $|n\rangle$ and $|m\rangle$ respectively. Through the prescriptions (8) and (9) it is seen that the adiabaticity parameter ξ_{nm} should be replaced by the symmetrized expression

$$\xi_{nm} = \frac{Z_1 Z_2 e^2}{\hbar}\left(\frac{1}{v_n} - \frac{1}{v_m}\right). \tag{10}$$

Since it would be rather complicated to use different collision functions for each pair of nuclear states, an important simplification of the symmetrization prescription is obtained if one renormalizes instead the nuclear matrix elements. All calculations of multiple Coulomb excitations have until now been performed with a symmetrization where one has used a common collision function but has substituted the ξ-parameters according to eq. (10) and has used renormalized expressions for the parameters $\chi_{nm}^{(\lambda)}$ in (1.7) according to the prescription

$$\chi_{nm}^{(\lambda)} = \frac{\sqrt{16\pi}(\lambda - 1)!\, Z_1 e \langle m \| \mathcal{M}(\mathrm{E}\lambda) \| n \rangle}{(2\lambda + 1)!!\, \hbar (v_n v_m)^{1/2} a_{nm}^{\lambda} \sqrt{2I_m + 1}}. \tag{11}$$

It is also seen directly that since these symmetrizations may be thought of as slight renormalizations of the nuclear energies and matrix elements, the symmetrized equations (1.2) will lead to solutions fulfilling unitarity. Furthermore, if we consider the problem of the reversed motion where the nucleus in the state $|\mathrm{f}\rangle$ is bombarded with projectiles of relative velocity v_f we find that

$$v_n^{(\mathrm{rev})} = v_n. \tag{12}$$

The differential equations which have to be solved are therefore the same for the two problems except for the initial conditions. Under these circumstances the amplitudes for the transitions from the state $|I_0 M_0\rangle$ to the state $|I_\mathrm{f} M_\mathrm{f}\rangle$ are equal to the amplitudes for the inverse transition according to eq. (II.8.22).

It thus follows that the quantities $P(0 \to \mathrm{f})$ for the two problems satisfy eq. (6) and that the following expression for the differential cross sections

$$\mathrm{d}\sigma(0 \to \mathrm{f}) = \frac{v_\mathrm{f}}{v_0} \tfrac{1}{4} |a_{\mathrm{f}0}|^2 P(0 \to \mathrm{f}, \chi_{nm}^{(\lambda)}, \xi_{nm}) \frac{1}{\sin^4(\frac{1}{2}\vartheta)}\, \mathrm{d}\Omega \tag{13}$$

satisfies condition (5).

CHAPTER VII

Multiple Excitation of Vibrational States

In this and the following chapter we shall discuss the multiple Coulomb excitation of collective states of vibrational and rotational character. As was shown in the preceding chapter the use of an explicit nuclear model allows one to compute the excitation amplitude in the sudden approximation in terms of a definite multiple integral, including in principle all the states of the model. The collective states that give rise to large electric multipole matrix elements are of special interest, since already bombardment with medium heavy charged projectiles leads to strong multiple excitation. It is known that in actual nuclei there are strong deviations from the idealized models which we are going to use. An example of the Coulomb excitation of a typical vibrational nucleus is given in figs. 1a-c. The advantage of using the pure models is that the integrals through which the multiple excitation amplitudes are expressed can be evaluated in a closed form. It is a special feature of the pure vibrational model that the excitation amplitude can be computed exactly, including finite excitation energies.

The simple results are well suited for a discussion of the dependence of multiple excitation processes on deflection angle and on the parameters χ and ξ, and we shall in the present chapter mainly discuss the vibrational model with this purpose in mind.

§ 1. *The vibrational model*

In the extreme vibrational model of an atomic nucleus the nuclear states can be thought of as eigenstates of a vibrational Hamiltonian of the form (see e.g. [BOH 75])

$$H_0 = \tfrac{1}{2} \sum_{\lambda\mu} \{D_\lambda |\dot{\alpha}_{\lambda\mu}|^2 + C_\lambda |\alpha_{\lambda\mu}|^2\}. \tag{1}$$

Here, $\alpha_{\lambda\mu}$ describes the vibrational amplitudes of multipolarity λ, μ, while D_λ and C_λ are parameters describing the inertia and the restoring force, respectively. The Hamiltonian (1) has been used to describe collective nuclear

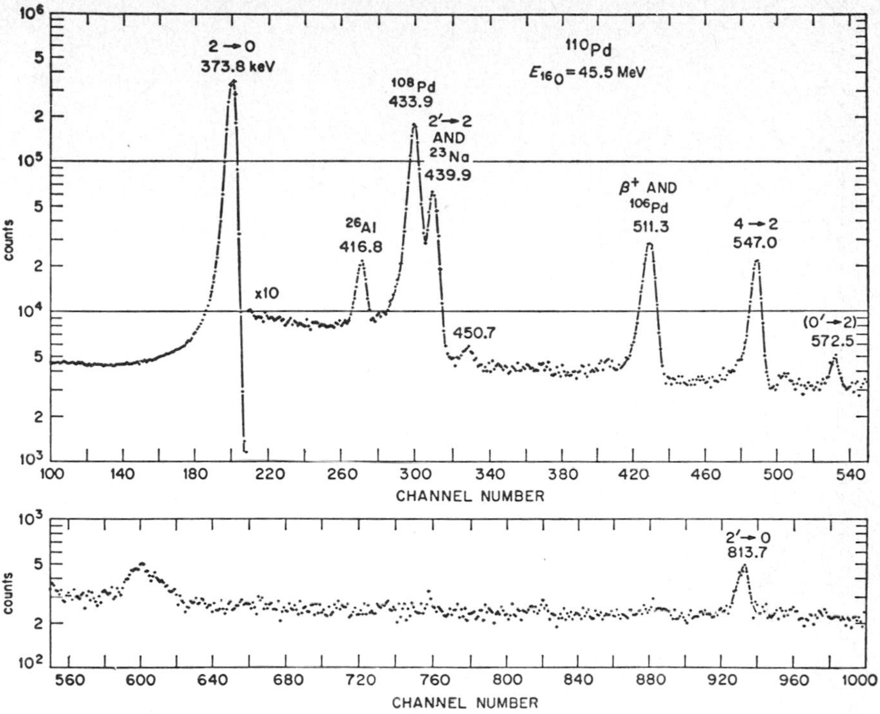

Fig. 1a. Gamma-ray spectrum resulting from bombardment of ^{110}Pd with 45.5 MeV oxygen ions. The gamma decays of the two-phonon triplet states as well as the crossover transition from the second 2^+ state can be discerned in the spectrum. (From Robinson, R. L. et al., 1969, Nucl. Phys. **A124**, 553.)

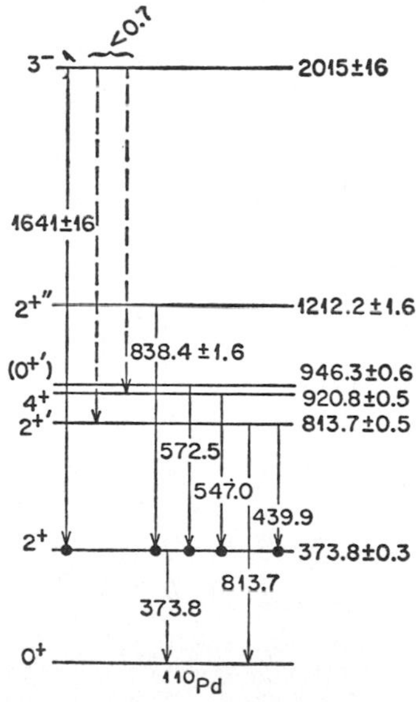

Fig. 1b. Coulomb excited levels in ^{110}Pd. (From Robinson, R. L. et al., 1969, Nucl. Phys. **A124**, 553.)

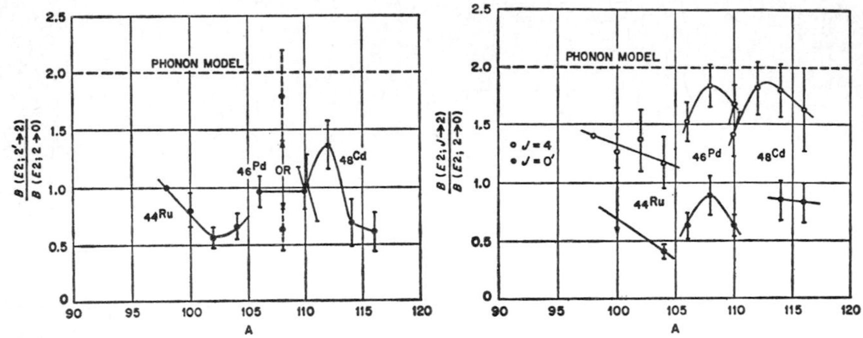

Fig. 1c. Results of the analysis of similar experiments with other vibrational nuclei. The ratio of B(E2) values are compared to the prediction of the pure phonon model. (From Robinson, R. L. et al., 1969, Nucl. Phys. **A124**, 553.)

dipole states (giant dipole resonance) as well as collective surface vibrations of quadrupole and higher multipole order. In the former case $\alpha_{\lambda\mu}$ describes the separation of the nuclear centres of mass and charge, while in the latter case $\alpha_{\lambda\mu}$ describes the shape of the nuclear surface, considering the equilibrium to be spherical. The description of vibrational states in deformed nuclei is somewhat more complicated and the excitation of such states will be discussed in §4.

It is well known that the main part of the dipole oscillator strength for nuclei lies at an energy of approximately 15–20 MeV. Since the minimum value of the parameter ξ for excitation of these states is of the order of 5, multiple excitation of these states is of no interest. In contrast, a large part of the quadrupole and octopole oscillator strength is found at low energies and a vibrational description may give a first approximation to many states in this energy region.

In the extreme vibrational model the eigenstates of the Hamiltonian (1) have the excitation energies

$$\Delta E_N = \hbar\omega_\lambda N = \hbar\sqrt{(C_\lambda/D_\lambda)}N, \qquad N = 0, 1, 2, \ldots \tag{2}$$

where N is the principal quantum number.

The eigenstates may be thought of as the ground state configuration plus a number of bosons, each carrying an angular momentum (λ, μ) corresponding to the multipole order. The bosons are conveniently described through the creation and annihilation operators a and $a^\dagger$, respectively, which fulfill the following commutation relations

$$\begin{aligned}[][a_{\lambda\mu}, a^\dagger_{\lambda'\mu'}] &= \delta_{\lambda\lambda'}\delta_{\mu\mu'} \\ [a_{\lambda\mu}, a_{\lambda'\mu'}] &= 0\end{aligned} \tag{3}$$

and which are defined by

$$\alpha_{\lambda\mu} = \sqrt{\frac{\hbar}{2\sqrt{C_\lambda D_\lambda}}}\,(a^\dagger_{\lambda\mu} + (-1)^\mu a_{\lambda-\mu}). \tag{4}$$

With the help of these operators the Hamiltonian (1) takes the simple form

$$H_0 = \sum_{\lambda\mu} \hbar\omega_\lambda [a^\dagger_{\lambda\mu} a_{\lambda\mu} + \tfrac{1}{2}]. \tag{5}$$

The first excited state is described by the state vector

$$|1, \lambda\mu\rangle = a^\dagger_{\lambda\mu}\,|0\rangle, \tag{6}$$

where the state $|0\rangle$ describes the ground state, which we consider to have angular momentum 0.

Due to the transformation properties of $\alpha_{\lambda\mu}$ under rotations it is seen that the state (6) has the angular momentum λ and the magnetic quantum number μ. The states containing more than one boson are strongly degenerate and may be decomposed into states of definite total angular momentum. For the two-quantum states

$$|2\rangle = a^\dagger_{\lambda\mu} a^\dagger_{\lambda\mu'}\,|0\rangle \tag{7}$$

the degeneracy $(\lambda + 1)(2\lambda + 1)$ corresponds to the total angular momenta, $0, 2, 4, \ldots, 2\lambda$.

The angular momentum eigenvector of the two-phonon states may be written

$$|2; IM\rangle = \frac{1}{\sqrt{2}} \sum_{\mu\mu'} \langle \lambda\mu\lambda\mu' | IM\rangle\, a^\dagger_{\lambda\mu} a^\dagger_{\lambda\mu'}\,|0\rangle, \tag{8}$$

where the factor $1/\sqrt{2}$ is due to the boson character of the phonons.

For higher excited states one may proceed in a similar manner. For an uncoupled state consisting of n_μ phonons with magnetic quantum number μ we shall use the notation

$$|n_{-\lambda}, \ldots, n_\mu, \ldots, n_\lambda\rangle = \frac{1}{\sqrt{n_{-\lambda}!\cdots n_\mu!\cdots n_\lambda!}} \times (a^\dagger_{\lambda-\lambda})^{n_{-\lambda}}\cdots(a^\dagger_{\lambda\mu})^{n_\mu}\cdots(a^\dagger_{\lambda\lambda})^{n_\lambda}\,|0\rangle. \tag{9}$$

From these states one may construct the states of total angular momenta, but in general the angular momentum quantum numbers are not sufficient to specify the state uniquely. The quantum numbers n_μ are connected with the principal quantum number introduced in (2) by

$$N = \sum_\mu n_\mu. \tag{10}$$

The transition probabilities between the vibrational states are determined by the matrix elements of the electric multipole moment of order λ. We shall assume that the multipole operator is linear in the amplitude $\alpha_{\lambda\mu}$, i.e. it can be written according to (4) as

$$\mathcal{M}(\mathrm{E}\lambda, \mu) = K_\lambda \sqrt{\frac{2\sqrt{C_\lambda D_\lambda}}{\hbar}}\, \alpha_{\lambda\mu}$$
$$= K_\lambda (a_{\lambda\mu}^\dagger + (-1)^\mu a_{\lambda-\mu}). \tag{11}$$

The coefficient K_λ may be expressed e.g. in terms of the reduced matrix element between the ground state and the first excited state. One finds

$$K_\lambda = \frac{(-1)^\lambda}{\sqrt{2\lambda+1}} \langle 0 \| \mathcal{M}(\mathrm{E}\lambda) \| \lambda \rangle. \tag{12}$$

A simple interpretation of the vibrational states can be made by assuming that the shape of the nuclear surface is given by

$$R(\theta, \phi) = R_0 \Big[1 + \sum_{\substack{\lambda\mu \\ \lambda > 1}} \alpha_{\lambda\mu} Y_{\lambda\mu}^*(\theta, \phi) \Big] \tag{13}$$

where R_0 is the equilibrium radius of the nucleus. If the charge density inside the nucleus is constant, the factor determining the multipole moment is given by

$$K_\lambda = \frac{3}{4\pi} Z_2 e R_0^\lambda \sqrt{\frac{\hbar}{2\sqrt{D_\lambda C_\lambda}}}, \tag{14}$$

i.e.

$$\mathcal{M}(\mathrm{E}\lambda, \mu) = \frac{3}{4\pi} Z_2 e R_0^\lambda \alpha_{\lambda\mu}. \tag{15}$$

In eqs. (14) and (15) $Z_2 e$ denotes the charge of the nucleus.

§ 2. *Evaluation of the excitation amplitude*

To compute the excitation amplitude we shall use the expression (II.3.30) for the *S*-matrix [ROB 61, 63]. Let us first consider the operator $\tilde{V}(t)$. Using eqs. (II.3.3) and (II.2.16) and inserting expressions (1.5) and (1.11) we obtain

$$\tilde{V}(t) = \exp\Big\{ \sum_{\lambda\mu} \mathrm{i}\omega_\lambda t a_{\lambda\mu}^\dagger a_{\lambda\mu} \Big\} \Big[\sum_{\lambda\mu} \frac{4\pi Z_1 e}{(2\lambda+1)} \bar{S}_{\mathrm{E}\lambda\mu}(t) K_\lambda [a_{\lambda\mu} + (-1)^\mu a_{\lambda-\mu}^\dagger] \Big]$$
$$\times \exp\Big\{ -\sum_{\lambda\mu} \mathrm{i}\omega_\lambda t a_{\lambda\mu}^\dagger a_{\lambda\mu} \Big\}. \tag{1}$$

This expression may be reduced considerably by utilizing the expression (VI.4.2) together with the commutation relations (1.3). One finds

$$\tilde{V}(t) = \sum_{\lambda\mu} \frac{4\pi Z_1 e}{2\lambda+1} \bar{S}_{\mathrm{E}\lambda\mu}(t) K_\lambda [a_{\lambda\mu} \exp(-\mathrm{i}\omega_\lambda t) + (-1)^\mu a^\dagger_{\lambda-\mu} \exp(\mathrm{i}\omega_\lambda t)]. \quad (2)$$

The operator R (see eq. (II.3.27)) is obtained by integrating eq. (2) over the time. Introducing the definitions (IV.1.4), (IV.2.2) and (IV.3.6) and utilizing eq. (IV.2.1) one obtains

$$R = -\sum_{\lambda\mu} (-1)^{\lambda+\mu} \chi^{(\lambda)}_{0\to\lambda} [R^*_{\lambda-\mu}(\vartheta, \xi) a_{\lambda\mu} + R_{\lambda\mu}(\vartheta, \xi) a^\dagger_{\lambda-\mu}], \quad (3)$$

where the parameters $\chi^{(\lambda)}$ and ξ refer to the transition from the ground state to the first excited state.

Next we consider the operator G, defined in eqs. (II.3.28) and (II.3.33). Using the expression (2), together with the commutation relation (1.3) one finds using the definitions (V.1.13–14)

$$\begin{aligned} G &= -\tfrac{1}{2} \sum_{\lambda\mu} |\chi^{(\lambda)}|^2 (-1)^\mu G^{\lambda\lambda}_{\mu-\mu}(\vartheta, \xi, -\xi) \\ &= -\tfrac{1}{2} \sum_{\lambda} |\chi^{(\lambda)}|^2 (-1)^\lambda \sqrt{2\lambda+1}\, G_{(\lambda\lambda)00}(\vartheta, \xi, -\xi). \end{aligned} \quad (4)$$

Since G is a constant c-number it contributes a constant, unobservable phase to all excitation amplitudes, and it can therefore be left out. From the fact that $[\tilde{V}(t), \tilde{V}(t')]$ is a c-number, it follows furthermore, according to eq. (II.3.34) and [BIA 69], that all higher terms in the series (II.3.30) containing multiple commutators vanish. We have thus obtained a closed expression for the Coulomb excitation S-matrix in the vibrational model, i.e.

$$\mathrm{e}^{2\mathrm{i}Q} = \exp\left\{-\mathrm{i} \sum_{\lambda\mu} (-1)^{\lambda+\mu} \chi^{(\lambda)} [R^*_{\lambda-\mu}(\vartheta, \xi) a_{\lambda\mu} + R_{\lambda-\mu}(\vartheta, \xi) a^\dagger_{\lambda\mu}]\right\}, \quad (5)$$

where we have neglected the phase G. The matrix elements of this operator between the ground state and the state (1.9) specified by the quantum numbers n_μ are most easily evaluated by using the formula

$$\exp(A+B) = \mathrm{e}^A \mathrm{e}^B \exp(-\tfrac{1}{2}[A, B]), \quad (6)$$

which holds for two operators, for which the commutator $[A, B]$ is a c-number. The excitation amplitude of the state (1.9) can thus be factorized according to

$$a_{n_{-\lambda}\cdots n_\mu \cdots n_\lambda} = a_{n_{-\lambda}} a_{n_\mu} a_{n_\lambda}, \quad (7)$$

where

$$a_{n_\mu} = \frac{1}{\sqrt{n_\mu!}} \langle 0| (a_{\lambda\mu})^{n_\mu} \exp\{-\mathrm{i}\chi^{(\lambda)}(-1)^{\lambda+\mu}[R^*_{\lambda-\mu}a_{\lambda\mu} + R_{\lambda-\mu}a^\dagger_{\lambda\mu}]\} |0\rangle$$

$$= \frac{1}{\sqrt{n_\mu!}} \langle 0| (a_{\lambda\mu})^{n_\mu} \exp\{-\mathrm{i}\chi^{(\lambda)}(-1)^{\lambda+\mu}R_{\lambda-\mu}a^\dagger_{\lambda\mu}\}$$

$$\times \exp\{\mathrm{i}\chi^{(\lambda)}(-1)^{\lambda+\mu}R^*_{\lambda-\mu}a_{\lambda\mu}\} \exp\{-\tfrac{1}{2}|\chi^{(\lambda)}|^2|R_{\lambda-\mu}|^2\} |0\rangle. \tag{8}$$

Expanding the exponential functions in eq. (8) in power series, only the first term of the second exponential contributes. In the first exponential only the term with n_μ creation operators contributes and one obtains the result

$$a_{n_\mu} = \frac{1}{\sqrt{n_\mu!}} [-\mathrm{i}\chi^{(\lambda)}(-1)^{\lambda+\mu}R_{\lambda-\mu}(\vartheta, \xi)]^{n_\mu} \exp\{-\tfrac{1}{2}|\chi^{(\lambda)}R_{\lambda-\mu}(\vartheta, \xi)|^2\}. \tag{9}$$

Inserting (9) into (7) we have obtained a relatively simple expression for the excitation amplitude of any state in the vibrational model. With this result one may compute all observable quantities in the Coulomb excitation process. Thus for the total excitation probability of the states with given principal quantum number N one finds

$$P_N = \sum_{n_\mu} \prod_{\mu=-\lambda}^{\lambda} \frac{1}{n_\mu!} [\chi^{(\lambda)}R_{\lambda-\mu}(\vartheta, \xi)]^{2n_\mu} \exp\{-|\chi^{(\lambda)}R_{\lambda-\mu}(\vartheta, \xi)|^2\}, \tag{10}$$

where the summation should be performed over all combinations n_μ with a fixed sum N according to eq. (1.10). Introducing the notation (IV.3.9) and performing the summation over n_μ according to the polynomial theorem, one obtains the following simple result

$$P_N = \frac{1}{N!} |\chi(\vartheta, \xi)|^{2N} \exp\{-|\chi(\vartheta, \xi)|^2\}. \tag{11}$$

It is noted that the excitation probability is given by a Poisson distribution (see fig. 2).

The average value of the principal quantum number after the collision is given by

$$\langle N\rangle = |\chi(\vartheta, \xi)|^2. \tag{12}$$

This number also describes the average number of elementary steps in the multiple excitation process. Comparing this to the estimate in § VI.1 it should be kept in mind that, within a vibrational band, the parameters χ connecting two states increase proportional to $\sqrt{N}$.

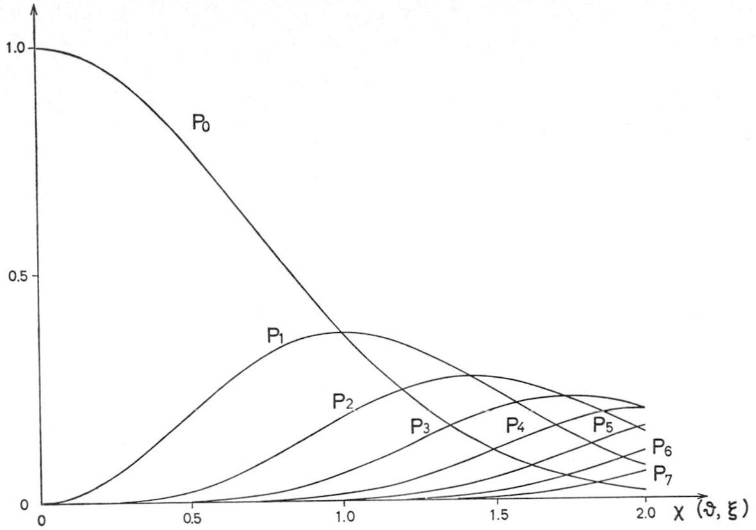

Fig. 2. Multiple Coulomb excitation of pure vibrational states. The excitation probability P_N of the state with principal quantum number N is given as a function of the parameter $\chi(\vartheta, \xi)$.

The result (11) can be interpreted in a very simple fashion by the following argument [ALD 60]. The excitation of a harmonic oscillator can be thought of as a collective excitation of a large number P of mutually uncoupled identical harmonic oscillators which are, each of them, only weakly excited. In this way we distribute the oscillator strength on a large number of degrees of freedom. The weak excitation of these oscillators can be treated by a perturbation calculation where the excitation probability is the same for all of them, viz. χ^2/P. Since the oscillators are mutually uncoupled, the probability that N oscillators are excited is given by a Poisson distribution where the average number N is P times the probability that each one is excited, i.e. χ^2. An alternative derivation of the results of the present section is given in § 5 below.

§ 3. *Vibrational excitation of spherical nuclei*

As an illustration of the results obtained in the previous section we consider the Coulomb excitation of pure vibrational states in spherical nuclei where one is interested in the cross section for the excitation of vibrational states of definite angular momentum. In even-even nuclei the excitation amplitude of the state $|N, IM\rangle$ with N phonons of angular momenta λ and total spin I, M is given by

$$a_{N,IM} = \langle N, IM|\, e^{2iQ}\, |0\rangle . \tag{1}$$

Expanding $|N, IM\rangle$ in terms of the states $|n_\mu\rangle$ according to eq. (1.8), the amplitude (1) can be expressed in terms of the elementary excitation amplitudes (2.9).

For the quadrupole vibrational states the results for the three lowest principal quantum numbers $N = 0$, 1 and 2 are given in table 1 in the coordinate system A. The total population of the states with $N = 0$, 1 and 2 is given by the Poisson distribution (2.11). For $N = 0$ and $N = 1$ only one spin state is present, while for $N = 2$ the probability is distributed over a triplet with $I = 0$, 2 and 4. The relative population of these three states can be found from table 1 to be

$$\frac{P_{N=2,I=0}}{P_{N=2}} = \tfrac{1}{5}\left|1 - \frac{|R^{\mathrm{A}}_{22} - R^{\mathrm{A}}_{2-2}|^2}{R^2_2}\right|^2, \tag{2}$$

$$\frac{P_{N=2,I=2}}{P_{N=2}} = \tfrac{2}{7}\left(1 - \frac{(R^{\mathrm{A}}_{22} - R^{\mathrm{A}}_{2-2})^2[(R^{\mathrm{A}}_{22} + R^{\mathrm{A}}_{2-2})^2 - 2R^{\mathrm{A}2}_{20}]}{R^4_2}\right), \tag{3}$$

$$\frac{P_{N=2,I=4}}{P_{N=2}} = \tfrac{18}{35}\left(1 + \frac{(R^{\mathrm{A}}_{22} - R^{\mathrm{A}}_{2-2})^2[17(R^{\mathrm{A}}_{22} + R^{\mathrm{A}}_{2-2})^2 - 6R^{\mathrm{A}2}_{20}]}{18R^4_2}\right). \tag{4}$$

TABLE 1

Excitation amplitudes of quadrupole vibrational states for a spherical even-even nucleus.

$N = 0$	$a_{0,00} = \exp(-\frac{1}{2}\chi(\vartheta, \xi)^2)$
$N = 1$	$a_{1,22} = -\mathrm{i}\chi R^{\mathrm{A}}_{2-2}\exp(-\frac{1}{2}\chi(\vartheta, \xi)^2)$
	$a_{1,20} = -\mathrm{i}\chi R^{\mathrm{A}}_{20}\exp(-\frac{1}{2}\chi(\vartheta, \xi)^2)$
	$a_{1,2-2} = -\mathrm{i}\chi R^{\mathrm{A}}_{22}\exp(-\frac{1}{2}\chi(\vartheta, \xi)^2)$
$N = 2$	$a_{2,00} = -\chi^2\sqrt{\frac{1}{10}}\exp(-\frac{1}{2}\chi(\vartheta, \xi)^2)\{2R^{\mathrm{A}}_{22}R^{\mathrm{A}}_{2-2} + \lvert R^{\mathrm{A}}_{20}\rvert^2\}$
	$a_{2,22} = -\chi^2\sqrt{\frac{1}{7}}\exp(-\frac{1}{2}\chi(\vartheta, \xi)^2)2R^{\mathrm{A}}_{2-2}R^{\mathrm{A}}_{20}$
	$a_{2,20} = -\chi^2\sqrt{\frac{1}{7}}\exp(-\frac{1}{2}\chi(\vartheta, \xi)^2)\{2R^{\mathrm{A}}_{22}R^{\mathrm{A}}_{2-2} - \lvert R^{\mathrm{A}}_{20}\rvert^2\}$
	$a_{2,2-2} = -\chi^2\sqrt{\frac{1}{7}}\exp(-\frac{1}{2}\chi(\vartheta, \xi)^2)2R^{\mathrm{A}}_{22}R^{\mathrm{A}}_{20}$
	$a_{2,44} = -\chi^2\sqrt{\frac{1}{2}}\exp(-\frac{1}{2}\chi(\vartheta, \xi)^2)\lvert R^{\mathrm{A}}_{2-2}\rvert^2$
	$a_{2,42} = -\chi^2\sqrt{\frac{3}{7}}\exp(-\frac{1}{2}\chi(\vartheta, \xi)^2)R^{\mathrm{A}}_{2-2}R^{\mathrm{A}}_{20}$
	$a_{2,40} = -\chi^2\sqrt{\frac{1}{35}}\exp(-\frac{1}{2}\chi(\vartheta, \xi)^2)\{R^{\mathrm{A}}_{22}R^{\mathrm{A}}_{2-2} + 3\lvert R^{\mathrm{A}}_{20}\rvert^2\}$
	$a_{2,4-2} = -\chi^2\sqrt{\frac{3}{7}}\exp(-\frac{1}{2}\chi(\vartheta, \xi)^2)R^{\mathrm{A}}_{22}R^{\mathrm{A}}_{20}$
	$a_{2,4-4} = -\chi^2\sqrt{\frac{1}{2}}\exp(-\frac{1}{2}\chi(\vartheta, \xi)^2)\lvert R^{\mathrm{A}}_{22}\rvert^2$

The amplitudes a are given for principal quantum numbers $N = 0$, 1, 2 and for a given spin and magnetic quantum number in coordinate system A.

It is noted that for $\xi = 0$ or for $\vartheta = \pi$ the three states are populated in the ratio 7:10:18. Since for forward scattering angles and finite values of ξ, R^{A}_{2-2} dominates over the other orbital integrals, the total excitation probability is, in this case, concentrated on the state with spin 4.

§ 4. *Vibrational excitation of deformed nuclei*

In deformed nuclei the expectation values of the parameters $\alpha_{\lambda\mu}$ in (1.13) are non-vanishing. If we assume that the shape of a nucleus can be described by quadrupole deformation, it is convenient to perform a coordinate transformation of the five parameters $\alpha_{2\mu}$ into the system of principal axes. The shape is then defined by two parameters β and γ while the orientation of the principal axes with respect to the fixed system is given by the three Eulerian angles. These five new parameters behave rather differently as compared to the original $\alpha_{2\mu}$. The three degrees of freedom of the Eulerian angles give rise to rotational spectra with fixed intrinsic structure and deformation. The degrees of freedom associated with β and γ give rise to vibrational spectra corresponding to oscillations of the intrinsic shape around equilibrium [BOH 74].

Since the deformation parameters $\alpha_{2\mu}$ transform as spherical tensors under rotation of the coordinate system we define the rotation to the intrinsic system by the condition that the intrinsic deformation parameters $\alpha'_{2\pm1}$ vanish and that $\alpha'_{22} = \alpha'_{2-2}$. The three Eulerian angles (ϕ, θ, ψ) together with the parameters β and γ defined by

$$\alpha'_{20} = \beta\cos\gamma, \qquad \alpha'_{22} = \sqrt{\tfrac{1}{2}}\,\beta\sin\gamma, \tag{1}$$

give an equivalent description of the five degrees of freedom associated with quadrupole deformations. We note that the quadrupole operator $\mathscr{M}(\mathrm{E}2, \mu)$ according to eq. (1.15) can be written in the new variables as

$$\mathscr{M}(\mathrm{E}2, \mu) = \sum_{\nu} \mathscr{M}_{\mathrm{int}}(\mathrm{E}2, \nu) D^2_{\mu\nu}(\phi, \theta, \psi), \tag{2}$$

where $\mathscr{M}_{\mathrm{int}}(\mathrm{E}2, \nu)$ has only two components proportional to the deformation parameters (1).

For a nucleus with an equilibrium deformation given by the values β_0 and γ_0 of the parameters β and γ we shall assume the Hamiltonian to be of the simple form

$$H = H_\beta + H_\gamma + H_\mathrm{r} + V_{\mathrm{E}2}(t), \tag{3}$$

where

$$H_\beta = \tfrac{1}{2}D_\beta\dot{\beta}^2 + \tfrac{1}{2}C_\beta(\beta - \beta_o)^2,$$
$$H_\gamma = \tfrac{1}{2}D_\gamma\dot{\gamma}^2 + \tfrac{1}{2}C_\gamma(\gamma - \gamma_o)^2, \tag{4}$$

and

$$V_{E2}(t) = \tfrac{4}{5}\pi\, Z_1 e \sum_\mu \mathscr{M}(E2, \mu)\bar{S}^*_{E2\mu}(t).$$

The Hamiltonian H_r is the kinetic energy of the rotational motion.

In the next chapter we shall consider the excitation of the rotational degrees of freedom especially in nuclei with axially symmetric equilibrium shape. Since the rotational mode of deformed nuclei is experimentally even more collective than the vibrational mode this degree of freedom is very strongly excited. In the limit where the rotational frequency is much smaller than the collision frequency one may in first order neglect the rotation of the nuclear axis during the collision and evaluate the excitation of the vibrational degrees of freedom β and γ as if the nuclear axis were fixed in space. We shall in the following furthermore assume an axial symmetric equilibrium deformation, i.e. $\gamma_o = 0$. The evaluation of the multiple excitation of the vibrational states then proceeds analogously to the treatment in the preceding section. We thus write the intrinsic quadrupole operator in the form

$$\mathscr{M}_{\text{int}}(E2, 0) = \frac{3}{4\pi} Z_2 e R_0^2 \left[(a_\beta^\dagger + a_\beta)\sqrt{\frac{\hbar}{2\sqrt{C_\beta D_\beta}}} + \beta_o\right],$$
$$\mathscr{M}_{\text{int}}(E2, \pm 2) = \frac{3}{4\pi} Z_2 e R_0^2 \frac{1}{\sqrt{2}} \beta_o (a_\gamma^\dagger + a_\gamma)\sqrt{\frac{\hbar}{2\sqrt{C_\gamma D_\gamma}}}, \tag{5}$$

where we have neglected second-order terms in $(\beta - \beta_o)$ and $(\gamma - \gamma_o)$.

The problem is now reduced to the excitation of two independent linear harmonic oscillators. The excitation probability of a state with oscillator quantum number N can therefore be written in the form

$$P_N = \frac{1}{N!}\,|\chi(\vartheta, \xi)|^{2N} \exp(-|\chi(\vartheta, \xi)|^2), \tag{6}$$

where $\chi(\vartheta, \xi)$ depends not only on ϑ and ξ but also on the Eulerian angles ϕ, θ, ψ through the relation

$$\chi(\vartheta, \xi) = \chi^\beta_{\text{int}} \sum_\mu (-1)^\mu D^2_{\mu 0}(\phi, \theta, \psi) R_{2-\mu}(\vartheta, \xi) \tag{7}$$

for β-vibrations, and

$$\chi(\vartheta, \xi) = \chi^{\gamma}_{\text{int}} \sum_{\mu} (-1)^{\mu} D^{2}_{\mu 2}(\phi, \theta, \psi) R_{2-\mu}(\vartheta, \xi) \tag{8}$$

for γ-vibrations. The intrinsic χ-parameters for the two types of vibration are conveniently defined by

$$\chi^{\beta}_{\text{int}} = \frac{\sqrt{16\pi}}{15} \frac{Z_1 e}{\hbar v} \frac{1}{a^2} \sqrt{5}\, \langle \phi^{\beta}_{0} | \mathcal{M}_{\text{int}}(\text{E2}, 0) | \phi_0 \rangle, \tag{9}$$

$$\chi^{\gamma}_{\text{int}} = \frac{\sqrt{16\pi}}{15} \frac{Z_1 e}{\hbar v} \frac{1}{a^2} \sqrt{5}\, \langle \phi^{\gamma}_{2} | \mathcal{M}_{\text{int}}(\text{E2}, 2) | \phi_0 \rangle, \tag{10}$$

where ϕ_0, ϕ^{β}_0 and ϕ^{γ}_2 are the intrinsic wave functions of the deformed nucleus in the ground state, in the first excited state of a β-vibration and of a γ-vibration, respectively. For small values of the intrinsic χ parameter we get the following expression of the excitation probability of the first excited state of the β- and γ-vibrations

$$P^{\beta} = |\chi^{\beta}_{\text{int}}|^2 |\sum_{\mu} D^{2*}_{\mu 0}(\phi, \theta, \psi)\, R_{2\mu}(\vartheta, \xi)|^2 \tag{11}$$

or

$$P^{\gamma} = |\chi^{\gamma}_{\text{int}}|^2 |\sum_{\mu} D^{2*}_{\mu 2}(\phi, \theta, \psi)\, R_{2\mu}(\vartheta, \xi)|^2 . \tag{12}$$

This result has to be averaged over all possible directions of the deformed nucleus. Assuming a random orientation (cf. § VIII. 3), we get then in both cases

$$P = |\chi_{\text{int}}|^2 \cdot \tfrac{1}{5} \sum_{\mu} |R_{2\mu}(\vartheta, \xi)|^2 = \tfrac{1}{5} |\chi_{\text{int}}|^2\, R^2_2(\vartheta, \xi). \tag{13}$$

From expression (13) it is seen that the χ-value computed from the known transition probabilities from the ground state to the 2^+ excited state of a β or γ vibrational state differs from χ_{int} by a factor of $\sqrt{5}$, i.e.

$$\chi_{\text{int}} = \sqrt{5} \chi^{(2)}_{0\to 2} . \tag{14}$$

This result can also easily be deduced from eqs. (9) and (10).

As a possible application of eq. (14) we mention the Coulomb fission process, where the deformed nucleus is excited by multiple excitation to the fission barrier. The fact that the effective χ parameter for the vibrational states is increased by a factor $\sqrt{5}$ is important in this connection. For a detailed discussion and for estimates of the cross sections see [BEY 69].

§ 5. *Classical treatment of vibrational excitations*

In this section we shall discuss the completely classical treatment of the Coulomb excitation of a system which can perform only pure harmonic vibrations. It is well known that this problem of forced harmonic motion can be solved explicitly and we shall compare the results of the classical calculation with the results obtained in § 2. It is instructive to see how the quantum mechanical results become indentical to the classical results for large values of χ.

The total classical Hamiltonian is given by

$$H(t) = \tfrac{1}{2} \sum_{\lambda\mu} [D_\lambda |\dot{\alpha}_{\lambda\mu}|^2 + C_\lambda |\alpha_{\lambda\mu}|^2] + 3Z_1Z_2e^2 \sum_{\lambda\mu} R_0^\lambda \frac{1}{2\lambda+1} \alpha^*_{\lambda\mu} \bar{S}_{\mathrm{E}\lambda\mu}(t), \tag{1}$$

where we inserted the multipole moment according to eq. (1.15). From this Hamiltonian one derives the following equation of motion

$$\ddot{\alpha}_{\lambda\mu} + \omega_\lambda^2 \alpha_{\lambda\mu} = -\frac{3Z_1Z_2e^2R_0^\lambda}{(2\lambda+1)D_\lambda} \bar{S}_{\mathrm{E}\lambda\mu}(t), \tag{2}$$

where we have introduced the frequencies (1.2) of the oscillators.

These equations can be solved by introducing the Green function corresponding to a given initial condition at time $t = -\infty$. One finds

$$\alpha_{\lambda\mu}(t) = \frac{3Z_1Z_2e^2R_0^\lambda}{2\pi(2\lambda+1)D_\lambda} \int_{-\infty}^{+\infty} \frac{\mathrm{d}\omega\, S_{\mathrm{E}\lambda\mu}(\omega)}{\omega^2 - \omega_\lambda^2 + \mathrm{i}\varepsilon\omega} \exp(-\mathrm{i}\omega t) + \alpha_{\lambda\mu}(\text{free}). \tag{3}$$

We have here introduced the orbital integrals $S_{\mathrm{E}\lambda\mu}(\omega)$ according to (IV.1.4). The term $\mathrm{i}\varepsilon\omega$ where ε is a small positive number ensures that for large negative times the amplitudes $\alpha_{\lambda\mu}(t)$ become equal to the free amplitude $\alpha_{\lambda\mu}(\text{free})$ before the collision. We are especially interested in the amplitude $\alpha_{\lambda\mu}(t)$ after the collision. This amplitude can be found from eq. (3) by closing the integration path in the lower half of the complex plane and by evaluating the contribution from the poles

$$\omega \approx \pm\, \omega_\lambda - \tfrac{1}{2}\mathrm{i}\varepsilon. \tag{4}$$

One obtains

$$\alpha_{\lambda\mu}(t) = \frac{-3\mathrm{i}Z_1Z_2e^2R_0^\lambda}{(2\lambda+1)D_\lambda 2\omega_\lambda}$$
$$\times \{S_{\mathrm{E}\lambda\mu}(\omega_\lambda)\exp(-\mathrm{i}\omega_\lambda t) - S_{\mathrm{E}\lambda\mu}(-\omega_\lambda)\exp(\mathrm{i}\omega_\lambda t)\} + \alpha_{\lambda\mu}(\text{free}). \tag{5}$$

In the simplest case which corresponds to the situation discussed in § 2, the amplitude $\alpha_{\lambda\mu}(\text{free})$ vanishes because the oscillator is at rest before the collision. In this case it is easy to evaluate the energy of the oscillator after the collision. One finds directly by inserting eq. (5) into (1)

$$\Delta E = \tfrac{9}{2} Z_1^2 Z_2^2 e^4 \sum_{\lambda\mu} \frac{R_0^{2\lambda}}{(2\lambda+1)^2 D_\lambda} |S_{\mathrm{E}\lambda\mu}(\omega_\lambda)|^2. \tag{6}$$

In order to compare this energy with the result of the quantum mechanical treatment we introduce the parameters $\chi^{(\lambda)}$ and the orbital integrals $R_\lambda(\vartheta, \xi)$ according to eqs. (IV.3.6), (1.12) and (1.14). For a definite multipole order λ we then obtain

$$\Delta E = |\chi^{(\lambda)}|^2 R_\lambda^2(\vartheta, \xi) \hbar\omega_\lambda, \tag{7}$$

where $R_\lambda^2(\vartheta, \xi)$ are the relative excitation probabilities defined by eq. (IV.3.3).

Comparing this result with the quantum mechanical result (2.11) for the excitation probability, we see that the quantum mechanical average excitation energy

$$\begin{aligned}\langle \Delta E \rangle &= \sum_N P_N \hbar\omega_\lambda N = \sum_N \exp\{-|\chi(\vartheta, \xi)|^2\} \frac{|\chi(\vartheta, \xi)|^{2N}}{N!} N\hbar\omega_\lambda \\ &= |\chi^{(\lambda)}(\vartheta, \xi)|^2 \hbar\omega_\lambda \end{aligned} \tag{8}$$

is equal to the classical result (7). It should be noted that it is also equal to the expectation value of the excitation energy in the perturbation theory.

The quantum mechanical excitation thus corresponds to a probability distribution around the classically well defined excitation energy (7). For large values of χ the Poisson distribution will approach a Gaussian distribution with a width $\chi(\vartheta, \xi)$ and the relative width will diminish as $\chi(\vartheta, \xi)^{-1}$ thus approaching the classical result. As an illustration we have plotted in fig. 3 the quantum mechanical energy distribution for $\chi(\vartheta, \xi) = 2$.

In § 2 the quantum mechanical problem of the Coulomb excitation of vibrational states was solved by computing the S-matrix elements. Actually, one may solve also the problem in a more elegant way which is completely analogous to the solution of the classical problem. This method is based on the Heisenberg picture, where the nuclear state is described by a time-independent state vector $|0\rangle$ and where the amplitudes $\alpha_{\lambda\mu}(t)$ are time dependent operators satisfying the equation of motion (2). Since the right-hand side is a c-number, the solution of eq. (2) is again given by eq. (3) where $\alpha_{\lambda\mu}(\text{free})$ is the operator corresponding to free harmonic vibrations, while the first term in eq. (3) is a c-number. In the Heisenberg picture the Hamiltonian after the collision is different from the Hamiltonian before the collision. The amplitudes of the state $|0\rangle$ on the eigenstates of the new Hamiltonian are the

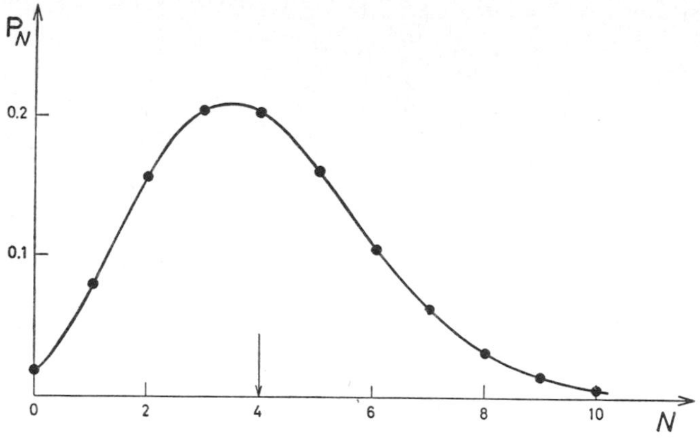

Fig. 3. The excitation probabilities for vibrational states with principal quantum numbers N for $\chi(\vartheta, \xi) = 2$. The excitation probabilities indicated by dots follow a Poisson distribution with average $N = 4$.

excitation amplitudes. To determine these amplitudes we note that the Hamiltonian after the collision can be written

$$H_f = \sum_{\lambda\mu} \hbar\omega_\lambda[(a^\dagger_{\lambda\mu} - \beta_{\lambda\mu})(a_{\lambda\mu} - \beta^*_{\lambda\mu}) + \tfrac{1}{2}], \tag{9}$$

where the c-number $\beta_{\lambda\mu}$ can be found by comparing eq. (1.4) with eq. (5) to be

$$\begin{aligned}\beta_{\lambda\mu} &= -\frac{3iZ_1Z_2e^2R_0^\lambda}{(2\lambda+1)}\sqrt{\frac{1}{2\hbar\sqrt{D_\lambda C_\lambda}}}\, S_{\lambda\mu}(-\omega_\lambda)\\ &= -i(-1)^{\lambda+\mu}\chi^{(\lambda)}R^*_{\lambda-\mu}(\vartheta, \xi).\end{aligned} \tag{10}$$

The eigenstates $|\tilde{n}\rangle$ of H_f for a definite degree of freedom λ, μ are given by

$$|\tilde{n}\rangle = \frac{1}{\sqrt{n!}}(a^\dagger_{\lambda\mu} - \beta_{\lambda\mu})^n\, |\tilde{0}\rangle, \tag{11}$$

where $|\tilde{0}\rangle$ is the ground state of (9). The excitation amplitudes a_n can thus be evaluated as

$$\begin{aligned}a_n &= \langle\tilde{n}|0\rangle = \langle\tilde{0}|\,\frac{1}{\sqrt{n!}}(a_{\lambda\mu} - \beta^*_{\lambda\mu})^n\,|0\rangle\\ &= \langle\tilde{0}|0\rangle\,\frac{1}{\sqrt{n!}}(-\beta^*_{\lambda\mu})^n.\end{aligned} \tag{12}$$

The scalar product $\langle\tilde{0}|0\rangle$ can be found from the condition that

$$\sum_n |a_n|^2 = 1. \tag{13}$$

One thus obtains up to an n-independent phase

$$a_n = \frac{1}{\sqrt{n!}}[-\mathrm{i}\chi^{(\lambda)}(-1)^{\lambda+\mu}R_{\lambda-\mu}(\vartheta, \xi)]^n \exp\{-\tfrac{1}{2}|\chi^{(\lambda)}|^2|R_{\lambda-\mu}(\vartheta, \xi)|^2\} \tag{14}$$

in accordance with eq. (2.9).

CHAPTER VIII

Multiple Excitation of Deformed Nuclei

In contrast to the excitation of vibrational states of spherical nuclei, the multiple excitation of rotational states in axially symmetric deformed nuclei cannot be expressed in a closed form. It is, however, possible to give rather simple expressions for the excitation amplitudes in the sudden approximation. These results are of considerable interest since the pure rotational model gives a reasonably good description of the rotational spectra of deformed nuclei and since the energies of these states are so low that the sudden approximation gives a reasonable first approximation. The deviations for the finite value of ξ can be included in an approximate way as was discussed in chapter VI.

It is also of considerable interest to study transitions from the ground state rotational band to rotational bands of different intrinsic structure. Such excitation of coupled rotational bands offer an instructive example of the methods which apply when the nuclear states can be decomposed into groups of strongly coupled states.

§ 1. *The rotational model*

In this chapter we study the excitation of nuclei with a non-spherical equilibrium shape. For such nuclei the Hamiltonian H_o is conveniently written in the form [BOH 75]

$$H_o = H_i + H_r + H_c. \tag{1}$$

The first term H_i is the intrinsic Hamiltonian that describes the motion of the nucleons in the intrinsic coordinate system defined by the equilibrium shape of the nucleus. The intrinsic motion includes vibrations of the nucleus around the equilibrium shape. The second term H_r represents the energy associated with the rotations of the intrinsic system and is an operator that only depends on the Eulerian angles which specify the orientation of the intrinsic coordinate system. The last term H_c represents the coupling between the intrinsic motion and the rotation.

We shall especially study nuclei with an axially symmetric shape, where the Hamiltonian H_r may be written in the form

$$H_r = \frac{\hbar^2}{2\mathscr{I}} \boldsymbol{R}^2, \tag{2}$$

where $\boldsymbol{R}^2$ denotes the square of the rotational angular momentum and $\mathscr{I}$ is the moment of inertia. The operator $\boldsymbol{R}^2$ can be expressed through the polar angles (θ, ϕ) of the symmetry axis in the following way

$$\boldsymbol{R}^2 = -\left\{\frac{\partial^2}{\partial\theta^2} + \cot\theta \frac{\partial}{\partial\theta} + \frac{1}{\sin^2\theta}\frac{\partial^2}{\partial\phi^2}\right\}. \tag{3}$$

In the pure rotational model one neglects the coupling term H_c and the eigenstates of (1) may then be written as a product of eigenstates of H_i and of H_r. The eigenstates of H_i can be classified according to the quantum number K, which denotes the component of the intrinsic angular momentum along the symmetry axis. The product wave function is thus given by

$$|IKM\rangle\,|NK\rangle = \sqrt{\frac{2I+1}{4\pi}}\, D^I_{MK}(\phi, \theta, 0)\,|NK\rangle, \tag{4}$$

where I and M are the total angular momentum and magnetic quantum numbers, respectively. The wave function (4) describes the rotation of the intrinsic state $|NK\rangle$ which is a function of the intrinsic coordinates only and which is specified by quantum numbers N and K.

The D-functions depending on the orientation of the intrinsic system are defined in appendix D which also contains explicit expressions for the lowest values of I. The eigenvalues of $H_i + H_r$, corresponding to the wave function (4) are given by

$$E_{I,N,K} = \frac{\hbar^2}{2\mathscr{I}} I(I+1) + E_{N,K}, \tag{5}$$

where $I = K, K+1\cdots(K > 0)$ and where $E_{N,K}$ is the eigenvalue of the intrinsic Hamiltonian. We include the possibility that the moment of inertia depends on the intrinsic configuration. The product wave function (4) would lead to a degeneracy of states with quantum numbers K and $-K$. It is, however, a consequence of the invariance principles that the true eigenstates $|IKMN\rangle$ are given by the symmetrized expressions

$$|IKMN\rangle = \begin{cases} |I0M\rangle\,|N0\rangle, & \text{for } K = 0 \\ \sqrt{\tfrac{1}{2}}\,\{|IKM\rangle\,|NK\rangle + (-1)^{I+K}|I-KM\rangle\,|N-K\rangle\}, & \\ & \text{for } K > 0. \end{cases} \quad (6)$$

The parity of the states is determined by the parity of the intrinsic wave function. For $K = 0$ only states with even I or only states with odd I occur.

It is well known from nuclear spectroscopy that even in strongly deformed nuclei there are deviations from the pure rotational model. These deviations can be ascribed to the presence of the coupling term H_c. While the explicit form of the Hamiltonian H_c depends on the details of the nuclear model one may argue that the most important part of H_c can be obtained from a series expansion in powers of the rotational angular momentum. Utilizing the invariance and symmetry principles appropriate for an axially symmetric nuclear shape, one obtains to second order

$$H_c = h_1 I_- + h_{-1} I_+ + h_2 I_-^2 + h_{-2} I_+^2 + h_0 \boldsymbol{I}^2. \quad (7)$$

In eq. (7) h_i denotes an operator in the intrinsic degrees of freedom which changes the K quantum number by i units. The operator $\boldsymbol{I}$ denotes the total angular momentum while I_+ and I_- are defined as the spherical components of this quantity in the intrinsic coordinate system, i.e.

$$I_\pm = (I_1 \pm \mathrm{i} I_2).$$

The effect of the interaction (7) may be taken into account by a perturbation treatment. For the perturbed energy states one finds to lowest order

$$E_{I,N,0} = \frac{\hbar^2}{2\mathscr{I}} I(I+1) + B(I(I+1))^2 + E_{N,0},$$

$$E_{I,N,1/2} = \frac{\hbar^2}{2\mathscr{I}} I(I+1) + a(-1)^{(I+1/2)}(I+\tfrac{1}{2}) + E_{N,1/2}, \quad (8)$$

$$E_{I,N,1} = \frac{\hbar^2}{2\mathscr{I}} I(I+1) + b(-1)^I(I-1)I + E_{N,1},$$

where the coefficients B, a and b can be expressed in terms of matrix elements of the operators h_i. For $K = \frac{1}{2}$ the so-called decoupling term proportional to a arises from the matrix elements of $h_{\pm 1}$ between the two terms in the symmetrized wave function.

The perturbed eigenstates $|IKMN\rangle_m$ can be written in the form

$$|IKMN\rangle_m = e^{iS}\,|IKMN\rangle, \tag{9}$$

where the Hermitian operator S causes mixing between the pure bands (6). The intrinsic matrix element of S may be expressed by the relations

$$\begin{aligned} \langle N' \pm (K+1)|\, S\, |N \pm K\rangle &= \pm i\varepsilon_1 I_{\mp}, \\ \langle N' \pm (K+2)|\, S\, |N \pm K\rangle &= i\varepsilon_2 I_{\mp}^2, \\ \langle N' \pm K|\, S\, |N \pm K\rangle &= i\varepsilon_0 \boldsymbol{I}^2, \end{aligned} \tag{10}$$

where the mixing coefficients ε_i are the matrix elements of h_i divided by the energy difference between the intrinsic states specified by N' and N.

The electric properties of the rotational states are given by the matrix elements of the electric multipole moment $\mathscr{M}(\mathrm{E}\lambda, \mu)$. It is convenient to express the multipole operator by the corresponding operator $\mathscr{M}_{\mathrm{int}}(\mathrm{E}\lambda, \mu)$ in the intrinsic system. We find

$$\mathscr{M}(\mathrm{E}\lambda, \mu) = \sum_{\nu} D^{\lambda}_{\mu\nu}(\phi, \theta, 0)\, \mathscr{M}_{\mathrm{int}}(\mathrm{E}\lambda, \nu), \tag{11}$$

where $\mathscr{M}_{\mathrm{int}}(\mathrm{E}\lambda, \nu)$ depends on the intrinsic coordinates only. The expectation values of $\mathscr{M}_{\mathrm{int}}(\mathrm{E}\lambda, 0)$ define the intrinsic multipole moments, while the non-diagonal matrix elements describe the transition between different rotational bands. For the case of $\lambda = 2$ one usually expresses the expectation value of $\mathscr{M}_{\mathrm{int}}(\mathrm{E}2, 0)$ by the intrinsic quadrupole moment Q_0 which is defined by

$$\langle N, \pm K|\, \mathscr{M}_{\mathrm{int}}(\mathrm{E}2, 0)\, |N, \pm K\rangle = \sqrt{\frac{5}{16\pi}}\, eQ_0. \tag{12}$$

The quadrupole moment Q_0 is characteristic of the band in question and independent of the sign of K.

For a pure rotational band one may express the intrinsic expectation value of $\mathscr{M}(\mathrm{E}2, \mu)$ in terms of Q_0 by inserting eq. (12) into eq. (11). One finds

$$\langle NK|\, \mathscr{M}(\mathrm{E}2, \mu)\, |NK\rangle = \tfrac{1}{2} Q_0 Y_{2\mu}(\theta, \phi). \tag{13}$$

The electric quadrupole matrix elements between the states in a rotational band are easily evaluated from this expression. For the reduced matrix elements one finds

$$\langle I_m \|\, \mathscr{M}(\mathrm{E}2)\, \| I_n\rangle = \sqrt{\frac{5}{16\pi}}\, (-1)^{I_m - K} \sqrt{(2I_m+1)(2I_n+1)} \begin{pmatrix} I_m & 2 & I_n \\ -K & 0 & K \end{pmatrix} eQ_0. \tag{14}$$

If deviations from the rotational model are taken into account by the perturbation treatment mentioned above, the electric multipole matrix elements take the form (see eq. (9))

$$ {}_m\langle IMKN|\,\mathcal{M}(\mathrm{E}\lambda,\mu)\,|I'M'K'N'\rangle_m = \langle IMKN|\,\mathrm{e}^{-\mathrm{i}S}\mathcal{M}(\mathrm{E}\lambda,\mu)\,\mathrm{e}^{\mathrm{i}S}\,|I'M'K'N'\rangle. \tag{15} $$

It is thus possible to introduce a renormalized electric multipole operator defined by

$$ \mathcal{M}_{\mathrm{r}}(\mathrm{E}\lambda,\mu) = \mathrm{e}^{-\mathrm{i}S}\mathcal{M}(\mathrm{E}\lambda,\mu)\,\mathrm{e}^{\mathrm{i}S}, \tag{16} $$

which to lowest order in S using eq. (11) is given by

$$ \mathcal{M}_{\mathrm{r}}(\mathrm{E}\lambda,\mu) = \mathcal{M}(\mathrm{E}\lambda,\mu) - \mathrm{i}\left[S, \sum_{\nu} D^{\lambda}_{\mu\nu}(\phi,\theta,0)\mathcal{M}_{\mathrm{int}}(\mathrm{E}\lambda,\nu)\right]. \tag{17} $$

The matrix elements of S are given by eq. (10). If one uses the renormalized multipole operator, one should use the pure rotational model wave function (6). The matrix elements of the electric quadrupole operator between different rotational states are given in tables 1 and 2, including deviations from the pure rotational model, according to eq. (17).

TABLE 1

Reduced quadrupole matrix elements within rotational bands.

$$ \langle K=0, I_2\|\,\mathcal{M}(\mathrm{E2})\,\|K=0, I_1\rangle = (2I_1+1)^{1/2}\langle I_1 020|I_2 0\rangle\sqrt{\frac{5}{16\pi}}\,eQ_0 $$
$$ \times\,[1+\zeta(I_2(I_2+1)+I_1(I_1+1))+\cdots] $$

$$ \langle KI_2\|\,\mathcal{M}(\mathrm{E2})\,\|KI_1\rangle = (2I_1+1)^{1/2}\langle I_1K20|I_2K\rangle\sqrt{\frac{5}{16\pi}}\,eQ_0 $$
$$ \times\left[1+\zeta\frac{1}{\langle I_1K20|I_2,K\rangle}\{\sqrt{(I_2+K)(I_2-K+1)}\,\langle I_1K2-1|I_2,K-1\rangle \right. $$
$$ \left. -\sqrt{(I_2-K)(I_2+K+1)}\,\langle I_1K21|I_2,K+1\rangle\}\right] $$

The first term is the value of the matrix element in the pure rotational model in terms of the intrinsic quadrupole moment Q_0, while the second term is due to effects of lowest order deviations from the pure model [BOH 75]. For the $K=0$ band, the lowest order correction is seen to be quadratic in I. In bands with $K\neq 0$ only deviations of first order in I have been included and terms arising from the interference between $|K\rangle$ and $|-K\rangle$ in the symmetrized wave function have been neglected.

TABLE 2

Reduced quadrupole matrix elements between two different rotational bands.

$$\langle KI_2 \| \mathcal{M}(\text{E2}) \| KI_1 \rangle = (2I_1+1)^{1/2} \langle I_1 K 2 0 | I_2 K \rangle \times [M_1 + M_2(I_2(I_2+1) - I_1(I_1+1))] + M_3(-1)^{I_1+1}(2I_1+1)^{1/2}\langle I_1 - 1 2 2 | I_2 1\rangle \delta(K,1)$$

$$\langle KI_2 \| \mathcal{M}(\text{E2}) \| K+2, I_1 \rangle = (2I_1+1)^{1/2}\langle I_1 K+2, 2-2 | I_2 K\rangle \times [M_1 + M_2(I_2(I_2+1) - I_1(I_1+1)) + \cdots]$$

$$\langle K+2+n, I_2 \| \mathcal{M}(\text{E2}) \| KI_1 \rangle = (2I_1+1)^{1/2}\langle I_1 K + n, 2 2 | I_2, K+n+2\rangle \times \left[\frac{(I_1-K)!\,(I_1+K+n)!}{(I_1-K-n)!\,(I_1+K)!}\right]^{1/2} \times \{M_1 + M_2[I_2(I_2+1) - I_1(I_1+1)] + M_3(-1)^{I_1+1/2}(I_1+\tfrac{1}{2})\delta(K,\tfrac{1}{2})\}$$

To leading order the matrix elements for $\Delta K \leqslant 2$ are proportional to a Clebsch-Gordan coefficient times an intrinsic quadrupole moment M_1. The deviations are partly caused by the coupling of the intrinsic motion to the rotation and partly by the interference between the terms $|K\rangle$ and $|-K\rangle$ in the symmetrized wave function [BOH 75].

§ 2. *Excitation in the sudden approximation*

Most of the calculations in the present chapter are based on the assumption that the excitation energies within a rotational band are small so that one can use the sudden approximation, while for the excitation of the intrinsic degrees of freedom the finite energies should be taken into account.

We may express this by assuming that the Hamiltonian H_r defined in eq. (1.2) commutes with the intrinsic Hamiltonian H_i as well as with the interaction $V_E(t)$. We write the interaction in the form (compare also eq. (VI.6.1))

$$V_E = V_i(\theta, \phi) + V_o(\theta, \phi), \tag{1}$$

where $V_o(\theta, \phi)$ indicates the interaction with a quadrupole moment Q_o which is assumed to be a constant, i.e.

$$V_o(\theta, \phi, t) = \tfrac{2}{5}\pi Z_1 e^2 Q_o \sum_{\mu} \bar{S}_{E2\mu}(t) Y^*_{2\mu}(\theta, \phi). \tag{2}$$

Here we have used eq. (II.1.12) and the definition (1.12). If the quadrupole moment Q_o is different for different intrinsic states, the difference should be included in the interaction V_i.

In the interaction representation (II.3.3) the potential (1) may thus be written

$$\tilde{V}(t) = \tilde{V}_i(\theta, \phi, t) + V_o(\theta, \phi, t), \tag{3}$$

where

$$\tilde{V}_i(\theta, \phi, t) = \exp(iH_i t/\hbar) V_i(\theta, \phi, t) \exp(-iH_i t/\hbar). \tag{4}$$

In the pure rotational model the two terms in eq. (3) commute and we may therefore write the S-matrix (II.3.17)

$$S = \mathscr{T} \exp\left\{-\frac{\mathrm{i}}{\hbar}\int_{-\infty}^{+\infty} \tilde{V}(t)\,\mathrm{d}t\right\}$$

$$= \exp\left\{-\frac{\mathrm{i}}{\hbar}\int_{-\infty}^{+\infty} V_{\mathrm{o}}(\theta, \phi, t)\,\mathrm{d}t\right\}\mathscr{T} \exp\left\{-\frac{\mathrm{i}}{\hbar}\int_{-\infty}^{+\infty} \tilde{V}_{\mathrm{i}}(\theta, \phi, t)\,\mathrm{d}t\right\}. \quad (5)$$

In this way we have achieved a separation between the intrinsic excitation and the excitation of rotational states built on these intrinsic states. Thus, for product states of the type (1.4) the excitation amplitude takes the form

$$a_{I_\mathrm{f}K_\mathrm{f}M_\mathrm{f}N_\mathrm{f},I_0K_0M_0N_0} = \langle I_\mathrm{f}K_\mathrm{f}M_\mathrm{f}|\exp\{-\mathrm{i}k_\mathrm{o}(\theta, \phi)\}a_{N_\mathrm{f}K_\mathrm{f},N_0K_0}(\theta, \phi)\,|I_0K_0M_0\rangle, \quad (6)$$

where

$$a_{N_\mathrm{f}K_\mathrm{f},N_0K_0}(\theta, \phi) = \langle N_\mathrm{f}K_\mathrm{f}|\,\mathscr{T} \exp\left\{-\frac{\mathrm{i}}{\hbar}\int_{-\infty}^{+\infty} \tilde{V}_{\mathrm{i}}(\theta, \phi, t)\,\mathrm{d}t\right\}|N_0K_0\rangle \quad (7)$$

indicates the intrinsic excitation amplitude for fixed orientation for the nuclear symmetry axis (see e.g. § VII.4). The quantity $k_\mathrm{o}(\theta, \phi)$ is given by

$$k_\mathrm{o}(\theta, \phi) = \frac{1}{\hbar}\int_{-\infty}^{+\infty} V_\mathrm{o}(\theta, \phi, t)\,\mathrm{d}t$$

$$= \tfrac{8}{3}\sqrt{\tfrac{1}{5}\pi}\, q \sum_{\mu} R_{2\mu}(\vartheta, 0)\, Y_{2\mu}^{*}(\theta, \phi), \quad (8)$$

where q defined by

$$q = \frac{Z_1 e^2 Q_\mathrm{o}}{4\hbar v a^2}, \quad (9)$$

plays the role of a common χ for the rotational model. Comparing the expression (9) with eq. (1.14) and eq. (IV.3.6) it is seen that q is connected with the χ for the transition from the ground state to the first rotational state in an even-even nucleus by the relation

$$q = \sqrt{\tfrac{45}{16}}\,\chi_{0\rightarrow 2}. \quad (10)$$

The factor $\exp(-\mathrm{i}k_\mathrm{o}(\theta, \phi))$ determines the amount of rotational angular momentum which is transferred and thus the probability distribution on the various rotational states within the band associated with the intrinsic quantum numbers N, K. Actually, the total excitation probability of this band

$P_{N_0K_0 \to NK}$ is given by the average value of $|a_{NK,N_0K_0}(\theta, \phi)|^2$. This can be seen directly from eq. (6), i.e.

$$P_{N_0K_0 \to N_fK_f} = \frac{1}{2I_0 + 1} \sum_{M_0 I_f M_f} |\langle I_f K_f M_f| \exp\{-ik_o(\theta, \phi)\}$$
$$\times\, a_{N_fK_f,N_0K_0}(\theta, \phi)\, |I_0 K_0 M_0\rangle|^2$$
$$= \frac{1}{2I_0 + 1} \sum_{M_0} \langle I_0 K_0 M_0|\, |a_{NK,N_0K_0}(\theta, \phi)|^2\, |I_0 K_0 M_0\rangle, \quad (11)$$

where we have utilized the fact that V_o commutes with $\tilde{V}_i$, and that the states $|I_f K_f M_f\rangle$ form a complete set in θ and ϕ for fixed K_f. The result (11) can, according to eq. (1.4) be written

$$P_{N_0K_0 \to N_fK_f} = \frac{1}{4\pi} \int |a_{N_fK_f,N_0K_0}(\theta, \phi)|^2\, d\Omega, \quad (12)$$

where we have used the unitarity of the D-function.

The amount of rotational angular momentum R which is transferred can be exhibited by expanding the function $\exp\{-ik_o(\theta, \phi)\}$ on spherical harmonics, i.e.

$$\exp\{-ik_o(\theta, \phi)\} = \sum_{RM} \sqrt{4\pi(2R + 1)}\, A_{RM}(\vartheta, q)\, Y_{RM}(\theta, \phi). \quad (13)$$

As will be shown in the following section the quantity $A_{RM}(\vartheta, q)\sqrt{2R + 1}$ is the amplitude for transferring the angular momentum R, M, to rotational motion. Since $k_o(\theta, \phi)$ has even parity, only even values of R occur in (13). Utilizing eq. (13), one may write the amplitude (6) in the form

$$a_{I_fK_fM_fN_f,I_0K_0M_0N_0} = \sum_{RM} \sqrt{4\pi(2R + 1)}\, A_{RM}(\vartheta, q)$$
$$\times\, \langle I_f K_f M_f|\, a_{N_fK_f,N_0K_0}(\theta, \phi)\, Y_{RM}(\theta, \phi)\, |I_0 K_0 M_0\rangle. \quad (14)$$

Above we have assumed that the rotational states are described by product wave functions. To obtain the actual excitation amplitude one should use the symmetrized wave function (1.6) and the amplitudes are then linear combinations of the expression (6) or (14). Since in most cases the interference terms between quantum numbers K and $-K$ can be neglected, the effects of symmetrization are rather trivial (see § 7 below.)

§ 3. *Excitation of the ground state band*

The main feature of the Coulomb excitation of deformed nuclei is the strong excitation of the ground state band. It is known that this excitation takes place mainly through the electric quadrupole interaction. In this section we shall limit ourselves to the computation of the quadrupole excitation in the ground state band. The effect of higher multipolarities as well as the excitation of other states can often be treated in a perturbation expansion and we shall defer the discussion of these effects until later.

In the sudden approximation the excitation amplitude is given by the matrix elements of $\exp(-ik_0)$ (see eq. (2.6)) between the symmetrized wave functions (1.6). In principle, for $K \neq 0$, matrix elements of $V_{\mathrm{E}}(t)$ occur between the states $|NK\rangle$ and $|N-K\rangle$. These matrix elements are expected to be small [BOH 75] and will be neglected. With the symmetrized wave function we are then left with two terms which are easily seen to be identical and for the computation of the excitation amplitude we may, therefore, use the product wave function (1.4).

To compute the excitation amplitude we may thus use eq. (2.14) with $a_{NK,NK} = 1$ and find the result

$$a_{I_{\mathrm{f}}M_{\mathrm{f}},I_0M_0} = \sum_{RM} \sqrt{(2I_0+1)(2I_{\mathrm{f}}+1)(2R+1)}(-1)^{M_{\mathrm{f}}-K} \times \begin{pmatrix} I_{\mathrm{f}} & I_0 & R \\ -M_{\mathrm{f}} & M_0 & M \end{pmatrix} \begin{pmatrix} I_{\mathrm{f}} & I_0 & R \\ -K & K & 0 \end{pmatrix} A_{RM}(\vartheta, q), \tag{1}$$

where the coefficients $A_{RM}(\vartheta, q)$ are explicitly given by

$$A_{RM}(\vartheta, q) = [4\pi(2R+1)]^{-1/2} \int_0^{2\pi} \mathrm{d}\phi \int_0^{\pi} \mathrm{d}\theta \sin\theta \, Y^*_{RM}(\theta, \phi) \times \exp\left\{ -\mathrm{i}\tfrac{8}{3}\sqrt{\tfrac{1}{5}\pi}\, q \sum_{\mu} R_{2\mu}(\vartheta, 0)\, Y^*_{2\mu}(\theta, \phi) \right\}. \tag{2}$$

The physical significance of A_{RM} can be seen explicitly by considering the special case of an even-even nucleus where the excitation amplitude according to eq. (1) is given by

$$a_{RM,00} = \sqrt{2R+1}\, A_{RM}(\vartheta, q), \tag{3}$$

which shows that $(2R+1)|A_{RM}|^2$ is the probability of transferring a rotational angular momentum R with component M. It is furthermore seen that the excitation amplitude (1) for an arbitrary rotational band is proportional to the excitation amplitude (3) multiplied by two vector addition coefficients which ensure that the final spin I_{f} is obtained by adding the initial angular momentum I_0 to the rotational angular momentum R. The triangular relations should be fulfilled, both in the laboratory system and in the intrinsic system, where the 3-component of R vanishes and the final amplitude is obtained by a summation over all possible transferred angular momenta.

An especially simple expression for A_{RM} is obtained for backwards scattering, where in coordinate system B only the terms with $\mu = 0$ in the exponent of (2) contribute. One may in this case perform the integration over ϕ, and the results may be written

$$A_{RM}(\pi, q) = \delta_{M0} \cdot \tfrac{1}{2} \exp(\mathrm{i}\tfrac{2}{3}q) \int_{-1}^{+1} \mathrm{d}x P_R(x) \exp(-2\mathrm{i}qx^2), \tag{4}$$

where $P_R(x)$ is the Legendre polynomial of order R and where we have used the explicit expression for P_2. The integral (4) may be evaluated in terms of a confluent hypergeometric function $_1F_1$ with the following result [ALD 60]

$$A_{R0}(\pi, q) = \frac{\Gamma(\frac{1}{2}(R+1))}{2\Gamma(\frac{1}{2}(2R+3))} \exp(-\mathrm{i}\tfrac{4}{3}q)(-2\mathrm{i}q)^{R/2}$$
$$\times\ {}_1F_1(\tfrac{1}{2}(R+2), \tfrac{1}{2}(2R+3); 2\mathrm{i}q), \tag{5}$$

which holds for even values of R.

The confluent hypergeometric function which appears in (5) can be expressed by means of Fresnel integrals. For $R = 0$ the expression (5) takes the simple form

$$A_{00}(\pi, q) = \sqrt{\frac{\pi}{4q}} \exp(\mathrm{i}\tfrac{2}{3}q)[C_2(2q) - \mathrm{i}S_2(2q)], \tag{6}$$

where $C_2(x)$ and $S_2(x)$ are the Fresnel integrals* tabulated e.g. in [ABR 64].

The functions A_{R0} for higher values of R are most easily obtained by means of recursion formulae. From the general recursion relations of $_1F_1$ one readily finds

$$(R+2)(2R-1)A_{R+2,0}(\pi, q) = \left[\frac{(2R+1)(2R-1)(2R+3)}{4\mathrm{i}q} + (2R+1)\right]$$
$$\times A_{R0}(\pi, q) + (R-1)(2R+3)A_{R-2,0}(\pi, q). \tag{7}$$

For the application of this formula one needs two consecutive A's. In the recursion relation it is convenient instead of A_{20} to use the non-physical A_{-20} which according to eq. (5) is a simple exponential function

$$A_{-20}(\pi, q) = -\frac{1}{4\mathrm{i}q} \exp(-\mathrm{i}\tfrac{4}{3}q). \tag{8}$$

* The notation of the Fresnel integrals differs from that used in [ALD 60] and in many other textbooks.

The functions $A_{R0}(\pi, q)$ in coordinate system B have been computed numerically in this way and the results are given in [ALD 60]. For other deflection angles one may evaluate the integral (2) in terms of a power series in the quantities $R^{\mathrm{B}}_{22}/R^{\mathrm{B}}_{20}$ (see [ALD 60] and [SIM 65]). As was discussed in § VI.3 one may use to a rather good approximation, the result for backwards scattering also for other deflection angles by means of the χ_{eff} or $\chi(\vartheta)$-approximation.

From the excitation amplitudes one may compute cross sections and angular distributions in the sudden approximation by means of the results of chapter III. Thus for the excitation probability of the state with spin I_{f} in a rotational band with ground state spin I_0 one finds, according to eqs. (III.1.2) and (1) the following result

$$P_{I_0\to I_{\mathrm{f}}}(\vartheta, q) = (2I_{\mathrm{f}}+1)\sum_{RM}(2R+1)\begin{pmatrix} I_{\mathrm{f}} & I_0 & R \\ -K & K & 0\end{pmatrix}^2 |A_{RM}(\vartheta, q)|^2$$

$$= (2I_{\mathrm{f}}+1)\sum_{R}\begin{pmatrix} I_{\mathrm{f}} & I_0 & R \\ -K & K & 0\end{pmatrix}^2 P_{0\to R}(\vartheta, q), \tag{9}$$

where $P_{0\to R}$ is the excitation probability for the special case of an even-even nucleus with ground state spin $I_0 = 0$, i.e.

$$P_{0\to I_{\mathrm{f}}}(\vartheta, q) = (2I_{\mathrm{f}}+1)\sum_{M}|A_{I_{\mathrm{f}}M}(\vartheta, q)|^2. \tag{10}$$

The expression (9) has a simple interpretation as the product of the probability $P_{0\to R}$ of transferring the rotational angular momentum R, times the square of the Clebsch-Gordan coefficient, coupling this angular momentum with the initial angular momentum to form the final angular momentum in the intrinsic system.

It is an interesting observation that the average transferred rotational energy $\langle \Delta E\rangle$ is identical to the one which would be obtained in first-order perturbation theory, even though the excitation probability in this description exceeds one. In order to prove this we consider an even-even nucleus with

$$\langle \Delta E\rangle = \sum_{I} P_{0\to I}(\vartheta, q) E_I, \tag{11}$$

where E_I is given by (1.5). Introducing the operator H_{r} from eqs. (1.2) and (1.3) we find

$$\langle \Delta E\rangle = \langle 0|\exp(\mathrm{i}k_{\mathrm{o}})H_{\mathrm{r}}\exp(-\mathrm{i}k_{\mathrm{o}})\,|0\rangle$$

$$= \frac{\hbar^2}{2\mathscr{I}}\langle 0|\left(\frac{\partial k_{\mathrm{o}}}{\partial\theta}\right)^2 + \frac{1}{\sin^2\theta}\left(\frac{\partial k_{\mathrm{o}}}{\partial\phi}\right)^2|0\rangle = E_2\,|\chi_{0\to 2}(\vartheta)|^2. \tag{12}$$

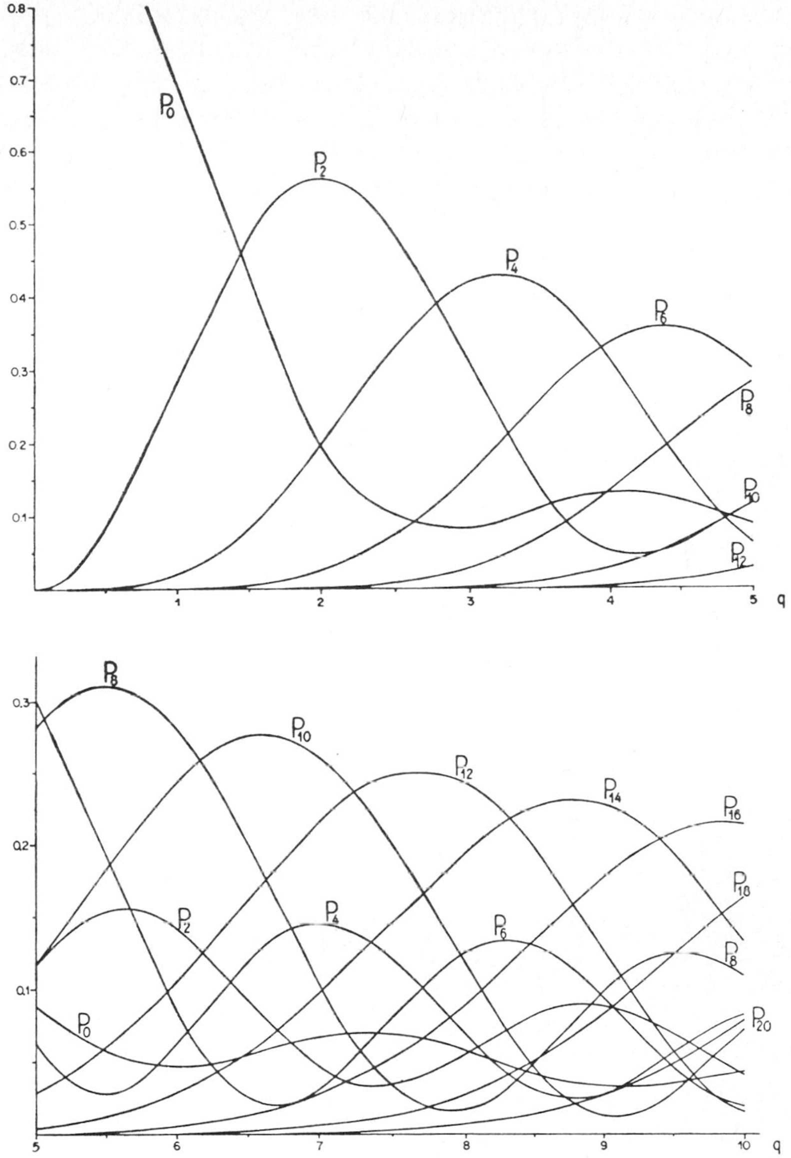

Fig. 1. The multiple Coulomb excitation of a pure rotational band in an even-even nucleus in the sudden approximation. The excitation probability P_I of the state with spin I is given as a function of the parameter q for backwards scattering. Note the change of scale between the values of P_I for $q < 5$ and $q > 5$.

We have here used the explicit form (2.8) for k_0 and the definition (IV.3.8).

The results (9) and (10) are especially simple to evaluate for backwards scattering where in the coordinate system B only the terms $M = 0$ contribute. The results for this special case are given in fig. 1. These results can be applied also for other deflection angles if one uses the $\chi(\vartheta)$-approximation. In this approximation the excitation probability is given by

$$P_{I_0 \to I_f}(\vartheta, q) \approx P_{I_0 \to I_f}(\pi, q(\vartheta)), \tag{13}$$

where $q(\vartheta)$ is defined by

$$q(\vartheta) = qR_2(\vartheta, 0). \tag{14}$$

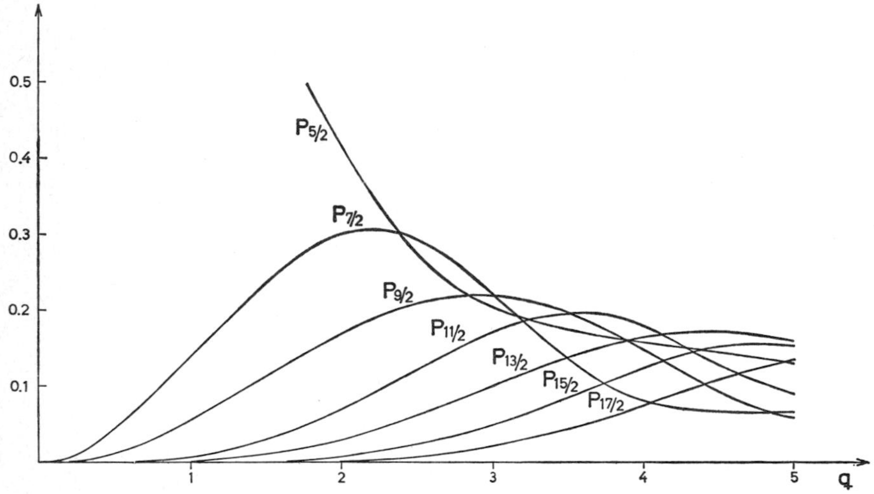

Fig. 2. The multiple Coulomb excitation of a pure rotational band in a nucleus with ground state spin $\frac{5}{2}$. The excitation probabilities P_I of the states with spin I are given in the sudden approximation as a function of the parameter q for backwards scattering.

As an application of eq. (9) we show in fig. 2 the excitation probabilities for a $K = \frac{5}{2}$ band.

§ 4. *Excitation of other rotational bands*

In evaluating the Coulomb excitation in a rotational band different from the ground state band, we first compute the excitation amplitudes with the product wave function (1.4) and discuss later the effect of the symmetrization. We consider the transition between different rotational bands to be small, so that the coupling can be treated as a perturbation.

It then follows from eq. (2.6) that the excitation amplitude of a state $|I_f M_f K_f\rangle$ can be written

$$b_{I_f K_f M_f, I_0 K_0 M_0} = -\mathrm{i} \sum_{\lambda} \langle I_f K_f M_f | \, k_\lambda \exp(-\mathrm{i}k_o) \, | I_0 K_0 M_0 \rangle, \tag{1}$$

with

$$k_o = \frac{1}{\hbar} \int_{-\infty}^{+\infty} V_o(t)\,\mathrm{d}t = \tfrac{4}{3} q \sum_{\mu} R_{2\mu}(\vartheta, 0)\, D^{2*}_{\mu 0}(\phi, \theta, 0), \tag{2}$$

and

$$\begin{aligned} k_\lambda &= \frac{1}{\hbar} \langle N_f K_f | \int_{-\infty}^{+\infty} \tilde{V}_i(t)\,\mathrm{d}t \, | N_0 K_0 \rangle \\ &= \chi^{(\lambda)}_{\mathrm{int}} \sqrt{2\lambda + 1} \sum_{\mu} (-1)^{\mu} R_{\lambda -\mu}(\vartheta, \xi)\, D^{\lambda}_{\mu K}(\phi, \theta, 0). \end{aligned} \tag{3}$$

In the expressions (1) to (3) we have used the notations

$$K = K_f - K_0 \tag{4}$$

and

$$\xi = \frac{E_f - E_0}{\hbar} \frac{a}{v}, \tag{5}$$

where $E_f - E_0$ is the energy difference between the intrinsic states $|N_f K_f\rangle$ and $|N_0 K_0\rangle$. Furthermore, we have used the expressions (2.8), (1.11) and (IV.2.2). The parameter q is defined in (2.9) while $\chi^{(\lambda)}_{\mathrm{int}}$ is given by

$$\chi^{(\lambda)}_{\mathrm{int}} = \frac{\sqrt{16\pi} Z_1 e(\lambda - 1)!}{\hbar v a^{\lambda}(2\lambda + 1)!!} \langle N_f K_f | \, \mathscr{M}_{\mathrm{int}}(\mathrm{E}\lambda, K) \, | N_0 K_0 \rangle. \tag{6}$$

This parameter which defines the strength of the transitions between the bands is defined in analogy to eqs. (IV.3.6) and (VII.4.9).

Utilizing the expansion (2.13) the problem of evaluating eq. (1) is reduced to an integral of the product of four D-functions. The result may e.g. be written in the form

$$\begin{aligned} b_{I_f M_f K_f, I_0 M_0 K_0} &= -\mathrm{i}\sqrt{(2I_0 + 1)(2I_f + 1)}(-1)^{M_f - K_f} \\ &\times \sum_{IM\lambda\mu} (-1)^{\mu} \sqrt{2I + 1} \begin{pmatrix} I_f & I_0 & I \\ -M_f & M_0 & M \end{pmatrix} \begin{pmatrix} I_f & I_0 & I \\ -K_f & K_0 & K \end{pmatrix} \chi^{(\lambda)}_{\mathrm{int}} \\ &\times R_{\lambda -\mu}(\vartheta, \xi) a_{IMK, \lambda\mu K}, \end{aligned} \tag{7}$$

where $a_{IMK,\lambda\mu K}$ is the excitation amplitude (3.1) in a ground state band of a nucleus with projection quantum number K for the transition from the state

$|\lambda, \mu\rangle$ to the state $|I, M\rangle$. The structure of this formula can be understood in a similar way as eq. (3.1) by interpreting the angular momentum I as the total transferred angular momentum, which must satisfy the triangular relations with the initial and final angular momenta I_0 and I_f, both in the intrinsic and the external coordinate systems. In turn, the explicit expression for $a_{IMK,\lambda\mu K}$ shows how the total transferred angular momentum is composed of the angular momenta λ transferred by the weak transition between the bands and the rotational angular momentum. The amplitude (7) is proportional to the product of the amplitudes for each of these steps, summed over all possible rotational and total angular momentum transfers.

In order to calculate the true excitation amplitudes one should consider that the nuclear wave functions in the rotational model are given by the symmetrized expressions (1.6) instead of the simple product (1.4).

For $K_0 = K_f = 0$, where both the initial and the final states are described by product wave functions, the Coulomb excitation amplitude is given directly by (7), i.e.

$$a_{I_f M_f 0, I_0 M_0 0} = b_{I_f M_f 0, I_0 M_0 0}. \tag{8}$$

It is noticed that the excitation amplitude in this case, vanishes unless $I_f + I_0 + \lambda$ is even, as is seen from eq. (7). Since, on the other hand, the symmetrization implies that in $K = 0$ bands only states of even or only states of odd spin occur, we obtain the following selection rule for the $K_0 = K_f = 0$ case

$$(-1)^{I_0 + I_f} = (-1)^{\lambda} = (-1)^{\Delta\pi}, \tag{9}$$

where $\Delta\pi$ is the relative parity of the two bands. If $K_0 = 0$ and $K_f > 0$ (or vice versa) we find that the excitation amplitude of the state $|I_f M_f K_f N_f\rangle$ is given by

$$a_{I_f M_f K_f, I_0 M_0 K_0} = \frac{1}{\sqrt{2}} [b_{I_f M_f K_f, I_0 M_0 0} + (-1)^{I_f + K_f} b_{I_f M_f - K_f, I_0 M_0 0}]. \tag{10}$$

Due to symmetry relations between matrix elements of the multipole moments, however, the two terms of (10) are equal and we get the result

$$a_{I_f M_f K_f, I_0 M_0 0} = \sqrt{2} b_{I_f M_f K_f, I_0 M_0 0}. \tag{11}$$

For the case where both K_0 and $K_f > 0$ the excitation amplitude contains four terms. Due to symmetry relations the expression reduces to

$$a_{I_f M_f K_f, I_0 M_0 K_0} = b_{I_f M_f K_f, I_0 M_0 K_0} + (-1)^{I_0 + K_0} b_{I_f M_f K_f, I_0 M_0 - K_0}. \tag{12}$$

The two terms are in (12) essentially different. The first term is proportional to the intrinsic matrix element between the states $|N_0 K_0\rangle$ and $|N_f K_f\rangle$ while the second is proportional to the matrix element between $|N_0, -K_0\rangle$ and $|N_f K_f\rangle$. In principle for $K_0 + K_f \leqslant \lambda$ both terms may contribute to the same transition.

From the above we may easily compute the excitation probabilities in the pure rotational model. For the excitation of a state of spin I_f in a $K = K_f$ band which is weakly coupled to the ground state band one finds

$$P_{I_f K_f} = \frac{1}{2I_0 + 1} \sum_{M_0 M_f} |a_{I_f M_f K_f, I_0 M_0 K_0}|^2, \tag{13}$$

where the excitation amplitude a is given by eqs. (8)–(12). For the evaluation of this expression we consider first the unsymmetrized expression (7). Inserting this in eq. (13) one finds

$$P_{I_f K_f} = (2I_f + 1) \sum_{\substack{I\lambda\lambda' \\ \mu\mu' M}} (-1)^{\mu+\mu'} \chi_{\text{int}}^{(\lambda)} \chi_{\text{int}}^{(\lambda')} R_{\lambda-\mu}(\vartheta, \xi) R^*_{\lambda'-\mu'}(\vartheta, \xi)$$
$$\times \begin{pmatrix} I_f & I_0 & I \\ -K_f & K_0 & K \end{pmatrix}^2 a_{IMK,\lambda\mu K}\, a^*_{IMK,\lambda'\mu' K}. \tag{14}$$

The total excitation probability (2.12) of the band with $K = K_f$ may either be evaluated directly from eq. (14) or one may use the expressions (2.12) and (3) to find

$$P_{K_f} = \sum_{\lambda\mu} |\chi_{\text{int}}^{(\lambda)}|^2 |R_{\lambda\mu}(\vartheta, \xi)|^2. \tag{15}$$

This is identical to the result one would obtain for the excitation probability from the intrinsic state $|N_0 K_0\rangle$ to the state $|N_f K_f\rangle$ in a perturbation treatment. The effect of the multiple excitation which takes place, is merely a redistribution of this probability on the various members of the $|N_f K_f\rangle$ band. This separation of the excitation probability becomes more evident if only one multipolarity occurs and if one uses the $q(\vartheta)$ approximation which ensures $\mu = \mu' = M$. One may then write the excitation probability

$$P_{I_f K_f} = |\chi_{\text{int}}^{(\lambda)}|^2 \sum_{\mu} |R_{\lambda-\mu}(\vartheta, \xi)|^2 |B_{I_f K_f}^{\lambda,\mu}(I_0 K_0)|^2 \tag{16}$$

with

$$|B_{I_f K_f}^{\lambda,\mu}(I_0 K_0)|^2 = (2I_f + 1) \sum_{I} \begin{pmatrix} I_f & I_0 & I \\ -K_f & K_0 & K \end{pmatrix}^2 |a_{I\mu K,\lambda\mu K}|^2. \tag{17}$$

This expression has essentially the form of a product of a transition probability (15) between the bands, times the excitation probabilities (17). These probability distributions are illustrated in fig. 3 for an even-even nucleus with $K_f = 0$, 1 and 2 and for $\vartheta = 180°$.

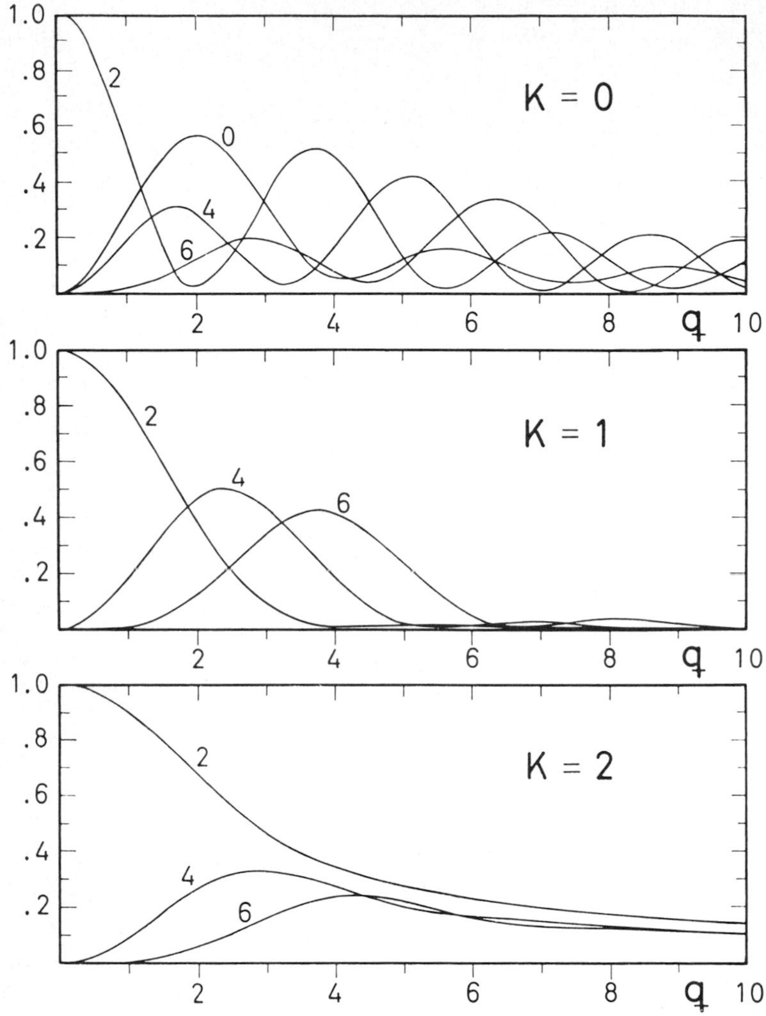

Fig. 3. The distribution coefficients $|B_{IK}^{\lambda,0}(q)|^2$ for $\lambda = 2$ and $I \leqslant 6$ as a function of q in an even-even nucleus. The coefficients which are defined in (4.17) indicate the relative excitation probabilities for various rotational states in a band with $K = 0$, 1 or 2. The functions are given separately for the three values of K while the spins I are indicated on the curves.

The effects of the symmetrization of the wave functions are easily included. For $K_0 = K_f = 0$ the results are unchanged. For $K_0 = 0$ and $K_f > 0$, or vice versa, there will appear an additional factor of 2 in the formulae (14)–(16). The total probability (15) will still have the same physical significance since the factor 2 appears also in the perturbation treatment. For this case the

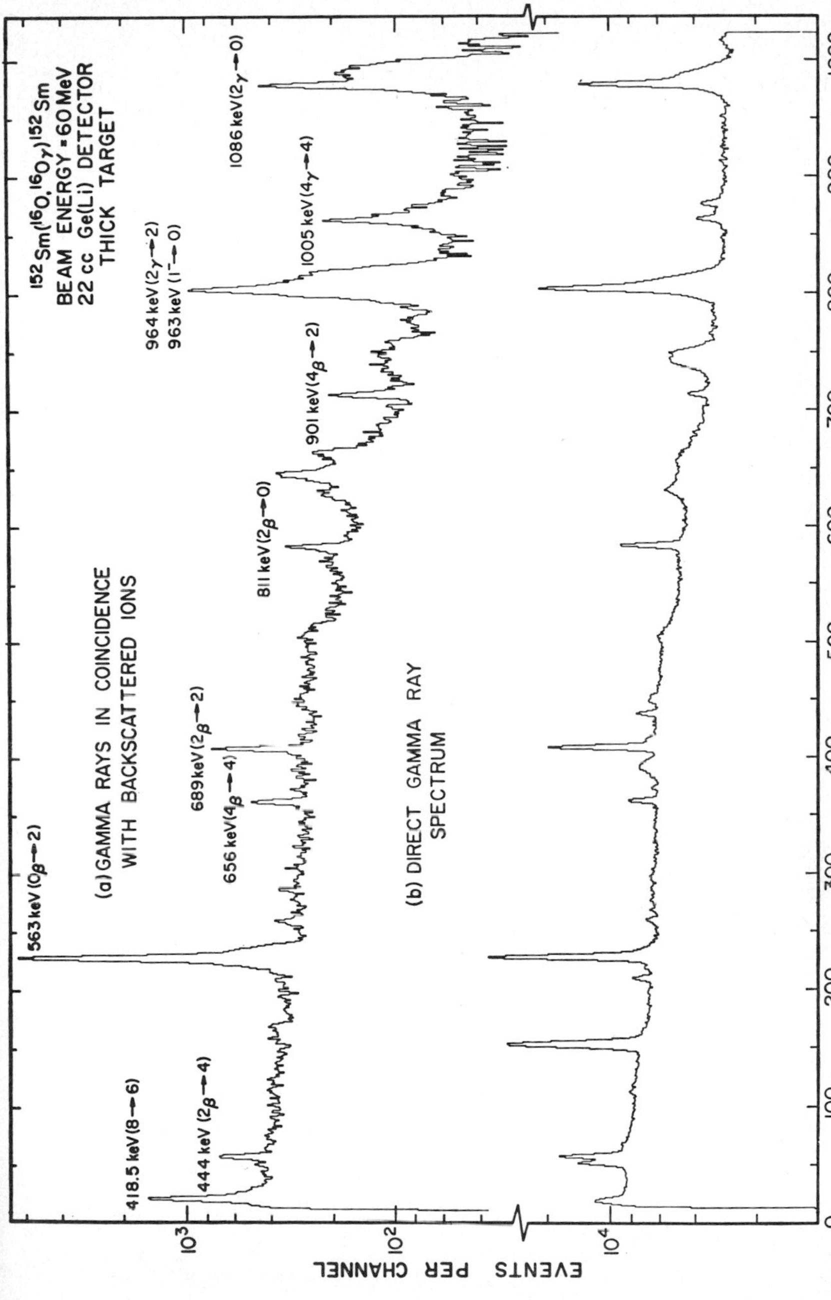

Fig. 4a. Gamma-ray spectra for ^{152}Sm(^{16}O, ^{16}Oγ)^{152}Sm. (From Greenberg, J. S., 1972, in: *Radioactivity in Nuclear Spectroscopy* (Gordon and Breach, New York) p. 1007.)

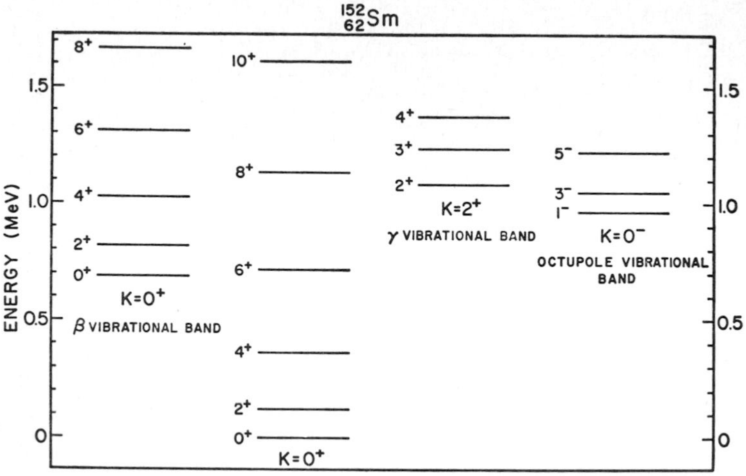

Fig. 4b. Partial energy level spectrum for ^{152}Sm. (From Greenberg, J. S., 1972, in: *Radioactivity in Nuclear Spectroscopy* (Gordon and Breach, New York) p. 1007.)

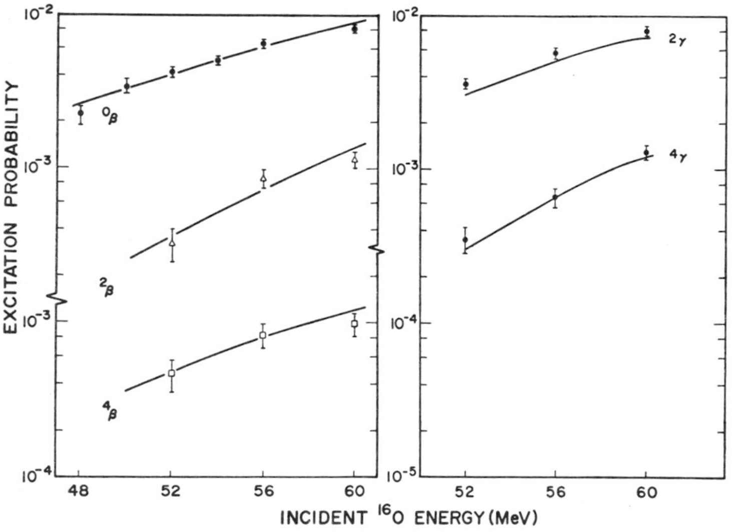

Fig. 4c. Measured average thick target excitation yields for the 0_β, 2_β, 4_β, 2_γ and 4_γ states in ^{152}Sm as a function ^{16}O ion energy. The solid lines are fits to the data using a multiple excitation calculation. (From Greenberg, J. S., 1972, in: *Radioactivity in Nuclear Spectroscopy* (Gordon and Breach, New York) p. 1007.)

factor 2 could be included in the definition of $\chi_{\text{int}}^{(\lambda)}$. Finally, for $K_0 > 0$ and $K_f > 0$ the excitation probabilities will in general contain three terms originating from the square of the two terms in eq. (12) plus an interference term. For the total excitation probability (15) one finds, however, that the interference term vanishes so that P_{K_f} becomes simply the sum of two expressions of the type (15), i.e.

$$P_{K_f} = \sum_{\lambda\mu} \{|\chi_{\text{int},K_0\to K_f}^{(\lambda)}|^2 + |\chi_{\text{int},-K_0\to K_f}^{(\lambda)}|^2\}|R_{\lambda\mu}(\vartheta, \xi)|^2. \tag{18}$$

While we considered above the Coulomb excitation of pure rotational bands, we shall finally mention the effect of deviations from the pure rotational model. These deviations are caused by the presence of the coupling term in the nuclear Hamiltonian which gives rise to admixtures in the pure rotational wave functions with different intrinsic structure.

Even small mixings may, however, in some cases have an important effect on the Coulomb excitation amplitude. This is the case e.g. if the excited band contains an admixture of the intrinsic wave function of the ground state band. The contribution to the Coulomb excitation amplitude from the admixture is proportional to the large matrix elements between identical intrinsic states, and even when the admixture is small this contribution may be comparable to the contribution from the main part of the wave functions.

Also mixing in the ground state band or in the excited band of states with K values between K_0 and K_f may be of importance. As an example we mention the excitation of a band with $K_f = K_0 + 3$, and with the same parity as the ground state. In this case, the direct transition is of E4 type. The mixing between the bands occurs only through a third-order term in H_c, and it may be expected that the admixture in the ground state band of states with $K = K_0 + 1$ as well as admixtures in the excited band with $K = K_0 + 2$ may lead to E2 transitions of comparable magnitude. A discussion of the effects associated with band mixing is given in [LÜT 64]. An experimental example of the excitation of the ground state and vibrational bands in a deformed nucleus is given in figs. 4a–c.

§ 5. *Deviation from the sudden approximation*

In § VI.4 it was outlined how one may calculate the deviation of the excitation amplitudes from the sudden approximation. The result was expressed in a power series in the ξ's which enter into the excitation process. For rotational bands, one may define a common ξ in terms of the moment of inertia in a similar way as we defined a common χ in terms of the intrinsic quadrupole moment. We shall use the notation

$$\xi = 3\hbar a/v\mathscr{I}\,, \tag{1}$$

where $\mathscr{I}$ is the moment of inertia entering in eq. (1.2). The quantity (1) is identical to the ξ corresponding to the excitation of the lowest rotational state in an even-even nucleus.

The excitation amplitudes which were evaluated in the previous paragraph are complex numbers. The first-order corrections must also be expected to be complex, and it follows therefore that the excitation probabilities have linear terms in ξ. This is in contrast to the first-order perturbation theory which is independent of ξ to first order in this quantity.

To first order, the excitation amplitude a_n may be written in the form

$$a_n = a_n^{(0)} + \Delta a_n\,, \tag{2}$$

where $a_n^{(0)}$ is the amplitude (3.1) in the sudden approximation. The first-order correction Δa_n is, according to eq. (VI.4.4) given by

$$\Delta a_n = \frac{\xi v}{6\hbar a} \langle n|\exp\left\{-\frac{\mathrm{i}}{\hbar}\int_{-\infty}^{+\infty} V_{\mathrm{o}}(t)\,\mathrm{d}t\right\}\int_{-\infty}^{+\infty}\mathrm{d}t\,\exp\left\{\frac{\mathrm{i}}{\hbar}\int_{-\infty}^{t} V_{\mathrm{o}}(t')\,\mathrm{d}t'\right\}$$

$$\times\, t[\boldsymbol{R}^2, V_{\mathrm{o}}(t)]\exp\left\{-\frac{\mathrm{i}}{\hbar}\int_{-\infty}^{t} V_{\mathrm{o}}(t')\,\mathrm{d}t'\right\}|0\rangle. \tag{3}$$

In eq. (3) we have suppressed the intrinsic degrees of freedom taking the expectation value of $V_{\mathrm{E}}(t)$ within the ground state band, i.e., according to eq. (2.2),

$$V_{\mathrm{o}}(t) = \langle NK|\, V_{\mathrm{E}}(t)\,|NK\rangle$$

$$= \frac{8\pi\hbar v a^2}{5}\, q\sum_{\mu} \bar{S}_{\mathrm{E}2\mu}(t)\, Y_{2\mu}^{*}(\theta, \phi). \tag{4}$$

Furthermore, we have inserted for the free Hamiltonian

$$H_{\mathrm{o}} = H_{\mathrm{r}} = \frac{\hbar^2}{2\mathscr{I}}\boldsymbol{R}^2 = \frac{\hbar v}{6a}\,\xi\,\boldsymbol{R}^2 \tag{5}$$

corresponding to a pure rotational model.

We shall only consider the case of even-even nuclei and shall furthermore use the $\chi(\vartheta)$-approximation where in coordinate system B we may neglect the terms with $\mu \neq 0$ in eq. (4). The more general case has been treated by [ALD 60]. In the $\chi(\vartheta)$-approximation we may easily evaluate the commutator in eq. (3) and we find

$$\int_{-\infty}^{+\infty} \mathrm{d}t \exp\left\{\frac{\mathrm{i}}{\hbar}\int_{-\infty}^{t} V_0(t')\,\mathrm{d}t'\right\}[\boldsymbol{R}^2, V_0(t)]t \exp\left\{-\frac{\mathrm{i}}{\hbar}\int_{-\infty}^{t} V_0(t')\,\mathrm{d}t'\right\}|0\rangle$$

$$= \mathrm{i}\,\frac{128\pi^2\hbar v^2 a^4}{25}\, q^2 \int_{-\infty}^{+\infty} t\bar{S}^{\mathrm{B}}_{\mathrm{E}20}(t) \int_{-\infty}^{t} \bar{S}^{\mathrm{B}}_{\mathrm{E}20}(t')\,\mathrm{d}t'\,\mathrm{d}t \left(\frac{\mathrm{d}}{\mathrm{d}\theta} Y_{20}(\theta, 0)\right)^2 |0\rangle. \quad (6)$$

We write the square of the derivative of $Y_{20}(\theta, 0)$ in the following way

$$\left(\frac{\mathrm{d}}{\mathrm{d}\theta} Y_{20}(\theta, 0)\right)^2 = \frac{3}{2\pi}\{-\tfrac{12}{7}P_4(\cos\theta) + \tfrac{5}{7}P_2(\cos\theta) + 1\}. \quad (7)$$

Inserting eqs. (7) and (6) into eq. (3) it is seen that the correction Δa_n can be expressed by the functions $A_{R0}(\pi, q)$ defined in eq. (3.4). One finds the following correction to the excitation amplitude of the state $|I_{\mathrm{f}}, 0\rangle$

$$\Delta a_{I_{\mathrm{f}}0} = \mathrm{i}\,\frac{32\pi}{25}\,\xi q^2 f_{00}\sqrt{2I_{\mathrm{f}}+1}$$
$$\times \sum_R (2R+1)\left[\begin{pmatrix} I_{\mathrm{f}} & 0 & R \\ 0 & 0 & 0\end{pmatrix}^2 + \tfrac{5}{7}\begin{pmatrix} I_{\mathrm{f}} & 2 & R \\ 0 & 0 & 0\end{pmatrix}^2 - \tfrac{12}{7}\begin{pmatrix} I_{\mathrm{f}} & 4 & R \\ 0 & 0 & 0\end{pmatrix}^2\right]$$
$$\times A_{R0}(\pi, q(\vartheta)), \quad (8)$$

where f_{00} is given by

$$f_{00} = v^3 a^3 \int_{-\infty}^{+\infty} \mathrm{d}t \cdot t\bar{S}_{\mathrm{E}20}(t) \int_{-\infty}^{t} S_{\mathrm{E}20}(t')\,\mathrm{d}t'. \quad (9)$$

This integral has been evaluated numerically for different values of the deflection angle and the results are given in table 3. While eq. (8) gives an approximate correction to the excitation amplitudes of states with $M_{\mathrm{f}} = 0$, there are quite important contributions also to the excitation probability for the states with $M_{\mathrm{f}} = \pm 1$ which are not included in the above treatment.

TABLE 3

The coefficient f_{00} for first-order corrections to the excitation of rotational states.

ϑ	f_{00}
180°	1.622×10^{-1}
150°	1.426×10^{-1}
120°	9.12×10^{-2}
90°	4.01×10^{-2}

The coefficient which is defined in eq. (5.9) is given for a few scattering angles.

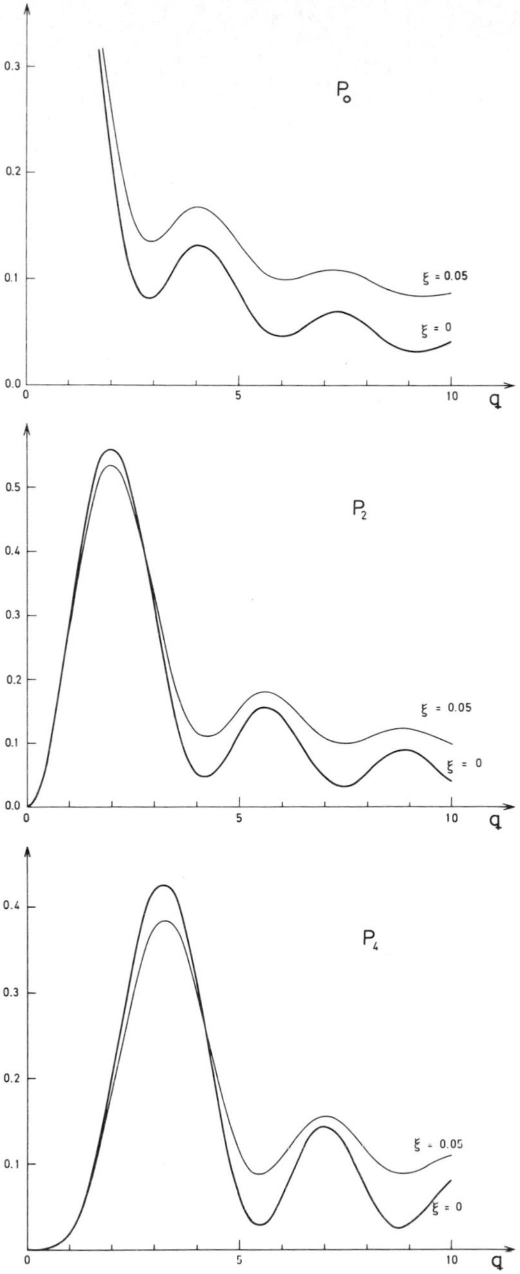

Fig. 5. The first-order correction in ξ for the excitation of a pure rotational band in an even-even nucleus. The curves show the excitation probabilities of the states with spin 0, 2 and 4 for backwards scattering as a function of q for the cases $\xi = 0$ and $\xi = 0.05$.

The result (8) can thus only be used for large deflection angles. A more complete treatment of the deviations from the sudden approximation, which is valid for all scattering angles, is given in [ALD 60].

From eq. (8) we may evaluate the correction to the excitation probability of the state of spin I. One finds the result

$$P_I(\vartheta, q) = P_I(\pi, q(\vartheta), \xi = 0) + \frac{f_{00}(\vartheta)}{f_{00}(\pi)} \Lambda_I(q(\vartheta)) \cdot \xi, \qquad (10)$$

where $P_I(\pi, q(\vartheta), \xi = 0)$ is the excitation probability (3.10) in the $q(\vartheta)$ approximation while the functions Λ_I are tabulated in [ALD 60] as functions of q for $I \leqslant 20$. They are oscillating functions of q which are of the order of magnitude one. The corrections for $\xi \neq 0$ are thus not dominated by the factor q^2 in eq. (8). The oscillations in Λ_I follow the oscillations of $P_I(\xi = 0)$ in such a way that the first maximum of P_I is cut down, while the excitation probability for larger values of q is increased. This increment is largest at the minima of P_I and the effect of $\xi \neq 0$ is thus essentially to smooth out the excitation curve. An indication of the accuracy of the result (10) can be obtained by comparing the limiting case of eq. (8) for $q \ll 1$ with the second-order perturbation calculation performed in chapter V. Such a comparison shows that eq. (8) should not be applied for angles less than 90 degrees. The effect is illustrated in fig. 5 for $\vartheta = 180°$, $\xi = 0.05$ and $I = 0$, 2 and 4.

§ 6. *Classical treatment*

It may be interesting to compare the results of the Coulomb excitation in rotational bands with the corresponding classical problem which should be applicable for large values of q. The classical treatment of a collision between a charged particle and a charged symmetrical top leads, as we shall see, to non-linear equations of motion. As in the quantum mechanical problem these equations can only be solved in a closed form in the limit where the collision time is short compared to the time of rotation of the top.

The classical Hamiltonian can be written in the form

$$H = \frac{v\xi}{6a\hbar}\left[p_\theta^2 + \frac{1}{\sin^2\theta} p_\phi^2\right] + \tfrac{8}{5}\pi\, \hbar v a^2 q \sum_\mu \bar{S}_{\mathrm{E}2\mu}(t)\, Y_{2\mu}^*(\theta, \phi), \qquad (1)$$

where ξ is defined by eq. (5.1) and q by eq. (2.9). The generalized momenta p_ϕ and p_θ are conjugate to the Eulerian angles ϕ and θ describing the orientation of the axis of the symmetric top.

The Hamiltonian equations of motion in coordinate system B are given by

$$\begin{aligned}\dot{\theta} &= \frac{v\xi}{3a\hbar}\,p_\theta = \frac{\partial H}{\partial p_\theta}\\ \dot{\phi} &= \frac{v\xi}{3a\hbar}\,\frac{p_\phi}{\sin^2\theta} = \frac{\partial H}{\partial p_\phi},\end{aligned} \tag{2}$$

and

$$\begin{aligned}\dot{p}_\phi &= 4\sqrt{\tfrac{6}{5}\pi}\,\hbar v a^2 q \sin\theta\,[\bar{S}^{\mathrm{B}}_{\mathrm{E22}}(t)\sin\theta\sin 2\phi - \mathrm{i}\bar{S}^{\mathrm{B}}_{\mathrm{E21}}(t)\cos\theta\cos\phi],\\ \dot{p}_\theta &= \frac{v\xi}{3a\hbar}\,\frac{p_\phi^2}{\sin^2\theta}\,\frac{\cos\theta}{\sin\theta} + 4\sqrt{\tfrac{1}{5}\pi}\,\hbar v a^2 q\{\bar{S}^{\mathrm{B}}_{\mathrm{E20}}(t)\cdot 3\cos\theta\sin\theta\\ &\quad - \mathrm{i}\bar{S}^{\mathrm{B}}_{\mathrm{E21}}(t)\sqrt{6}(\cos^2\theta - \sin^2\theta)\sin\phi - \bar{S}^{\mathrm{B}}_{\mathrm{E22}}(t)\sqrt{6}\sin\theta\cos\theta\cos 2\phi\}.\end{aligned} \tag{3}$$

From these equations one obtains the following set of coupled equations of motion

$$\begin{aligned}\ddot{\theta} &= \tfrac{1}{2}\dot{\phi}^2\sin 2\theta + 2\sqrt{\tfrac{1}{5}\pi}\,v^2 a q\xi\\ &\quad\times[\bar{S}^{\mathrm{B}}_{\mathrm{E20}}(t)\sin 2\theta - 2\mathrm{i}\sqrt{\tfrac{2}{3}}\bar{S}^{\mathrm{B}}_{\mathrm{E21}}(t)\cos 2\theta\sin\phi - \sqrt{\tfrac{2}{3}}\bar{S}^{\mathrm{B}}_{\mathrm{E22}}(t)\sin 2\theta\sin 2\phi],\\ \sin\theta\,\ddot{\phi} &= -2\cos\theta\,\dot{\phi}\dot{\theta} + 4\sqrt{\tfrac{2}{15}\pi}\,v^2 a q\xi\\ &\quad\times[\bar{S}^{\mathrm{B}}_{\mathrm{E22}}(t)\sin\theta\sin 2\phi - \mathrm{i}\bar{S}^{\mathrm{B}}_{\mathrm{E21}}(t)\cos\theta\cos\phi].\end{aligned} \tag{4}$$

These non-linear equations can be solved in the limiting case where the collision is so sudden that one can neglect the motion of the top during the collision. If we assume that the top is at rest before the collision, the generalized momenta after the collision $(p_\phi)_{\mathrm{f}}$ and $(p_\theta)_{\mathrm{f}}$ can then be found directly from (3). By integration over time one finds

$$\begin{aligned}(p_\phi)_{\mathrm{f}} &= 4\sqrt{\tfrac{2}{3}}\,\hbar q\sin^2\theta_{\mathrm{i}}\sin 2\phi_{\mathrm{i}}\,R^{\mathrm{B}}_{22}(\vartheta, 0),\\ (p_\theta)_{\mathrm{f}} &= 4\hbar q\cos\theta_{\mathrm{i}}\sin\theta_{\mathrm{i}}\,[R^{\mathrm{B}}_{20}(\vartheta, 0) - \sqrt{\tfrac{2}{3}}\cos 2\phi_{\mathrm{i}}\,R^{\mathrm{B}}_{22}(\vartheta, 0)],\end{aligned} \tag{5}$$

where ϕ_{i} and θ_{i} denote the orientation of the top before the collision in coordinate system B.

According to (1) the energy which has been transferred to the top is given by

$$\Delta E = \frac{8}{3}\frac{v\xi\hbar q^2}{a}\sin^2\theta_{\mathrm{i}}\,\{\cos^2\theta_{\mathrm{i}}\,[R^{\mathrm{B}}_{20} - \sqrt{\tfrac{2}{3}}R^{\mathrm{B}}_{22}\cos 2\phi_{\mathrm{i}}]^2 + \tfrac{2}{3}\sin^2 2\phi_{\mathrm{i}}\,(R^{\mathrm{B}}_{22})^2\}. \tag{6}$$

It is seen that the energy transfer depends strongly on the initial orientation of the top and that a maximum energy transfer is obtained for

$$\theta_i = \tfrac{1}{4}\pi \qquad \text{and} \qquad \phi_i = 0, \tag{7}$$

whereby

$$\Delta E_{\text{max}} = \tfrac{2}{3}\Delta E_2 \cdot q^2[R_{20}^{\text{B}} - \sqrt{\tfrac{2}{3}}R_{22}^{\text{B}}]^2, \tag{8}$$

and where ΔE_2 is the energy of the first excited state

$$\Delta E_2 = 6\hbar^2/2\mathscr{I} = v\xi\hbar/a. \tag{9}$$

It is noted that this maximum energy transfer corresponds to a maximum angular momentum transfer of

$$L_{\text{max}} = 2q\hbar[R_{20}^{\text{B}}(\vartheta, 0) - \sqrt{\tfrac{2}{3}}R_{22}^{\text{B}}(\vartheta, 0)]. \tag{10}$$

Since the probability distribution in ϕ_i and θ_i is given by

$$P(\phi_i, \theta_i)\,\mathrm{d}\phi_i\,\mathrm{d}\theta_i = \frac{1}{4\pi}\sin\theta_i\,\mathrm{d}\theta_i\,\mathrm{d}\phi_i, \tag{11}$$

corresponding to an isotropic probability distribution of the orientation of the top before the collision, one may calculate the classical probability distribution of the energy of the top after the collision from eq. (6). Also one may evaluate the average value of the energy after the collision, i.e.

$$\begin{aligned}\langle \Delta E \rangle &= \frac{1}{4\pi}\int_0^{\pi}\int_0^{2\pi} \sin\theta_i\,\mathrm{d}\theta_i\,\mathrm{d}\phi_i\,\Delta E(\phi_i, \theta_i) \\ &= \tfrac{16}{45}\frac{v\xi\hbar q^2}{a}[(R_{20}^{\text{B}})^2 + 2(R_{22}^{\text{B}})^2] = \Delta E_2 \cdot |\chi(\vartheta)|^2.\end{aligned} \tag{12}$$

It is noted that the average excitation energy is exactly equal to the corresponding quantum mechanical result (3.12).

§ 7. *Effects of higher multipole moments*

In the sudden approximation it is easy to include the effect of higher multipole moments on the excitation of pure rotational states. If we consider e.g. a hexadecapole interaction (II.1.12)

$$V_{\text{E}4}(t) = \frac{5\pi Z_1 e}{9}\sum_{\mu}\mathscr{M}(\text{E}4, -\mu)(-1)^{\mu}\frac{1}{r^5(t)}\,Y_{4\mu}(\theta(t), \phi(t)) \tag{1}$$

and assume that this interaction commutes with the quadrupole interaction, we find to lowest order in V_{E4} the following excitation amplitude

$$a_{I_fM_f,I_0M_0} = \langle I_fM_f| \exp\{-ik_o(\theta, \phi)\}\left\{1 - \frac{i}{\hbar}\int_{-\infty}^{+\infty} V_{E4}(t)\,dt\right\}|I_0M_0\rangle, \quad (2)$$

where $k_o(\theta, \phi)$ is given by eq. (2.8).

The excitation amplitude (2) thus consists of two terms,

$$a_{I_fM_f,I_0M_0} = a^{rot}_{I_fM_f,I_0M_0} + \Delta a_{I_fM_f,I_0M_0}, \quad (3)$$

i.e. the unperturbed excitation amplitude a^{rot} given by (3.1) and a correction proportional to the matrix element of the E4 moment. This correction is given by

$$\Delta a_{I_fM_f,I_0M_0} = \sum_{I_zM_z} a^{rot}_{I_fM_f,I_zM_z} a^{(1)}_{I_zM_z,I_0M_0}. \quad (4)$$

The first-order amplitude $a^{(1)}$ is given by eq. (IV.3.7), i.e.

$$a^{(1)}_{I_zM_z,I_0M_0} = -i\chi^{(4)}_{I_0\to I_z}(-1)^{I_0-M_0}\cdot 3\sqrt{2I_0+1}\begin{pmatrix} I_0 & 4 & I_z \\ -M_0 & \mu & M_z \end{pmatrix} R_{4\mu}(\vartheta, 0). \quad (5)$$

As an illustration we consider the case of an even-even nucleus where eq. (3) may be written as

$$a_{IM,00} = a^{rot}_{IM,00} - i\chi^{(4)}_{0\to 4}\sum_{\mu}(-1)^{\mu}R_{4-\mu}(\vartheta, 0)a^{rot}_{IM,4\mu}. \quad (6)$$

This equation may serve as a qualitative basis for a discussion of the optimum conditions for the observation of a hexadecapole interaction in a rotational band.

CHAPTER IX

Quantum Mechanical Treatment

The description of the Coulomb excitation process that has been given in the preceding chapters was based on a classical description of the motion of the projectile. As was discussed especially in § II.2 this approximation is very accurate as long as the parameter η is large compared to unity. For the collision between heavy ions below the Coulomb barrier η is larger than 20 if projectiles heavier than oxygen are used. For bombardments with α-particles η is of the order of 10 or less and for protons it may become of the order of two to three, still without strong penetration (see § X.4).

For the cases with light projectiles where quantal effects are most important one may usually treat the Coulomb excitation by first-order perturbation theory. The quantum mechanical first-order theory has been given by [BIE 55], [ALD 55] and [ALD 56]. These calculations have shown that the major part of the quantum mechanical corrections to the classical results can be incorporated in the classical description by a proper symmetrization of the parameters ξ and χ with respect to the relative velocities. For the total cross section the additional corrections are very small indeed and actually quite unimportant. For quantities like angular distributions of γ-rays, which are sensitive to phases, the corrections should be taken into account and it is expected that they behave essentially as $1/\eta$ as a function of η (see § 4 below). The quantum mechanical corrections to second-order effects like reorientation are also expected to be significant since they depend on phases ([SMY 68], [ALD 69]).

In the present chapter we shall derive the differential equations which determine the multiple Coulomb excitation quantum mechanically and study the transition via the WKB approximation to the classical limit in order to justify the symmetrization procedure which was given in § VI.8.

§ 1. *Coupled radial equations*

In the quantum mechanical treatment of Coulomb excitation we consider a stationary situation where the excitation amplitudes are determined by the radial wave functions in the inelastic channels. These radial wave functions are solutions of coupled radial differential equations of second order with boundary conditions corresponding to outgoing waves in all the inelastic channels. We consider only the monopole-multipole terms in the interaction (II.1.7) and assume that only the target nucleus can be excited. The Hamiltonian in the centre-of-mass system is then of the form

$$H = H_o(2) + T_{\mathrm{rel}} + V_{\mathrm{E}}(2, \boldsymbol{r}). \tag{1}$$

Here $H_o(2)$ is the Hamiltonian of the free target nucleus, T_{rel} is the relative kinetic energy

$$T_{\mathrm{rel}} = -\frac{\hbar^2}{2m_o} \Delta_{\mathrm{rel}}, \tag{2}$$

where m_o is the reduced mass of target and projectile, and $V_{\mathrm{E}}(2, \boldsymbol{r})$ is the monopole-multipole electric interaction (see eq. (II.1.12))

$$V_{\mathrm{E}}(2, \boldsymbol{r}) = \sum_{\lambda\mu} \frac{4\pi Z_1 e}{2\lambda+1} r^{-(\lambda+1)} Y_{\lambda\mu}(\theta, \phi) \mathcal{M}(\mathrm{E}\lambda, -\mu)(-1)^{\mu}. \tag{3}$$

We shall try to find the stationary solutions of the equation

$$H\,|\Psi\rangle = E\,|\Psi\rangle, \tag{4}$$

which behave asymptotically as

$$\begin{aligned} |\Psi(\boldsymbol{k}_0)\rangle \approx{}& |I_0 M_0\rangle \exp\{\mathrm{i}[\boldsymbol{k}_0\boldsymbol{r} + \eta_0 \log(k_0 r - \boldsymbol{k}_0\boldsymbol{r})]\} \\ &+ \sum_n f_n(\vartheta, \varphi)\,|I_n M_n\rangle \frac{1}{r} \exp\{\mathrm{i}(k_n r - \eta_n \log 2k_n r)\}, \end{aligned} \tag{5}$$

corresponding to an incoming plane wave with relative wave number $\boldsymbol{k}_0$ on the nucleus in the ground state $|I_0 M_0\rangle$ giving rise to outgoing spherical waves in the elastic and inelastic channels n. It follows directly from eq. (4) that the wave number k_n is determined by

$$E = E_0 + \frac{\hbar^2 k_0^2}{2m_o} = E_n + \frac{\hbar^2 k_n^2}{2m_o}. \tag{6}$$

The quantity $f_n(\vartheta, \varphi)$ is the scattering amplitude from which the inelastic differential centre-of-mass cross section is determined by

$$\left(\frac{\mathrm{d}\sigma}{\mathrm{d}\Omega}\right)_n = \frac{k_n}{k_0} |f(\vartheta, \varphi)|^2. \tag{7}$$

In the solution of this problem we shall use a partial wave expansion into eigenstates of the relative orbital angular momenta. Since the total relative angular momentum is conserved with the interaction (1)–(3) it is convenient to introduce basis functions with definite total angular momentum J and component N along the z-axis. These channel-spin wave functions may thus be written

$$|(lI_n)JN\rangle = \sum_{m\,M_n} \langle lmI_nM_n|JN\rangle\, Y_{lm}(\vartheta, \varphi)\, |I_nM_n\rangle . \tag{8}$$

It follows from eqs. (4)–(6) that any eigenstate $|\Psi\rangle$ of eq. (4) can be expanded in the form

$$|\Psi\rangle = \sum_{lI_nJN} \frac{\alpha_{JN}}{r}\, g_{(lI_n)J}(r)\, |(lI_n)JN\rangle , \tag{9}$$

where α_{JN} denote arbitrary coefficients while $g_{(lI_n)J}(r)$ must satisfy the following real coupled differential equations

$$\left(\frac{\mathrm{d}^2}{\mathrm{d}r^2} + k_n^2 - \frac{l(l+1)}{r^2} - \frac{2\eta_n k_n}{r}\right) g_{(lI_n)J}(r) = \sum_{l'I_m} V^J_{lI_n,l'I_m}(r)\, g_{(l'I_m)J}(r) . \tag{10}$$

The coupling $V^J_{lI_n,l'I_m}(r)$ is given by

$$V^J_{lI_n,l'I_m}(r) = \frac{4\sqrt{\pi}m_0Z_1e}{\hbar^2} \sum_{\lambda} \sqrt{\frac{(2l+1)(2l'+1)}{2\lambda+1}}\, (-1)^{J+I_m+\lambda}$$
$$\times \langle I_m\|\mathscr{M}(\mathrm{E}\lambda)\|I_n\rangle \begin{pmatrix} l & l' & \lambda \\ 0 & 0 & 0 \end{pmatrix} \begin{Bmatrix} J & l' & I_m \\ \lambda & I_n & l \end{Bmatrix} \frac{1}{r^{\lambda+1}} . \tag{11}$$

It is seen that $V^J(r)$ is symmetric under interchange of $l'I_m$ with lI_n and is independent of the magnetic quantum number N. The index N on $g(r)$ has therefore been left out. If we choose the z-axis along the direction $\boldsymbol{k}_0$ of the incoming projectile the index N is fixed to be equal to the magnetic quantum number M_0 of the nucleus in the initial state. It is noted that through the expression (11) the parity conservation ($l + l' + \lambda =$ even) is ensured. The solutions of eq. (10) must be carried out by starting the integration at the origin where the functions $g(r)$ are zero. There is a large number of independent solutions which can be found by using different initial conditions. In fact, there are as many linearly independent finite real solutions as there are functions $g(r)$ for a given J and we shall denote the various solutions by an index q and write the solutions $g^{(q)}_{(lI_n)J}(r)$. From these solutions one can construct the most general solution by a linear superposition, and to determine the scattering state $|\Psi(\boldsymbol{k}_0)\rangle$ one must choose the linear combination in such a way that eq. (9) has the asymptotic behaviour (5).

We do this in two stages. First we determine special solutions of eq. (10) which have the asymptotic form

$$g^{(l_0I_0)}_{(lI_n)J}(r) \approx \delta_{I_nI_0}\delta_{ll_0}\exp\{-\mathrm{i}\varphi_{I_0l}(r)\} - \sqrt{\frac{k_0}{k_n}}\, r^J_{lI_n,l_0I_0}\exp\{\mathrm{i}\varphi_{I_nl}(r)\} \tag{12}$$

for $r \to \infty$. The phases $\varphi_{I_nl}(r)$ are defined by

$$\varphi_{I_nl}(r) = k_nr - \eta_n \log 2k_nr - \tfrac{1}{2}\pi l + \sigma_l(\eta_n), \tag{13}$$

where $\sigma_l(\eta)$ is the Coulomb phase shift

$$\sigma_l(\eta) = \arg \Gamma(l + 1 + \mathrm{i}\eta). \tag{14}$$

The asymptotic condition (12) corresponds to incoming waves for $I = I_0$ and for a definite orbital angular momentum $l = l_0$ only. Solutions with all possible values of l_0 consistent with given J and I_0 should be constructed.

In practice the solution (12) may be constructed from the arbitrary real solution $g^{(q)}_{(lI_n)J}(r)$ by considering the asymptotic form of the linear combinations

$$\sum_q \beta_q g^{(q)}_{(lI_n)J}(r) \approx \sum_q \beta_q C^{(q)}_{(lI_n)J} \sin[k_nr - \eta_n \log 2k_nr - \tfrac{1}{2}l\pi + \sigma_l(\eta_n) + \delta^{(q)}_{(lI_n)J}], \tag{15}$$

where the real coefficients C and the real phase shifts δ express the information about the scattering process which is contained in the radial functions $g^{(q)}(r)$. Comparison with eq. (12) shows that $g^{(l_0I_0)}(r)$ is obtained if β_q satisfies the equations

$$\sum_q \tfrac{1}{2}\mathrm{i}C^{(q)}_{(lI_n)J}\beta_q \exp\{-\mathrm{i}\delta^{(q)}_{(lI_n)J}\} = \delta_{II_0}\delta_{ll_0}, \tag{16}$$

from which one finds

$$r^J_{lI_n,l_0I_0} = \sqrt{(k_n/k_0)}\,\tfrac{1}{2}\mathrm{i}\sum_q \beta_q C^{(q)}_{(lI_n)J}\exp\{\mathrm{i}\delta^{(q)}_{(lI_n)J}\}. \tag{17}$$

from the functions (12) it is easy to construct the scattering state $|\Psi(\boldsymbol{k}_0)\rangle$. According to (5) the coefficients α_{JN} in eq. (9) must satisfy the equations

$$\sum_{\substack{lI_nJN\\ l_0M_nm}} \frac{\alpha^{l_0}_{JN}}{r}\left[\delta_{I_nI_0}\delta_{ll_0}\exp\{-\mathrm{i}\varphi_{I_0l_0}(r)\} - \sqrt{\frac{k_0}{k_n}}\, r^J_{lI_n,l_0I_0}\exp\{\mathrm{i}\varphi_{I_nl}(r)\}\right]$$

$$\times \langle lmI_nM_n|JN\rangle\, Y_{lm}(\vartheta, \varphi)\,|I_nM_n\rangle$$

$$= \sum_{l_0}(2l_0 + 1)\mathrm{i}^{l_0}\frac{\sin(k_0r - \eta_0 \log 2k_0r - \tfrac{1}{2}\pi l_0)}{k_0r}P_{l_0}(\cos\vartheta)\,|I_0M_0\rangle$$

$$+ \sum_{I_nM_n} f_{I_0M_0\to I_nM_n}(\vartheta, \varphi)\frac{\exp(k_nr - \eta_n \log 2k_nr)}{r}|I_nM_n\rangle, \tag{18}$$

which shows that $\alpha_{JN}^{l_0}$ is given by

$$\alpha_{JN}^{l_0} = \frac{\sqrt{\pi(2l_0+1)}}{k_0}\, \mathrm{i}^{l_0+1} \exp\{\mathrm{i}\sigma_{l_0}(\eta_0)\} \langle l_0 0 I_0 M_0 | JN \rangle . \tag{19}$$

The scattering amplitude can thus be determined by

$$f_{I_0M_0\to I_nM_n}(\vartheta,\varphi) = \frac{\mathrm{i}\sqrt{\pi}}{\sqrt{k_0k_n}} \sum_{l_0lJ} \sqrt{2l_0+1}\, \langle l_0 0 I_0 M_0 | JM_0\rangle \langle lmI_nM_n|JM_0\rangle\, \mathrm{i}^{l_0-l}\, Y_{lm}(\vartheta,\varphi)$$
$$\times\, [\delta_{I_0I_n}\delta_{ll_0} - \exp\{\mathrm{i}(\sigma_{l_0}(\eta_0)+\sigma_l(\eta_n))\} r^J_{lI_n,l_0I_0}] \tag{20}$$

in terms of the quantities $r^J_{lI_n,l_0I_0}$ in eq. (12).

We note that the matrix $r^J_{lI_n,l_0I_0}$ is connected with the S-matrix in the channel spin representation [BLA 52] by

$$r^J_{lI_n,l_0I_0} = \exp\{-\mathrm{i}[\sigma_l(\eta_n)+\sigma_{l_0}(\eta_0)]\} \langle (lI_n)J|\, S\, |(l_0I_0)J\rangle . \tag{21}$$

The T-matrix which was introduced in (III.1.4) is proportional to the scattering amplitude (20), i.e.

$$\langle I_nM_n\boldsymbol{k}_n|\, T\, |I_0M_0\boldsymbol{k}_0\rangle = -\frac{2\pi\hbar^2}{m_0} f_{I_0M_0\to I_nM_n}(\vartheta,\varphi) . \tag{22}$$

§ 2. *Perturbation theory*

As was mentioned earlier, the quantal corrections to the classical results are expected to be especially important for light projectiles, such as protons, deuterons or α-particles. For these projectiles the Coulomb interaction is so weak ($\chi^{(\lambda)} \lesssim 1$) that first- or second-order perturbation theory gives an accurate description of the excitation process. In the perturbation expansion the general theory presented in the preceding section is considerably simplified so that explicit results can be obtained. In this section we consider this limit of the quantal treatment. The results can also be used for heavy ions for those partial waves of large angular momenta where perturbation theory applies.

To obtain the perturbation expansion it is convenient to introduce the Green function belonging to the coupled differential equations (1.10). The Green function belonging to the boundary condition of outgoing waves may be written [GOL 64]

$$G(r,r') = -\frac{1}{k_n} h_l^+(k_nr_>)F_l(k_nr_<)$$
$$= \frac{2}{\pi}\int_0^\infty \frac{F_l(kr)\,F_l(kr')\,\mathrm{d}k}{k_n^2-k^2+\mathrm{i}\varepsilon} , \tag{1}$$

where $r_>$ and $r_<$ are the larger and the smaller, respectively, of the quantities r and r' while

$$h_l^{\pm}(kr) = G_l(kr) \pm \mathrm{i}F_l(kr). \tag{2}$$

The symbols $F_l(kr)$ and $G_l(kr)$ denote the regular and the irregular Coulomb wave functions [MOT 49] which behave asymptotically as

$$\begin{aligned} F_l(kr) &\approx \sin \varphi_l(r), \\ G_l(kr) &\approx \cos \varphi_l(r), \end{aligned} \tag{3}$$

where $\varphi_l(r)$ is given by eq. (1.13).

With the help of the Green function (1) the radial equations (1.10) with the boundary condition (1.12) can be written

$$\begin{aligned} g_{(lI_n)J}(r) &= \frac{2}{\mathrm{i}}\, \delta_{I_nI_0}\delta_{ll_0}F_{l_0}(k_0r) + \frac{2}{\pi}\int_0^\infty \mathrm{d}k \int_0^\infty \mathrm{d}r'\, \frac{F_l(kr')\, F_l(kr)}{k_n^2 - k^2 + \mathrm{i}\varepsilon} \\ &\quad \times \sum_{l'I_m} V^J_{lI_n,l'I_m}(r')\, g_{(l'I_m)J}(r') \\ &= \frac{2}{\mathrm{i}}\, \delta_{I_nI_0}\delta_{ll_0}F_{l_0}(k_0r) - \frac{1}{k_n}\int_0^\infty \mathrm{d}r'\, h_l^+(k_nr_>)F_l(k_nr_<) \\ &\quad \times \sum_{l'I_m} V^J_{lI_n,l'I_m}(r') g_{(l'I_m)J}(r'). \end{aligned} \tag{4}$$

Here and in the following we delete the indices l_0 and I_0 on $g(r)$. By studying the asymptotic form of eq. (4) and comparing with the definition (1.12) one may write the reaction matrix in the form

$$r^J_{lI_n,l_0I_0} = \delta_{I_nI_0}\delta_{ll_0} + \frac{1}{\sqrt{k_0k_n}}\sum_{I_ml'}\int_0^\infty \mathrm{d}r'\, F_l(k_nr')V^J_{lI_n,l'I_m}(r')g_{(l'I_m)J}(r'). \tag{5}$$

Perturbation expansions can be obtained by iteration from the expressions (4) and (5) using as zeroth-order approximation the first term in eq. (4). It is noted that already the zeroth-order approximation for $g(r)$ leads to the first-order perturbation result for r^J.

Let us now introduce the strength parameter χ in the definition of the coupling potential (1.11). One finds

$$\begin{aligned} V^J_{lI_n,l'I_m}(r) &= \sum_\lambda \chi^{(\lambda)}_{I_m\to I_n}(-1)^{I_m+J+\lambda}\sqrt{\frac{(2l+1)(2l'+1)(2I_m+1)}{2\lambda+1}} \\ &\quad \times \frac{(2\lambda+1)!!}{(\lambda-1)!}\begin{pmatrix} l & l' & \lambda \\ 0 & 0 & 0\end{pmatrix}\begin{Bmatrix} J & l' & I_m \\ \lambda & I_n & l\end{Bmatrix}\left(\frac{a_{nm}}{r}\right)^{\lambda+1}\frac{\sqrt{k_nk_m}}{a_{nm}}, \end{aligned} \tag{6}$$

where we have used the symmetrized definition (VI.8.11), i.e.

$$a_{nm} = Z_1Z_2e^2/m_0v_nv_m. \tag{7}$$

Since the main contribution to the integral (5) comes from distances $r \sim 2a$ the last term in (5) is of the order of magnitude of the dominating χ-values.

The perturbation expansion for r^J can be written in the form

$$r^J_{lI_n,l_0I_0} = \delta_{I_nI_0}\delta_{ll_0} + r^J_{lI_n,l_0I_0}(1) + r^J_{lI_n,l_0I_0}(2) + \cdots, \tag{8}$$

where

$$r^J_{lI_n,l_0I_0}(1) = -\mathrm{i}\sum_{\lambda} \chi^{(\lambda)}_{I_0\to I_n} W^{(J,\lambda)}_{lI_n,l_0I_0} I^{\lambda}_{l_0l}(\eta_0, \eta_n). \tag{9}$$

The geometrical factor W is defined by

$$W^{(J,\lambda)}_{lI_n,l_0I_0} = \tfrac{1}{2}\sqrt{2I_0+1}\,\frac{(2\lambda+1)!!}{(\lambda-1)!}\sqrt{\frac{(2l_0+1)(2l+1)}{2\lambda+1}}\,(-1)^{J+I_0+\lambda}$$
$$\times \begin{pmatrix} l_0 & l & \lambda \\ 0 & 0 & 0 \end{pmatrix} \begin{Bmatrix} J & l_0 & I_0 \\ \lambda & I_n & l \end{Bmatrix}, \tag{10}$$

while $I^{\lambda}_{l_0l}(\eta_0, \eta_n)$ denotes the radial matrix element

$$I^{\lambda}_{l_0l}(\eta_0, \eta_n) = 4a^{\lambda}_{0n}\int_0^{\infty} F_{l_0}(k_0r)\frac{1}{r^{\lambda+1}}F_l(k_nr)\,\mathrm{d}r. \tag{11}$$

In the definition (11) we have introduced the factor a^{λ}_{0n} such that the matrix element I is dimensionless and a function of η_0 and η_n only. It should be noted that through this definition the radial integral (11) approaches the classical orbital integral $I_{\lambda\mu}(\vartheta, \xi)$ in the limit of $\eta_0, \eta_n \gg 1$ with a fixed difference $\eta_n - \eta_0 = \xi$ (see § 3 below). These matrix elements $I^{\lambda}_{l_0l}$ can be evaluated in terms of generalized hypergeometric functions of two variables (see [ALD 56]). The result is

$$I^{\lambda}_{l_0l}(\eta_0, \eta_n) = \frac{|\Gamma(l_0+1+\mathrm{i}\eta_0)|\,|\Gamma(l+1+\mathrm{i}\eta_n)|}{(2l_0+1)!\,(2l+1)!}(l_0+l-\lambda+1)!$$
$$\times\, \mathrm{i}^{l_0+l-\lambda+2}x^{l_0+1}(-y)^{l+1}\exp\{-\tfrac{1}{2}\pi(\eta_0+\eta_n)\}(\eta_n-\eta_0)^{\lambda}$$
$$\times\, F_2(l_0+l-\lambda+2, l_0+1+\mathrm{i}\eta_0, l+1-\mathrm{i}\eta_n, 2l_0+2, 2l+2; x, y) \tag{12}$$

where

$$x = \frac{2\eta_n}{\eta_n-\eta_0} \quad \text{and} \quad y = -\frac{2\eta_0}{\eta_n-\eta_0}. \tag{13}$$

Since the Appell function F_2 is usually defined in terms of a series expansion which does not converge in the point (13), which is of interest, it is necessary to study the analytic continuation. We shall not discuss this problem here, but only refer to the literature [ALD 55], [BIE 55]. For the evaluation of $I^{\lambda}_{l_0 l}$ it is of importance that there exists a large number of recursion relations which connect matrix elements with different values of l_0, l and λ. This means that in practice only the matrix elements for the lowest values of l have to be evaluated directly from the formula (12).

For the limiting cases where $\eta_n = \eta_0$, i.e. $\xi = 0$, the radial matrix elements (12) reduce to elementary functions. The results for $\lambda = 1$, 2 and 3 are collected in table 1. For the special case of dipole excitation ($\lambda = 1$) the Appell function F_2 can be written in terms of ordinary hypergeometric functions. One finds the result

$$I^{1}_{l,l+1}(\eta_0, \eta_f)$$
$$= 2\left(\frac{\eta_f - \eta_0}{\eta_f + \eta_0}\right)^{i(\eta_f+\eta_0)} \exp(-\tfrac{1}{2}\pi\xi)\, \frac{(-x_o)^{l+1}}{(2l+2)!}\, |\Gamma(l+1+i\eta_0)|\, |\Gamma(l+2+i\eta_f)|$$
$$\times \Big\{ \eta_0\, {}_2F_1(l+1-i\eta_0, l+1-i\eta_f, 2l+2; x_o)$$
$$- \eta_f \frac{|l+1+i\eta_0|^2}{(2l+2)(2l+3)} (-x_o)\, {}_2F_1(l+2-i\eta_0, l+2-i\eta_f, 2l+4; x_o) \Big\}, \tag{14}$$

TABLE 1

The radial integral $I^{\lambda}_{ll'}(\eta, \eta)$ for $\xi = 0$.

$$I^{1}_{l,l+1}(\eta, \eta) = I^{1}_{l+1,l}(\eta, \eta) = \frac{2\eta}{|l+1+i\eta|}$$

$$I^{2}_{l,l+2}(\eta, \eta) = I^{2}_{l+2,l}(\eta, \eta) = \frac{2}{3}\, \frac{\eta^2}{|l+1+i\eta|\, |l+2+i\eta|}$$

$$I^{2}_{ll}(\eta, \eta) = \frac{2\eta^2}{l(l+1)(2l+1)} [2l+1-\pi\eta+2\eta \operatorname{Im} \psi(l+1+i\eta)]$$

$$I^{3}_{l,l+3}(\eta, \eta) = I^{3}_{l+3,l}(\eta, \eta) = \frac{4}{15}\, \frac{\eta^3}{|l+1+i\eta|\, |l+2+i\eta|\, |l+3+i\eta|}$$

$$I^{3}_{l,l+1}(\eta, \eta) = I^{3}_{l+1,l}(\eta, \eta) = \frac{4\eta^3}{3l(l+1)(l+2)(2l+1)(2l+3)|l+1+i\eta|}$$
$$\times \{3|l+1+i\eta|^2[2l+1-\pi\eta+2\eta \operatorname{Im} \psi(l+1+i\eta)] - l(l+1)(2l+1)\}$$

The integrals which are defined in (2.11) are given for $\lambda = 1$, 2, and 3 as functions of η. It is noted that the imaginary part of the logarithmic derivative ψ of the Γ-function can be expressed by elementary functions through the relation

$$\operatorname{Im} \psi(l+1+i\eta) = \tfrac{1}{2}\pi \coth \pi\eta + 1/2\eta - \eta \textstyle\sum_{n=0}^{l} 1/(n^2+\eta^2).$$

where

$$x_o = -4\eta_0\eta_f/(\eta_f - \eta_0)^2. \tag{15}$$

The analytic continuation of this function in terms of the variable $1/x_o$ is well known and the result may be written

$$\begin{aligned}I^1_{l,l+1}(\eta_0, \eta_f) = \mathrm{Re}\bigg\{&\frac{2\exp(-\frac{1}{2}\pi\xi)}{l+1}\,\Gamma(-\mathrm{i}\xi)\left(\frac{4\eta_0\eta_f}{(\eta_0+\eta_f)^2}\right)^{\mathrm{i}(\eta_0+\eta_f)/2}\left(\frac{4\eta_0\eta_f}{\xi^2}\right)^{-\mathrm{i}\xi/2}\\ &\times\Big[|l+1+\mathrm{i}\eta_f|\eta_0\exp\{\mathrm{i}[\sigma_l(\eta_f)-\sigma_l(\eta_0)]\}\\ &\times {}_2F_1\left(l+1-\mathrm{i}\eta_0, -l-\mathrm{i}\eta_0, 1+\mathrm{i}\xi; \frac{1}{x_o}\right)\\ &-|l+1+\mathrm{i}\eta_0|\eta_f\exp\{\mathrm{i}[\sigma_{l+1}(\eta_f)-\sigma_{l+1}(\eta_0)]\}\\ &\times {}_2F_1\left(l+2-\mathrm{i}\eta_0, -l-1-\mathrm{i}\eta_0, 1+\mathrm{i}\xi; \frac{1}{x_o}\right)\Big]\bigg\}.\end{aligned} \tag{16}$$

From eq. (9) one can easily obtain the scattering amplitude (1.20) or the T-matrix (1.22). Since the summation over J can be performed according to

$$\begin{aligned}&\sum_{j_3}(2j_3+1)\begin{pmatrix} j_1 & j_2 & j_3\\ m_1 & m_2 & m_3\end{pmatrix}\begin{pmatrix} l_1 & l_2 & j_3\\ n_1 & n_2 & m_3\end{pmatrix}\begin{Bmatrix} j_1 & j_2 & j_3\\ l_1 & l_2 & l_3\end{Bmatrix}\\ &= (-1)^{j_1+m_1+l_1+n_1}\begin{pmatrix} j_1 & l_2 & l_3\\ -m_1 & n_2 & m_1-n_2\end{pmatrix}\begin{pmatrix} l_1 & j_2 & l_3\\ -n_1 & m_2 & n_1-m_2\end{pmatrix}\end{aligned} \tag{17}$$

we obtain for the T-matrix

$$\begin{aligned}&\langle \boldsymbol{k}_n I_n M_n|\, T\,|\boldsymbol{k}_0 I_0 M_0\rangle\\ &= \frac{\mathrm{i}\pi\hbar^2}{m_0 k_0}\delta_{I_0I_n}\delta_{M_0M_n}\sum_{l_0}(2l_0+1)[\exp\{\mathrm{i}2\sigma_{l_0}(\eta_0)\}-1]P_{l_0}(\cos\vartheta)\\ &+\frac{\pi^{3/2}\hbar^2}{m_o}\frac{1}{\sqrt{k_0k_n}}\sum_{\lambda l l_0}\chi^{(\lambda)}_{I_0\to I_n}I^{\lambda}_{l_0l}(\eta_0,\eta_n)\exp\{\mathrm{i}[\sigma_{l_0}(\eta_0)+\sigma_l(\eta_n)]\}\\ &\times(2l_0+1)\mathrm{i}^{l_0-l}(-1)^{l+l_0+I_n+M_n}\sqrt{\frac{(2l+1)(2I_0+1)}{2\lambda+1}\frac{(2\lambda+1)!!}{(\lambda-1)!}}\\ &\times\begin{pmatrix} I_n & I_0 & \lambda\\ -M_n & M_0 & -m\end{pmatrix}\begin{pmatrix} l_0 & l & \lambda\\ 0 & -m & m\end{pmatrix}\begin{pmatrix} l_0 & l & \lambda\\ 0 & 0 & 0\end{pmatrix}Y_{lm}(\vartheta,\varphi).\end{aligned} \tag{18}$$

The first term arises from the elastic Coulomb scattering while the second term contains elastic and inelastic scattering due to multipole interaction. From the T-matrix (18) one may evaluate the differential and total cross section as well as the particle parameters for γ-emission, according to the formula of chapter III.

The cases of electric dipole excitation ($\lambda = 1$) is especially simple, since it is possible in this case to give a closed expression for the differential and total cross sections in first-order perturbation theory [SOM 39]. One finds

$$\frac{\mathrm{d}\sigma_{\mathrm{E1}}}{\mathrm{d}\Omega} = \frac{v_{\mathrm{f}}}{v_0} |a_{\mathrm{f0}}|^2 |\chi^{(1)}_{0\to\mathrm{f}}|^2 \cdot \frac{2\pi^2\eta_0\eta_{\mathrm{f}}}{\xi^2} \frac{\exp(2\pi\eta_0)}{(\exp(2\pi\eta_0)-1)(\exp(2\pi\eta_{\mathrm{f}})-1)}$$

$$\times \frac{\mathrm{d}}{\mathrm{d}x}\left\{-x\frac{\mathrm{d}}{\mathrm{d}x}|{}_2F_1(-\mathrm{i}\eta_0, -\mathrm{i}\eta_{\mathrm{f}}, 1; x)|^2\right\}, \tag{19}$$

where F is the hypergeometric function of the variable

$$x = -\frac{4\eta_0\eta_{\mathrm{f}}}{\xi^2}\sin^2(\tfrac{1}{2}\vartheta). \tag{20}$$

Similarly the total dipole cross section is given by

$$\sigma_{\mathrm{E1}} = \frac{v_{\mathrm{f}}}{v_0}|a_{\mathrm{f0}}|^2|\chi^{(1)}_{0\to\mathrm{f}}|^2 \frac{2\pi^3\exp(2\pi\eta_0)}{(\exp(2\pi\eta_0)-1)(\exp(2\pi\eta_{\mathrm{f}})-1)}$$

$$\times\, x_{\mathrm{o}}\frac{\mathrm{d}}{\mathrm{d}x_{\mathrm{o}}}|{}_2F_1(-\mathrm{i}\eta_0, -\mathrm{i}\eta_{\mathrm{f}}, 1; x_{\mathrm{o}})|^2 \tag{21}$$

where x_{o} is given by eq. (15).

In general, the calculation of the cross sections and particle parameters must be based on the T-matrix (18) and the result is therefore obtained as a sum over partial waves. The cross section can be written in a form similar to the symmetrized classical expression (IV.7.4) and one may thus define differential and total quantum mechanical cross-section functions $\mathrm{d}f_{\mathrm{E}\lambda}(\eta_0, \vartheta, \xi)$ and $f_{\mathrm{E}\lambda}(\eta_0, \xi)$ which approach the corresponding classical quantities $\mathrm{d}f_{\mathrm{E}\lambda}(\vartheta, \xi)$ and $f_{\mathrm{E}\lambda}(\xi)$ for $\eta_0 \gg 1$, i.e.

$$\mathrm{d}\sigma_{\mathrm{E}\lambda} = \left(\frac{Z_1 e}{\hbar v_0}\right)^2 a_{0\mathrm{f}}^{-2\lambda+2} B(\mathrm{E}\lambda, I_0 \to I_{\mathrm{f}})\, \mathrm{d}f_{\mathrm{E}\lambda}(\eta_0, \vartheta, \xi). \tag{22}$$

Similarly, one may define a quantum mechanical particle parameter analogous to (IV.4.25) and (IV.4.27).

The explicit expression for all these quantities are simplified somewhat when integrations over the scattering angles are performed, since one may then utilize the orthogonality properties of the spherical harmonics appearing in (18). We shall here quote the result for the total cross section function

$f_{E\lambda}(\eta, \xi)$ and the particle parameter $a_k^{E\lambda}(\eta, \xi)$ defined by (IV.4.27). One finds [ALD 56]

$$f_{E\lambda}(\eta_0, \xi) = \frac{4\pi^2}{\eta_0\eta_f} \frac{1}{(2\lambda+1)^2} \sum_{l_0 l_f} (2l_0+1)(2l_f+1) \begin{pmatrix} l_0 & l_f & \lambda \\ 0 & 0 & 0 \end{pmatrix}^2 |I^{\lambda}_{l_0 l_f}(\eta_0, \eta_f)|^2, \tag{23}$$

and

$$a_k^{E\lambda}(\eta_0, \xi) = b_k^{(\lambda)}(\eta_0, \xi)/b_0^{(\lambda)}(\eta_0, \xi), \tag{24}$$

where

$$\begin{aligned} b_k^{(\lambda)}(\eta_0, \xi) = \begin{pmatrix} \lambda & \lambda & k \\ 1 & -1 & 0 \end{pmatrix}^{-1} & \sum_{l_0 l_0' l_f} (-1)^{l_f+1}(2l_0+1)(2l_0'+1)(2l_f+1) \\ & \times \mathrm{i}^{l_0-l_0'} \exp\{\mathrm{i}[\sigma_{l_0}(\eta_0) - \sigma_{l_0'}(\eta_0)]\} \begin{Bmatrix} \lambda & \lambda & k \\ l_0 & l_0' & l_f \end{Bmatrix} \begin{pmatrix} \lambda & l_0 & l_f \\ 0 & 0 & 0 \end{pmatrix} \\ & \times \begin{pmatrix} \lambda & l_0' & l_f \\ 0 & 0 & 0 \end{pmatrix} \begin{pmatrix} l_0 & l_0' & k \\ 0 & 0 & 0 \end{pmatrix} I^{\lambda}_{l_0 l_f}(\eta_0, \eta_f) I^{\lambda}_{l_0' l_f}(\eta_0, \eta_f). \end{aligned} \tag{25}$$

Some numerical results for $\lambda = 1$ and 2 are given in § 4 below.

The quantal corrections to the second-order Coulomb excitation are generally expected to be somewhat larger than the corrections to the first-order theory, because the total cross section will contain interferences between different matrix elements.

The formal result for the r-matrix to second order is obtained by evaluating $g_{(lI_n)J}(r)$ to first order, inserting the result into eq. (5). One obtains [ALD 72] with the notation (8) and with $r^J_{lI_n, l_0 I_0}(1)$ given by (9)

$$\begin{aligned} r^J_{lI_n, l_0 I_0}(2) = -\tfrac{1}{2} \sum_{I_z \lambda \lambda' l'} \chi^{(\lambda)}_{0\to z} \chi^{(\lambda')}_{z\to n} W^{(J\lambda)}_{l'I_z, l_0 I_0} W^{(J\lambda')}_{lI_n, l'I_z} \Big\{ & I^{\lambda}_{l_0 l'}(\eta_0, \eta_z) I^{\lambda'}_{l'l}(\eta_z, \eta_n) \\ & + \frac{2\mathrm{i}}{\pi} \mathscr{P} \int_0^{\infty} \frac{I^{\lambda}_{l_0 l'}(\eta_0, \eta) I^{\lambda'}_{l'l}(\eta, \eta_n)}{\eta^2 - \eta_z^2} \frac{\eta_z^{\lambda+\lambda'+1}}{\eta^{\lambda+\lambda'}} \mathrm{d}\eta \Big\}. \end{aligned} \tag{26}$$

We have changed the integration over k in eq. (4) into an integral over η through the relation

$$\eta \cdot k = \text{const.} \tag{27}$$

Furthermore, we have separated the integral over η into real and imaginary parts corresponding to a pole contribution and the principal part integral, so that the result appears in a form similar to the one used in the classical theory (see eq. (V.1.11)).

It is noted that the first term can be written as half the product of the first-order r-matrices corresponding to the transitions $0 \to z$ and $z \to n$ summed over intermediate states in complete analogy to the classical result (V.1.8). In contrast to the classical result, however, the principal part integral does not vanish in the limit of the sudden approximation, i.e. $k_0 = k_z = k_n$.

§ 3. *The WKB approximation*

In this section we shall study the transition from the quantum mechanical to the classical description of Coulomb excitation. This transition is especially simple in the first-order perturbation theory, where the use of the WKB approximation for the radial wave functions rather directly leads to a result which can be compared with the classical theory ([ALD 54], [BRE 54], [BEN 56], [BRE 55], [LAZ 55]).

The comparison between the two treatments is somewhat masked by the fact that the quantum mechanical expressions are based on an expansion in partial waves. For a given deflection angle the scattering amplitude (1.20) receives contributions from many angular momenta both in the incoming and outgoing channels. In the classical limit, however, there is a unique connection between the scattering angle ϑ and the angular momentum $\boldsymbol{l}$, (cf. (II.9.7)), i.e.

$$|\boldsymbol{l}| = am_0 v \cot(\tfrac{1}{2}\vartheta), \tag{1}$$

or since $|\boldsymbol{l}| \approx (l + \frac{1}{2})\hbar$

$$l + \tfrac{1}{2} = \eta \cot(\tfrac{1}{2}\vartheta). \tag{2}$$

In a quantum mechanical formulation this must mean that for large values of η the terms in the sum (1.20) over l and l_0 interfere destructively except for l values in the neighbourhood of the value (2). To prove this in detail we consider the asymptotic form of the scattering amplitude (1.20) for large values of η and l where

$$Y_{lm}(\vartheta, \varphi) \sim \frac{1}{\pi} \mathrm{e}^{\mathrm{i}m\varphi}(\sin\vartheta)^{-1/2} \cos\{(l + \tfrac{1}{2})\vartheta + \tfrac{1}{2}(m - \tfrac{1}{2})\pi\}, \tag{3}$$

and

$$\exp\{\mathrm{i}\sigma_l(\eta)\} \sim \exp \mathrm{i}\left\{(l + \tfrac{1}{2}) \arctan \frac{\eta}{l + \frac{1}{2}} - \eta + \tfrac{1}{2}\eta \log[\eta^2 + (l + \tfrac{1}{2})^2]\right\} \tag{4}$$

neglecting terms of order $1/l$ and $1/\eta$.

In order to study the limiting process of (1.20) in detail it is necessary to investigate the asymptotic behaviour of the Clebsch-Gordan coefficient and r-matrix for large values of l. We shall here only note that both are smooth functions of l. Since the phases (3) and (4) vary rapidly as functions of l, the partial wave contribution to the scattering amplitude will cancel, except from the region of l values where the phases proportional to $(l + \frac{1}{2})$ cancel. This condition leads to the relation (2), where one should use average values of l and η corresponding to initial and final states. It follows from this argument that the quantum mechanical scattering amplitude for large values of η depends essentially only on the r-matrix for those l values which satisfy (2). Since the corresponding classical result is proportional to the excitation amplitude a_{IM,I_0M_0} there must exist a simple relation between these two quantities. In fact we shall prove

$$r^J_{lI,l_0I_0} \approx a_{IM,I_0M_0}(-1)^{l-l_0}, \tag{5}$$

where a_{IM,I_0M_0} is to be evaluated in the coordinate system A with a scattering angle given by (2) and with

$$M_0 = J - l_0$$

and (6)

$$M = J - l.$$

Let us first study the relations (5) and (6) in first-order perturbation theory where there exists the following simple relation

$$I^{\lambda}_{l_0l_f}(\eta_0, \eta_f) \approx I_{\lambda,l_f-l_0}(\vartheta, \xi) \tag{7}$$

between the radial integral (2.11) and the Coulomb excitation function (IV.2.7) for $\eta \gg 1$.

Qualitatively this relation can be understood for $l_0 = l_f$ in the limit $\xi \ll 1$ by noting that the radial matrix element is of the type

$$I^{\lambda}_{ll}(\eta, \eta) \approx 4a^{\lambda} \int_0^{\infty} |F_l(kr)|^2 \frac{1}{r^{\lambda+1}} \, \mathrm{d}r. \tag{8}$$

The radial density $|F_l(kr)|^2$ is approximately half of the relative density in phase space $v/\dot{r}$ outside the classical turning point r_0, while it is zero inside the region $r < r_0$. We thus obtain

$$I^{\lambda}_{ll}(\eta, \eta) \approx 2a^{\lambda}v \int_{r_0}^{\infty} \frac{\mathrm{d}r}{\dot{r}} \frac{1}{r^{\lambda+1}}. \tag{9}$$

Inserting the parametrization (II.9.10) for $r(t)$ it is seen that the right-hand side of eq. (9) is identical to the Coulomb excitation function $I_{\lambda 0}(\vartheta, 0)$ defined in eq. (IV.2.7).

The more accurate proof of the relation (7) may be based on the WKB approximation*. Thus we insert in the radial integral

$$I^{\lambda}_{l_0 l_f}(\eta_0, \eta_f) = 4a^{\lambda}_{0f} \int_0^{\infty} F_{l_0}(k_0 r) \frac{1}{r^{\lambda+1}} F_{l_f}(k_f r)\, dr \tag{10}$$

the approximate wave function

$$F_l(kr) \approx [f(r)/k^2]^{-1/4} \sin \varphi, \tag{11}$$

where

$$\varphi = \tfrac{1}{4}\pi + \int_{r_0}^{r} \sqrt{f(r)}\, dr, \tag{12}$$

and

$$f(r) = k^2 - 2\eta k/r - l(l+1)/r^2. \tag{13}$$

The expression (11) holds outside the classical turning point r_o defined by $f(r_o) = 0$. The contribution to the radial matrix element (10) from the region $r < r_o$ is of lower order in η and is neglected in the present approximation.

Inserting the expression (11) in (10) one finds that the radial matrix element involves two terms, the first containing the sum of the phases φ_0 and φ_f of the initial and final wave functions and the second containing the difference. The first is of similar order of magnitude as the contribution inside the turning point and shall therefore be neglected. In the second term the phase difference may be expanded as follows

$$\varphi_0 - \varphi_f \approx \Big\{ (k_0 - k_f) \int_{r_0}^{r} [f(r)]^{-1/2} k \, dr - [l_0(l_0+1) - l_f(l_f+1)] \int_{r_0}^{r} [f(r)]^{-1/2} \frac{dr}{2r^2} \Big\}, \tag{14}$$

where we have used the relation $k_0\eta_0 = k_f\eta_f$. In the functions $f(r)$ the parameters k, η and l refer to average values for initial and final states. The differences between the turning points for the initial and final states have been neglected since the contributions are again of higher order.

Evaluating the integrals in (14) and introducing the substitution

$$kr = [\eta^2 + (l + \tfrac{1}{2})^2]^{1/2} \cosh w + \eta \tag{15}$$

* The relation (7) can also be obtained by expressing the radial integral (2.12) in a suitable form and performing the limit $|l + i\eta| \to \infty$ for fixed ξ (cf. [BIE 55]).

one obtains

$$\varphi_0 - \varphi_f \approx \xi[\varepsilon \sinh w + w] + \mu \arccos \frac{\varepsilon + \cosh w}{1 + \varepsilon \cosh w}, \tag{16}$$

where

$$\xi = \eta_f - \eta_0, \tag{17}$$

while

$$\varepsilon = \frac{[\eta^2 + (l + \frac{1}{2})^2]^{1/2}}{\eta} = \left\{1 + \frac{(l_0 + \frac{1}{2}(\mu + 1))^2}{\eta^2}\right\}^{1/2} \tag{18}$$

and

$$\mu = l_f - l_0. \tag{19}$$

Inserting this result in (10) we have proved eq. (7), i.e.

$$I^{\lambda}_{l_0 l_f}(\eta_0, \eta_f) \approx I_{\lambda\mu}(\vartheta, \xi). \tag{20}$$

It is difficult to estimate the accuracy of the approximate relation (20) since the WKB wave functions deviate considerably from the true wave function, especially in the neighbourhood of the classical turning point. In general one must expect that the deviations are of the order of magnitude $1/\eta$. It is noted, however, that there is a freedom in the choice of the averaging in the parameters η and l which may also lead to inaccuracies of the order $1/\eta$. From a numerical comparison of the exact radial integrals and the classical Coulomb excitation functions it seems that by the choice (15–19) the first-order terms vanish such that the relation (20) holds up to terms of order $1/\eta^2$ (c.f. discussion in [BRE 59] p. 540 ff.).

To obtain the relation (5) in first-order perturbation theory we must study the geometrical factors (2.10) in the limit of large values of l_0, l and J. Inserting the following asymptotic formulae for the $3j$- and $6j$-symbols

$$\begin{pmatrix} l_1 & l_2 & \lambda \\ m_1 & m_2 & \mu \end{pmatrix} \approx \frac{(-1)^{l_1 - \lambda - m_2}}{\sqrt{2l_1 + 1}} D^{\lambda}_{l_1 - l_2, \mu}(0, \vartheta, 0), \tag{21}$$

with

$$\cos\vartheta = m_2/(l_2 + \tfrac{1}{2}) \qquad (0 \leqslant \vartheta \leqslant \pi) \tag{22}$$

and

$$\begin{Bmatrix} \lambda_1 & \lambda_2 & \lambda_3 \\ l_1 & l_2 & l_3 \end{Bmatrix} \approx \frac{(-1)^{2l_1 - 2\lambda_1}}{\sqrt{2l_1 + 1}} \begin{pmatrix} \lambda_1 & \lambda_2 & \lambda_3 \\ l_3 - l_2 & l_1 - l_3 & l_2 - l_1 \end{pmatrix}, \tag{23}$$

we find

$$W^{(J,\lambda)}_{lI_n,l_0I_0} = \sqrt{\pi}(2I_0+1)^{1/2}\frac{(2\lambda-1)!!}{(\lambda-1)!}(-1)^{I_0-M_0+\mu}$$

$$\times\, Y_{\lambda\mu}(\tfrac{1}{2}\pi, 0)\begin{pmatrix} I_0 & \lambda & I_n \\ -M_0 & \mu & M \end{pmatrix}, \tag{24}$$

where M_0 and M are defined by (6). Introducing this result in eq. (2.9) together with the result (20) we obtain the right-hand side of eq. (IV.3.7) with $R^{A}_{\lambda\mu}$ given by eq. (IV.2.6) which proves (5) in first-order perturbation theory.

From this result it is possible to show that the relation (5) holds in general, because one may reduce the evaluation of the r-matrix in higher-order perturbation theory to integrals of products of first-order matrix elements (see e.g. eq. (2.26)). We shall, however, prove the relation (5) directly by showing that it is possible to formulate the radial equation (1.10) in such a way that in the limit $\eta \gg 1$ it becomes equivalent to the coupled time dependent equations (VI.1.2) for the excitation amplitude.

To do this we introduce the functions

$$g^{\pm}_{(lI)J}(r) = a^{J\pm}_{lI}(r)\, h^{\pm}_{l}(k_I r)\sqrt{k_0/k_I}, \tag{25}$$

where $h^{\pm}(kr)$ is defined by eq. (2.2). We require that the functions $g^{\pm}_{(lI)J}(r)$ are solutions of the radial equations (1.10) and that the radial wave function $g_{(lI)J}(r)$ that has the asymptotic behaviour (1.12) is given by

$$g^{l_0I_0}_{(lI)J}(r) = g^{-}_{(lI)J}(r) - g^{+}_{(lI)J}(r). \tag{26}$$

From these definitions it is seen that we may choose the solution $g^{-}(r)$ such that at infinity it only contains incoming waves in the entrance channel and no outgoing waves in any channel. We normalize $g^{-}(r)$ such that the coefficient a^{J-}_{lI} has the asymptotic form

$$a^{J-}_{lI,l_0I_0}(\infty) = \delta_{II_0}\delta_{ll_0}. \tag{27}$$

From the asymptotic condition (1.12) and the definitions (25) and (26) it then follows

$$a^{J+}_{lI,l_0I_0}(\infty) = r^{J}_{lI,l_0I_0}. \tag{28}$$

Furthermore, the regularity of (26) at $r=0$ leads to the following condition

$$\lim_{r\to 0}\{a^{J+}_{lI,l_0I_0}(r) - a^{J-}_{lI,l_0I_0}(r)\} = 0. \tag{29}$$

It is convenient to formulate the equations for a^+ and a^- in terms of integral equations of the type (2.4). Writing this equation in terms of an integral with the lower limit r utilizing (2.5) it is seen that the equation can be split in an equation for g^- and an equation for g^+, which, in terms of the functions $a^\pm$, take the form

$$a^{J-}_{lI,l_0I_0}(r) = \delta_{II_0}\delta_{ll_0} - \frac{1}{2\mathrm{i}\sqrt{k_I}\,h_l^-(k_I r)} \times \int_r^\infty \mathrm{d}r'\{h_l^+(k_I r)\,h_l^-(k_I r') - h_l^-(k_I r)\,h_l^+(k_I r')\} \times \sum_{l'I'} \frac{1}{\sqrt{k_{I'}}} V^J_{lI,l'I'}(r')\,h_{l'}^-(k_{I'}r')\,a^{J-}_{l'I',l_0I_0}(r'), \tag{30}$$

and

$$a^{J+}_{lI,l_0I_0}(r) = r^J_{lI,l_0I_0} + \frac{1}{2\mathrm{i}\sqrt{k_I}\,h_l^+(k_I r)} \times \int_r^\infty \mathrm{d}r'\{h_l^-(k_I r)\,h_l^+(k_I r') - h_l^+(k_I r)\,h_l^-(k_I r')\} \times \sum_{l'I'} \frac{1}{\sqrt{k_{I'}}} V^J_{lI,l'I'}(r')\,h_{l'}^+(k_{I'}r')\,a^{J+}_{l'I',l_0I_0}(r'). \tag{31}$$

In equation (31) the r-matrix should be determined in such a way that the condition (29) is fulfilled.

As a rather trivial application of this method we may reconsider the first-order perturbation theory in which one should substitute for $a^\pm(r')$ under the integral in (30) and (31) the zeroth-order approximation,

$$a^{J-(0)}_{lI,l_0I_0}(r) = a^{J+(0)}_{lI,l_0I_0}(r) = \delta_{II_0}\delta_{ll_0}. \tag{32}$$

Equating the resulting first-order functions $a^{\pm(1)}(r)$ at $r = 0$ and utilizing the fact that

$$\lim_{r\to 0} \frac{h_l^+(kr)}{h_l^-(kr)} = 1 \tag{33}$$

one obtains an equation for r^J_{lI,l_0I_0} that is identical to the first-order solution of (2.5).

We shall now show that the functions $a^\pm(r)$ are the quantum mechanical analogues of the excitation amplitudes as functions of $r(t)$. To see this we notice first that the first terms in the integrals (30) and (31) contain the product $h_l^\pm(k_I r')\,h_{l'}^\mp(k_{I'}r')$ which oscillates very rapidly for large values of η.

In the classical limit these terms may thus be neglected, and the integral equations reduce to

$$a^{J-}_{lI,l_0I_0}(r) = \delta_{II_0}\delta_{ll_0} + \frac{1}{2\mathrm{i}\sqrt{k_I}}\int_r^\infty \mathrm{d}r'\, h_l^+(k_I r')$$

$$\times \sum_{l'I'} \frac{1}{\sqrt{k_{I'}}} V^J_{lI,l'I'}(r')\, h_{l'}^-(k_{I'}r')\, a^{J-}_{l'I',l_0I_0}(r'),$$

$$a^{J+}_{lI,l_0I_0}(r) = r^J_{lI,l_0I_0} - \frac{1}{2\mathrm{i}\sqrt{k_I}}\int_r^\infty \mathrm{d}r'\, h_l^-(k_I r')$$

$$\times \sum_{l'I'} \frac{1}{\sqrt{k_{I'}}} V^J_{lI,l'I'}(r')\, h_{l'}^+(k_{I'}r')\, a^{J+}_{l'I',l_0I_0}(r'). \tag{34}$$

These equations are equivalent to two sets of first-order differential equations, i.e.

$$\frac{\mathrm{d}}{\mathrm{d}r}\, a^{J\pm}_{lI,l_0I_0}(r) = \pm\frac{1}{2\mathrm{i}} \sum_{l'I'} \frac{1}{\sqrt{k_I k_{I'}}}\, h_l^\mp(k_I r) V^J_{lI,l'I'}(r)\, h_{l'}^\pm(k_{I'}r)\, a^{J\pm}_{l'I',l_0I_0}(r), \tag{35}$$

with the initial condition (27) and the connecting formulae (29.) To show that these equations are equivalent to the coupled equations (VI.1.2) we insert the WKB approximation for the radial function $h_l^\pm(r)$ and neglect the contribution inside the classical turning points. Thus, according to eqs. (11)–(13) we write

$$h_l^\pm(k_I r) = [f(r)/k_I^2]^{-1/4} \exp(\pm \mathrm{i}\varphi_{Il}). \tag{36}$$

In evaluating the product of the two functions h^+ and h^- in eq. (35) one introduces the parametrization (15) leading to the result (16) for the phase difference $\varphi_{Il}(r) - \varphi_{I'l'}(r)$. The errors in this substitution are of the order of $1/\eta^2$ or $1/l^2$. We shall furthermore assume that we may choose the same parameter w for all terms in the sum in eq. (35). This actually implies that one uses different values of r for each term in such a way that $w = 0$ corresponds to the classical turning point for all terms. This approximation is less serious than it seems offhand, because, as we shall see below, the important values of l and l' are correlated with the η values in such a way that the excentricity ε defined by eq. (18) is approximately the same for each pair of states.

Finally, we shall assume that the matching condition (29) is already fulfilled at the classical turning points, i.e. for $w = 0$, and we introduce the definition

$$a^J_{lI,l_0I_0}(w) = \begin{cases} a^{J-}_{lI,l_0I_0}(r(w)), & \text{for } w < 0 \\ a^{J+}_{lI,l_0I_0}(r(w)), & \text{for } w > 0. \end{cases} \tag{37}$$

In this way the two equations (35) can be combined into the equation

$$\frac{\mathrm{d}}{\mathrm{d}w}(a^J_{lI,l_0I_0}(w)) = -\mathrm{i}\sum_{\lambda l'I'} \chi^{(\lambda)}_{I'\to I} W^{(J,\lambda)}_{lI,l'I'} \exp\{\mathrm{i}\xi_{II'}(\varepsilon \sinh w + w)\} \times \frac{[\cosh w + \varepsilon + \mathrm{i}\sqrt{\varepsilon^2 - 1}\sinh w]^{l-l'}}{(\varepsilon \cosh w + 1)^{\lambda + l - l'}} a^J_{l'I',l_0I_0}(w), \tag{38}$$

with the initial condition

$$a^J_{lI,l_0I_0}(-\infty) = \delta_{II_0}\delta_{ll_0}. \tag{39}$$

We have introduced the notation (2.10) and the symmetrized parameters (17) and (18). Introducing finally the asymptotic expressions (21) and (23) for the 3-*j* and 6-*j* symbols in W one obtains directly the coupled equations (VI.1.2) for the classical excitation amplitudes if one makes the identification

$$a^J_{lI,l_0I_0}(w) = a^{\mathrm{A}}_{IM,I_0M_0}(w)(-1)^{l-l_0}, \tag{40}$$

where the classical excitation amplitude is evaluated in coordinate system A. The connection between J, l, l_0, M and M_0 is given by eq. (6). For $w = \infty$ this offers the final proof of the relation (5).

Finally, we investigate the scattering amplitude in the classical limit in detail in order to prove the connection with the classical excitation amplitudes. We thus insert in the expression (1.20) for the scattering amplitude the asymptotic expressions (3) and (4) for the spherical harmonics and the Coulomb phase shift. Furthermore, we replace the Clebsch-Gordan coefficients with their asymptotic form (21). We then get

$$\begin{aligned} f_{I_0M_0\to I_nM_n}(\vartheta,\varphi) = {} & \frac{\mathrm{i}}{\sqrt{\pi k_0 k_n}} \mathrm{e}^{\mathrm{i}m\varphi}(\sin\vartheta)^{-1/2} \sum_{l_0 l J} \mathrm{i}^{l_0 - l}\sqrt{2l_0 + 1} \\ & \times D^{I_0}_{M_0 J - l_0}(0, \tfrac{1}{2}\pi, 0) D^{I_n}_{M_n J - l}(0, \tfrac{1}{2}\pi, 0)(-1)^{l - l_0} \\ & \times [\delta_{I_nI_0}\delta_{ll_0} - \exp\{\mathrm{i}[\sigma_l(\eta_n) + \sigma_{l_0}(\eta_0)]\} a^{\mathrm{A}}_{IJ-l, I_0J-l_0}] \\ & \times \cos\{(l + \tfrac{1}{2})\vartheta + (m - \tfrac{1}{2})\tfrac{1}{2}\pi\}, \end{aligned} \tag{41}$$

where it may be noted that $\sigma_l(\eta_n)$ according to eq. (4) can be written

$$\sigma_l(\eta_n) = \int_0^{l+1/2} \operatorname{arccot}(l'/\eta_n)\,\mathrm{d}l' + \eta_n \log \eta_n. \tag{42}$$

As was mentioned at the beginning of this section the summation in (41) can be evaluated approximately by the method of stationary phases. In fact, the summation rule may, in general, be written

$$\sum_l A(l)\exp\{\mathrm{i}B(l)\} \approx A(\bar{l})\sqrt{\frac{\pi}{|\beta|}} \exp\left\{\mathrm{i}\left[B(\bar{l}) + \tfrac{1}{4}\pi\frac{\beta}{|\beta|}\right]\right\}, \tag{43}$$

where $\bar{l}$ is determined by the condition of a stationary phase (see e.g. [MOT 49] p. 123)

$$\mathrm{d}B(\bar{l})/\mathrm{d}l = 0 \tag{44}$$

while the quantity β is defined by

$$\beta = \tfrac{1}{2}\mathrm{d}^2B(\bar{l})/\mathrm{d}l^2. \tag{45}$$

Since the scattering angle is defined to be positive, only one of the exponential functions in the expression $\cos[(l+\frac{1}{2})\vartheta + \frac{1}{2}(m-\frac{1}{2})\pi]$ can fulfill the condition (44) of a stationary phase. For this term one thus obtains

$$B(l) = \int_0^{l_0+1/2} \operatorname{arccot}\frac{l'}{\eta_0}\,\mathrm{d}l' + \int_0^{l+1/2} \operatorname{arccot}\frac{l'}{\eta_n}\,\mathrm{d}l'$$
$$+\ \eta_0 \log \eta_0 + \eta_n \log \eta_n - (l+\tfrac{1}{2})\vartheta - \tfrac{1}{2}(m-\tfrac{1}{2})\pi. \tag{46}$$

The condition (44) leads to the relation

$$\bar{\eta}/(\bar{l}+\tfrac{1}{2}) = \tan(\tfrac{1}{2}\vartheta), \tag{47}$$

where $\bar{\eta}$ and $\bar{l}$ are averages of η_0 and η_n or l_0 and l, respectively. Utilizing the expression (43) we thus obtain

$$f_{I_0M_0\to I_nM_n}(\vartheta,\varphi) = -\tfrac{1}{2}a_{0n}\,\frac{1}{\sin^2(\frac{1}{2}\vartheta)}\exp\{\mathrm{i}[-\bar{\eta}\log\sin^2(\tfrac{1}{2}\vartheta) + 2\sigma_0(\bar{\eta})]\}$$
$$\times \sum_{MM'} D^{I_0*}_{M'M_0}(\tfrac{1}{2}(\vartheta+\pi),\tfrac{1}{2}\pi,-\varphi-\tfrac{1}{2}\pi)\,D^{I_n}_{MM_n}(\tfrac{1}{2}(\vartheta+\pi),\tfrac{1}{2}\pi,-\varphi-\tfrac{1}{2}\pi)\,a^{\mathrm{A}}_{IM,I_0M'}, \tag{48}$$

where we have used the abbreviations

$$M = J - l, \qquad M' = J - l_0. \tag{49}$$

It is seen that the Eulerian angles in the D-functions correspond to a rotation from the coordinate system A in which the classical amplitudes $a^{\mathrm{A}}_{IM,I_0M'}$ are defined to a coordinate system in which the z-axis is along the direction of the incoming projectile and in which the scattering amplitude $f_{I_0M_0\to I_nM_n}$ is defined.

Introducing the excitation amplitudes a^{C} in the coordinate system C (see e.g. eq. (III.3.4)) it is seen that we may write eq. (48) in the form

$$f^{\mathrm{C}}_{I_0M_0\to I_nM_n} = -\tfrac{1}{2}a_{0n}\,\frac{1}{\sin^2(\frac{1}{2}\vartheta)}\exp\{\mathrm{i}[-\bar{\eta}\log\sin^2(\tfrac{1}{2}\vartheta) + 2\sigma_0(\bar{\eta})]\}a^{\mathrm{C}}_{I_nM_n,I_0M_0}. \tag{50}$$

If one utilizes the connection (1.22) between the T-matrix and the scattering amplitude we obtain the formula quoted in eq. (III.1.5) which obviously holds independently of the choice of coordinate system.

§ 4. *Discussion of quantum mechanical corrections*

In this section we shall discuss the results that we have obtained in the preceding sections, and study especially the accuracy of the semiclassical approximation.

Through the WKB treatment we proved in the last section the validity of the symmetrization procedure which was derived in a heuristic way earlier in § VI.8. Besides offering an elegant way of bringing contact between the classical and the quantal treatments, the WKB approximation provides a method for evaluating Coulomb excitation amplitudes in situations where the quantal corrections are small. It is also possible to combine the various methods and thereby obtain new ways of evaluating Coulomb excitation more accurately than it is possible in the simple classical theory. Thus one method would be to evaluate the r-matrix elements by the connection formula (3.5) to the classical excitation amplitudes and insert these numbers in the exact quantum mechanical formulae for the excitation amplitudes, where summations are performed over all angular momenta.

As was stated earlier it is quite difficult to estimate analytically the accuracy of the WKB approximation as well as of the classical treatment. Generally, one would expect both methods to deviate from the exact calculations by terms of the order $1/\eta$.

This general rule has been verified numerically in a number of cases [ALD 69]. In the first-order perturbation theory, where explicit formulae can be given, one can study the accuracy of the classical approximation in more detail. A simple comparison of the radial matrix elements for $\xi = 0$ (see table 1) with the Coulomb excitation functions for $\xi = 0$ (see table H.1), shows that

$$I_{ll'}^{\lambda}(\eta, \eta) = I_{\lambda, l'-l}(\vartheta, 0)[1 + o(1/\eta^2)], \tag{1}$$

where ϑ is defined by

$$(l + l' + 1)/2\eta = \cot(\tfrac{1}{2}\vartheta). \tag{2}$$

In fact, it turns out to be a general rule that by the choice of the relation (3.47) for the connection of ϑ with l, l', η_0 and η_f the first-order correction terms $1/\eta$ vanish. For $\xi \neq 0$ this can be proved explicitly for the case of dipole excitation by performing an asymptotic expansion of eq. (2.16) for $\eta_0 \gg 1$ with fixed ξ. One obtains the result (IV.2.8) with corrections of the order $1/\eta^2$. This interesting observation which was discussed also in connection with the WKB approximation in the previous section gives rise to an efficient approximation method by which the quantum mechanical expressions of § 2, e.g. (2.22)–(2.25) are used, but the radial matrix elements are substituted by

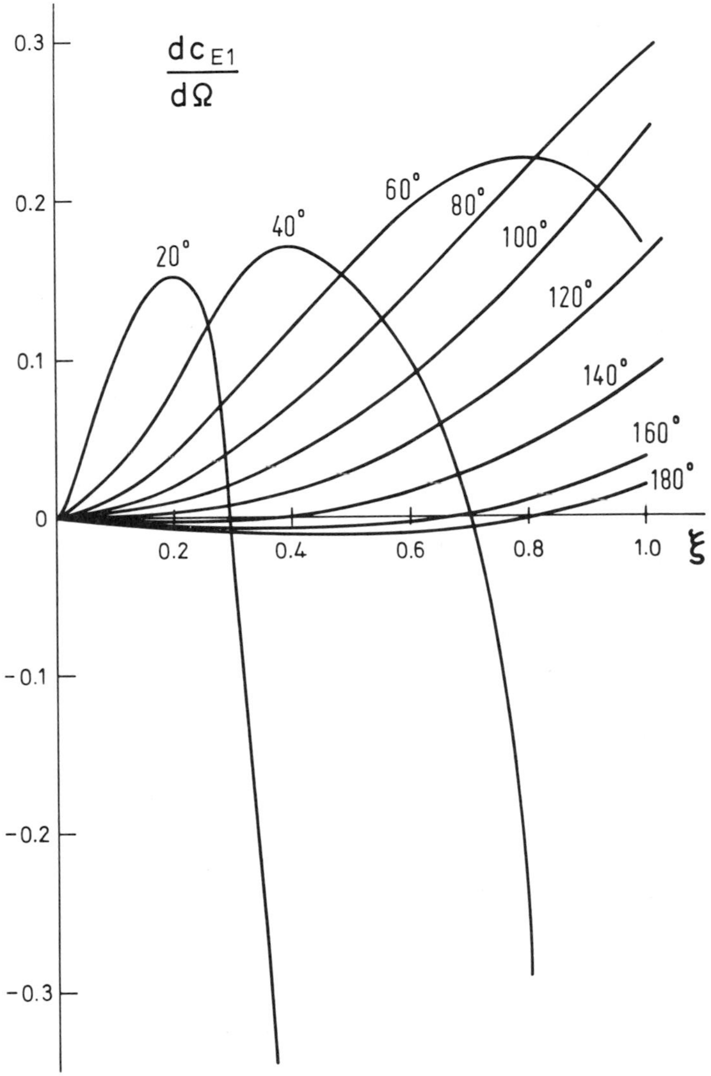

Fig. 1. The quantum mechanical corrections to the differential dipole cross section in first-order Coulomb excitation. The coefficient $dc_{E1}/d\Omega$ defined in eq. (4.3) has been plotted as a function of ξ for a few values of ϑ indicated on the curves.

the classical functions according to eq. (3.20). In this way one obtains in first-order perturbation theory results which are correct to terms of the order $1/\eta^2$.

This method is, however, of little importance for the differential and total cross section where it turns out that even the classical limit is correct to terms of order $1/\eta^2$. For the differential cross section one thus finds

$$\frac{\mathrm{d}f_{\mathrm{E}\lambda}(\eta, \vartheta, \xi)}{\mathrm{d}\Omega} = \frac{\mathrm{d}f_{\mathrm{E}\lambda}(\vartheta, \xi)}{\mathrm{d}\Omega} - \frac{1}{\bar{\eta}^2}\frac{\mathrm{d}c_{\mathrm{E}\lambda}}{\mathrm{d}\Omega} \tag{3}$$

where the first term is defined in eq. (IV.4.5). The quantity $\mathrm{d}c_{\mathrm{E}\lambda}/\mathrm{d}\Omega$ is illustrated in figs. 1 and 2 for $\lambda = 1$ and 2 as a function of ξ for different ϑ.

For the total cross sections one can easily prove that the accuracy of the classical expression is again of order $1/\eta^2$ by considering the expression (2.23) in the limit of large values of η and l. The results may be written analogous to

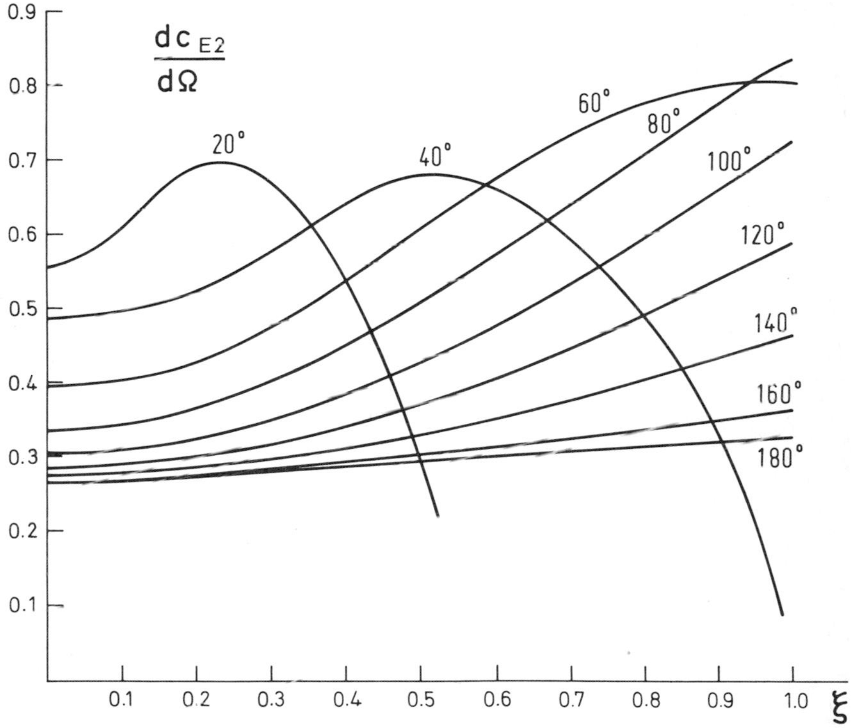

Fig. 2. The quantum mechanical corrections to the differential quadrupole cross section in first-order Coulomb excitation. The coefficient $\mathrm{d}c_{\mathrm{E}2}/\mathrm{d}\Omega$ defined in eq. (4.3) has been plotted as a function of ξ for a few values of ϑ indicated on the curves.

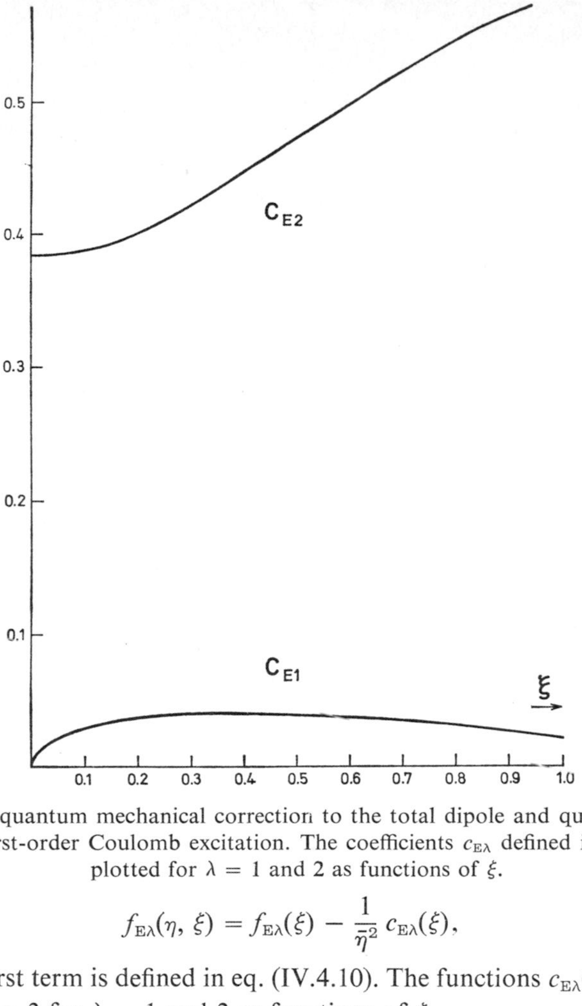

Fig. 3. The quantum mechanical correction to the total dipole and quadrupole cross sections in first-order Coulomb excitation. The coefficients $c_{E\lambda}$ defined in eq. (4.4) are plotted for $\lambda = 1$ and 2 as functions of ξ.

(3), i.e.

$$f_{E\lambda}(\eta, \xi) = f_{E\lambda}(\xi) - \frac{1}{\bar{\eta}^2} c_{E\lambda}(\xi), \tag{4}$$

where the first term is defined in eq. (IV.4.10). The functions $c_{E\lambda}(\xi)$ have been plotted in fig. 3 for $\lambda = 1$ and 2 as functions of ξ.

For the particle parameters it turns out that the classical limit cannot be performed without introducing errors of the order $1/\eta$. This can be seen explicitly from expression (2.25) for the integrated particle parameters describing the angular distribution of the deexcitation γ-rays. These quantities have been evaluated numerically in [ALD 56] and the results can be represented in the following form

$$a_k^{E\lambda}(\eta_0, \xi) = a_k^{E\lambda}(\xi) - \frac{1}{\bar{\eta}} d_k^{E\lambda}(\xi), \tag{5}$$

where the first term is defined by eq. (IV.4.27) and plotted in figs. IV.18–19. The quantum corrections $d_k^{E\lambda}(\xi)$ can be evaluated by inserting the classical integrals in eq. (2.25) and the result is illustrated for $\lambda = 1$ and 2 in figs. 4 and 5.

The high accuracy of the classical first-order theory for the differential and total cross sections was a consequence of the coincidence of the first-order radial matrix elements and the classical Coulomb excitation functions. In second-order perturbation theory the accuracy of the relation (3.5) is not as high, the correction being of the order $1/\eta$. This can be seen directly from the expression (2.26) by performing the limiting process $\eta \to \infty$. Correction terms of this order of magnitude are to be expected for all classical higher-order Coulomb excitation calculations including the procedure (3.38) for solving the coupled radial equation.

As an example of the quantum mechanical corrections to higher-order Coulomb excitation we mention the effect of static moments. In a quantal treatment the cross section for the excitation of a state of spin 2 in an even-even nucleus may be written in a form similar to eq. (V.6.1), i.e.

$$\mathrm{d}\sigma_2 = \mathrm{d}\sigma_2(\text{first order})[1 + c(\eta_0, \vartheta, \xi)\chi^{(2)}_{2\to 2}]. \tag{6}$$

For finite values of η the coefficient c may be written in the form

$$c(\eta_0, \vartheta, \xi) = c(\vartheta, \xi) + \frac{1}{\eta} e(\vartheta, \xi). \tag{7}$$

The classical limit $c(\vartheta, \xi)$ is given in fig. V.7 where one should use $s = 1$. The function $e(\vartheta, \xi)$ is illustrated in fig. 6 as a function of ξ for a few values of ϑ.

It is the experience that in Coulomb excitation the expansion to lowest order in $1/\eta$ is sufficiently accurate for all practical cases.

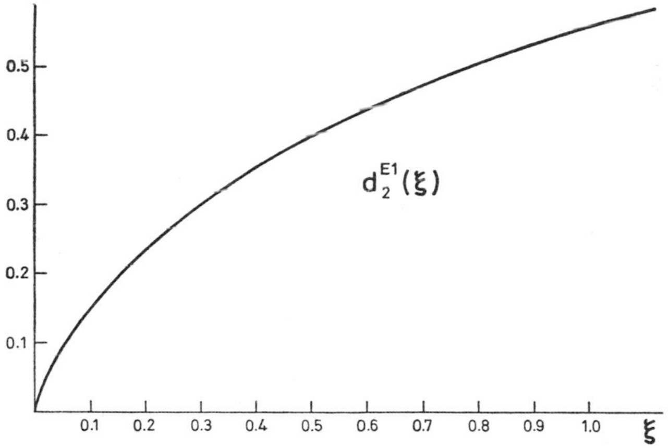

Fig. 4. The quantum mechanical correction to the particle parameter (IV.4.27) in first-order dipole excitation. The coefficient $d_2^{E1}(\xi)$ defined in eq. (4.5) is plotted as a function of ξ.

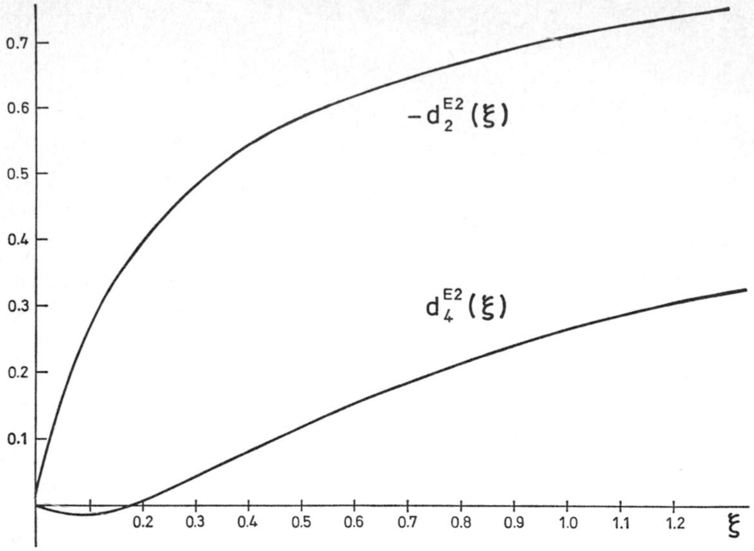

Fig. 5. The quantum mechanical correction to the particle parameter (IV.4.27) in first-order quadrupole excitation. The coefficients $d_k^{E2}(\xi)$ defined in eq. (4.5) are plotted for $k = 2$ and 4 as functions of ξ. Note that the negative value of $d_2^{E2}(\xi)$ is given.

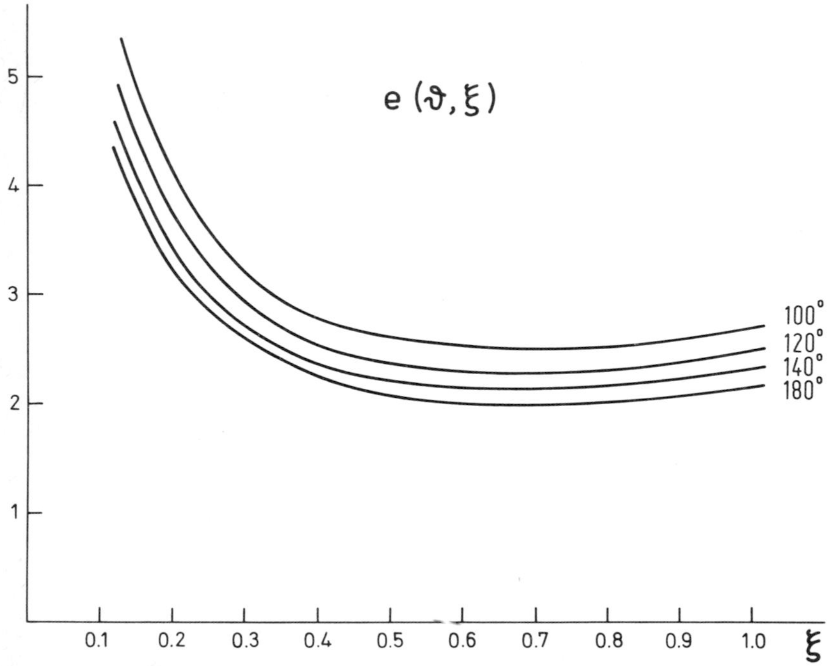

Fig. 6. The quantal correction to the reorientation coefficient (V.6.1) in second-order Coulomb excitation. The coefficient $e(\vartheta, \xi)$ defined in eq. (4.7) has been plotted as a function of ξ for a few values of ϑ.

CHAPTER X

Application to Experiments

In this chapter we collect a number of formulae which are useful for the experimental application of the theory of Coulomb excitation. We shall also quote the results from related fields which are of importance in designing experiments using heavy ion Coulomb excitation.

§ 1. *Transformation to centre-of-mass system*

In the preceding chapters we have separated the centre-of-mass motion from the relative motion such that there is a complete symmetry between projectile and target and such that no explicit distinction between projectile excitation and target excitation was required. Since we realize that it is easier to translate all formulae to the laboratory system than to translate an experimental physicist into the centre-of-mass system we shall here quote the necessary transformation formulae. In these formulae we shall neglect relativistic effects, since the projectile velocities in Coulomb excitation experiments are usually quite small (cf. § 7 below).

The centre-of-mass transformation is illustrated in fig. 1, where we have indicated the projectile by P and the target by T. The scattering angle of the projectile in the centre-of-mass system has been denoted by ϑ_P and the corresponding target deflection angle by $\vartheta_T = \pi - \vartheta_P$, both measured with respect to the projectile velocity. The velocities of the projectile and target in the centre-of-mass system before the collision have been denoted by $\boldsymbol{u}_P$ and $\boldsymbol{u}_T$, and after the collision by $\boldsymbol{u}'_P$ and $\boldsymbol{u}'_T$, respectively. The final velocities in the laboratory system where the target is initially at rest have been denoted by $\boldsymbol{w}'_P$ and $\boldsymbol{w}'_T$, respectively, while the deflection angles in the laboratory system are θ_P and θ_T. A survey of this notation is given in table 1. With a bombarding energy E_P the relative velocity v_i before the collision is given by

$$v_i = \sqrt{2E_P/m_P}, \tag{1}$$

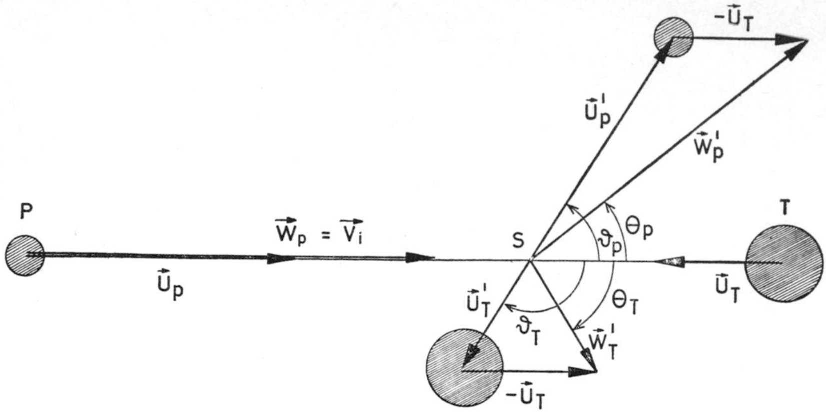

Fig. 1. The velocities of target and projectile before and after the collision. The velocities in the centre-of-mass system are denoted by $\boldsymbol{u}$ and $\boldsymbol{u}'$, respectively, while the velocities in the laboratory system are denoted by $\boldsymbol{w}$ and $\boldsymbol{w}'$, respectively. The indices P and T refer to projectile and target. The notation is summarized in table 1.

where m_P is the projectile mass. In the centre-of-mass system we find the following velocities of projectile and target

$$u_\text{P} = v_\text{i} \frac{m_\text{T}}{m_\text{P} + m_\text{T}} \tag{2}$$

and

$$u_\text{T} = v_\text{i} \frac{m_\text{P}}{m_\text{P} + m_\text{T}}, \tag{3}$$

where m_T is the target mass. For an inelastic collision with total excitation energy ΔE the velocities in the centre-of-mass system after the collision are found to be

$$u'_\text{P} = \frac{m_\text{T}}{m_\text{P} + m_\text{T}} \sqrt{\frac{2}{m_\text{P}} \tilde{E}'} \tag{4}$$

and

$$u'_\text{T} = \frac{m_\text{P}}{m_\text{P} + m_\text{T}} \sqrt{\frac{2}{m_\text{P}} \tilde{E}'}, \tag{5}$$

where we have used the abbreviation

$$\tilde{E}' = E_\text{P} - \Delta E' = E_\text{P} - \Delta E(1 + m_\text{P}/m_\text{T}), \tag{6}$$

which is proportional to the total energy in the centre-of-mass system after the collision (see table 1). The relative velocity v_f after the collision is thus given by

$$v_\text{f} = u'_\text{P} + u'_\text{T} = \sqrt{2\tilde{E}'/m_\text{P}}. \tag{7}$$

TABLE 1

A survey of our notation for the kinematic quantities of projectile and target in the centre-of-mass and the laboratory system.

	Projectile		Target	
mass	m_P		m_T	
	laboratory system	centre-of-mass system	laboratory system	centre-of-mass system
scattering angle	θ_P	$\vartheta_P = \vartheta$	θ_T	$\vartheta_T = \pi - \vartheta$
solid angle	$d\omega_P$	$d\Omega_P = d\Omega$	$d\omega_T$	$d\Omega_T = d\Omega$
initial velocities	$w_P = v_i$	u_P	$w_T = 0$	u_T
final velocities	w'_P	u'_P	w'_T	u'_T
initial energies	E_P	—	$E_T = 0$	—
final energies	E'_P	—	E'_T	—

Total kinetic energy

	laboratory system	centre-of-mass system
initial	E_P	$E = E_P(1 + m_P/m_T)^{-1}$
final	$E_P - \Delta E$	$E' = E - \Delta E = \tilde{E}'(1 + m_P/m_T)^{-1}$

The dashed quantities (as e.g. w'_P) depend on the excitation energy of the nuclear state N. This index number is sometimes specified explicitly (e.g. $w'_P(N)$). The same is true for the centre-of-mass scattering angles.

From fig. 1 it is seen that the deflection angles ϑ_P and θ_P are connected by the relation

$$\frac{\sin(\vartheta_P - \theta_P)}{\sin \theta_P} = \frac{m_P}{m_T}\sqrt{\frac{E_P}{\tilde{E}'}} = \tau, \tag{8}$$

where the quantity τ lies in the range $0 \leqslant \tau \leqslant \infty$. The corresponding relation between ϑ_T and θ_T is

$$\frac{\sin(\vartheta_T - \theta_T)}{\sin \theta_T} = \sqrt{\frac{E_P}{\tilde{E}'}} = \tilde{\tau}, \tag{9}$$

where the quantity $\tilde{\tau}$ is always larger than or equal to unity. Note that for elastic scattering where $\tilde{\tau} = 1$, one finds $\theta_T = \frac{1}{2}(\pi - \vartheta)$.

From the solution of (8) and (9) one may determine the final velocities in the laboratory system. We obtain

$$w'_{\mathrm{P}} = \frac{m_{\mathrm{T}}}{m_{\mathrm{P}} + m_{\mathrm{T}}} \frac{\sin \vartheta_{\mathrm{P}}}{\sin \theta_{\mathrm{P}}} v_{\mathrm{f}}, \tag{10}$$

and

$$w'_{\mathrm{T}} = \frac{m_{\mathrm{P}}}{m_{\mathrm{P}} + m_{\mathrm{T}}} \frac{\sin \vartheta_{\mathrm{T}}}{\sin \theta_{\mathrm{T}}} v_{\mathrm{f}}. \tag{11}$$

The relations (1)–(11) give a complete kinematic connection between the quantities in the laboratory system and the centre-of-mass system. In practical applications of these formulae various situations may arise which we shall briefly discuss.

We notice first that the Coulomb excitation amplitudes are calculated from the knowledge of the initial relative velocity v_{i}, the final relative velocities v_{f} in the various inelastic channels, and the scattering angle ϑ_{P} in the centre-of-mass system. The resulting excitation amplitudes apply to a coordinate system moving with the final velocity of the excited projectile or target nucleus.

The most common situation will be that of target excitation where the scattering angle of the inelastically scattered projectile θ_{P} is observed. In the solution of eq. (8) for the determination of $\vartheta = \vartheta_{\mathrm{P}}$ there may occur an ambiguity if the quantity τ exceeds unity, i.e. if

$$\tau = \frac{m_{\mathrm{P}}}{m_{\mathrm{T}}} \sqrt{\frac{E_{\mathrm{P}}}{\tilde{E}'}} > 1. \tag{12}$$

This ambiguity is illustrated by rewriting eq. (8) in the form

$$\tan \theta_{\mathrm{P}} = \frac{\sin \vartheta_{\mathrm{P}}}{\cos \vartheta_{\mathrm{P}} + \tau}. \tag{13}$$

The connection (13) between θ_{P} and ϑ_{P} is illustrated in fig. 2 for the three cases $\tau < 1$, $\tau = 1$ and $\tau > 1$. It is seen that for $\tau > 1$ there is a maximum laboratory scattering angle $\theta_{\max}$ given by $\sin \theta_{\max} = 1/\tau$. Thus for given θ_{P} there will be either two solutions for ϑ_{P} or none at all according to whether

$$\sin \theta_{\mathrm{P}} \begin{cases} > 1/\tau, & \text{no solution} \\ = 1/\tau, & \vartheta_{\mathrm{P}} = \frac{1}{2}\pi + \theta_{\mathrm{P}} \\ < 1/\tau, & \text{two solutions.} \end{cases} \tag{14}$$

Since $\vartheta_{\mathrm{T}} = \pi - \vartheta_{\mathrm{P}}$ one may determine the recoil angle θ_{T} of the target from (9) and the final velocities w'_{P} and w'_{T} in the laboratory system from (10) and

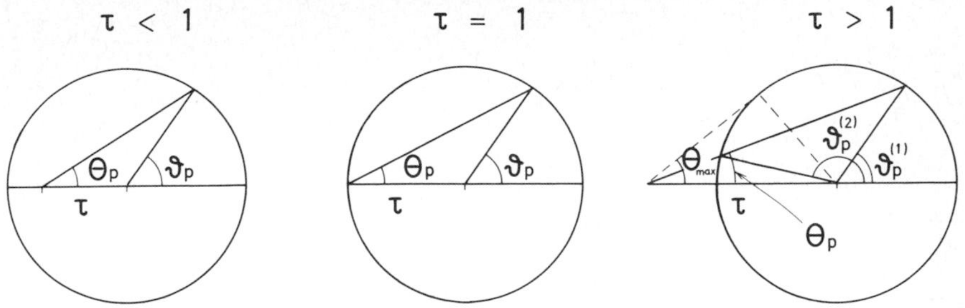

Fig. 2. The connection between the laboratory scattering angle θ_P and the centre-of-mass scattering angle ϑ_P. The trigonometric relation (1.13) is illustrated for different values of the parameter τ defined by (1.12). It is especially noted that for $\tau > 1$ and $\theta_P < \theta_{max}$ there are two solutions $\vartheta_P^{(1)}$ and $\vartheta_P^{(2)}$ for the centre-of-mass scattering angle.

(11). It is noted that there is always an ambiguity in ϑ for given θ_T, since $\tilde{\tau}$ in (9) is always larger than unity.

To determine the differential cross section when the solid angle of the scattered projectile $d\omega_P$ is known in the laboratory system one must know the ratio of this solid angle to the centre-of-mass solid angle $d\Omega_P$. One finds according to eq. (8)

$$d\omega_P/d\Omega_P = \sin^2\theta_P\,|\cos(\vartheta_P - \theta_P)|/\sin^2\vartheta_P. \tag{15}$$

One may similarly evaluate the relative solid angle $d\omega_T/d\Omega_T$ for the detection of the recoil. It is noted that $d\Omega_T = d\Omega_P$. According to eq. (9) one finds furthermore

$$d\omega_T/d\Omega_T = \sin^2\theta_T\,|\cos(\vartheta_T - \theta_T)|/\sin^2\vartheta_T. \tag{16}$$

The expressions (15) and (16) become singular in the limit of backwards scattering. One finds in this limit

$$d\omega_P/d\Omega_P \approx 1/(1-\tau)^2 \tag{17}$$

and

$$d\omega_T/d\Omega_T \approx 1/(1+\tilde{\tau})^2 \tag{18}$$

for $\vartheta_P \approx \pi$, both when $\theta_P \approx \pi$ and when $\theta_P \approx 0$ ($\tau > 1$).

We note that for a given value of the laboratory angle θ_P the scattering angle $\vartheta = \vartheta_P$ depends, in principle, on the excitation energy ΔE. This effect is mostly so small that it may be neglected as was discussed in chapter VI.

Finally we quote the laboratory energies E'_P and E'_T of the projectile and the target after collision. One finds from (10) and (11)

$$E'_P = \left(\frac{m_T}{m_P + m_T}\right)^2 (1 + \tau^2 + 2\tau\cos\vartheta_P)\tilde{E}' \tag{19}$$

and

$$E_{\mathrm{T}}' = \frac{m_{\mathrm{P}} m_{\mathrm{T}}}{(m_{\mathrm{P}} + m_{\mathrm{T}})^2} (1 + \tilde{\tau}^2 + 2\tilde{\tau} \cos \vartheta_{\mathrm{T}}) \tilde{E}' . \tag{20}$$

For backwards scattering the projectile energy is given by

$$E_{\mathrm{P}}'(\vartheta = \pi) = \left(\frac{m_{\mathrm{T}} - m_{\mathrm{P}} \tilde{\tau}}{m_{\mathrm{P}} + m_{\mathrm{T}}} \right)^2 \tilde{E}' . \tag{21}$$

All formulae in the present section are independent of whether the projectile or the target (or both) are excited. The quantity ΔE is the total excitation energy.

It should be emphasized that if one detects the excitation energy by means of an energy measurement of the subsequent γ-quanta, the frequency of the γ-rays may be changed by a Doppler shift. This effect is discussed in § 6 below.

§ 2. *TORBEN units*

Although internationally recognized systems of units exist (SI), it is a well known fact that nuclear physicists rarely measure the projectile energy in joule and projectile mass in kilogram, but use instead the unrational units of MeV, atomic mass units, etc. In this section we shall collect some of the most important formulae in such TORBEN*-units, which give a more direct feeling for the order of magnitudes involved. Although the measurements and the theory of Coulomb excitation can rarely be made with an accuracy better than one percent, we quote the expressions for relevant measurable quantities [TAY 69] to much higher accuracy mainly in order to facilitate the comparison between different computations made e.g. with different computer programs. The units which have been used in the following are compiled in table 2.

We consider first the kinematic quantities given in the preceding section. The initial velocity in units of the velocity of light c is according to (1.1) given by

$$v_{\mathrm{i}}/c = 0.0463370 \sqrt{E_{\mathrm{P,MeV}}/A_{\mathrm{P}}} . \tag{1}$$

It is noted that the mass number of the projectile A_{P} according to the definition is only approximately an integer.

* We propose to name these practical units after Torben Huus who always insists that all theoretical formulae should be translated into these units. The name TORBEN may also be thought of as an abbreviation of "Theoretical Observables Rewritten for the Benefit of Experimental Nucleists".

TABLE 2
The units and magnitudes of various quantities used in the text. (The numbers are taken from [TAY 69].)

Quantity	Measured in units of	Numerical value in CGS units
Mass	atomic mass unit ($=\frac{1}{12}\,{}^{12}$C)	$M = 1.66053 \times 10^{-24}$ g
Length	fm	10^{-13} cm
Charge	elementary charge	$e = 4.80325 \times 10^{-10}$ esu
Energy	MeV	1.60219×10^{-6} erg
$\hbar$	—	$\hbar = 1.05459 \times 10^{-27}$ erg·s
c	—	$c = 2.997925 \times 10^{10}$ cm/s
Cross section	barn	10^{-24} cm^2
$B(\mathrm{E}\lambda)$	$e^2(\mathrm{fm})^{2\lambda}$	—
$B(\mathrm{M}\lambda)$	$(e\hbar/2M_{\mathrm{P}}c)^2(\mathrm{fm})^{2\lambda-2}$ $= (0.105157)^2e^2\mathrm{fm}^{2\lambda}$	—

The final relative velocity v_{f} is according to eq. (1.7) given by

$$v_{\mathrm{f}}/c = 0.0463370\sqrt{\tilde{E}'_{\mathrm{MeV}}/A_{\mathrm{P}}}, \tag{2}$$

where

$$\tilde{E}'_{\mathrm{MeV}} = E_{\mathrm{P,MeV}} - \Delta E_{\mathrm{MeV}}(1 + A_{\mathrm{P}}/A_{\mathrm{T}}). \tag{3}$$

The laboratory velocity of the scattered projectile and of the recoiling target are given in terms of v_{f} by the formulae (1.10)–(1.11).

The laboratory energy of the projectile after the collision is found from (1.19), i.e.

$$E'_{\mathrm{P,MeV}} = \left(1 + \frac{A_{\mathrm{P}}}{A_{\mathrm{T}}}\right)^{-2} \frac{\sin^2 \vartheta_{\mathrm{P}}}{\sin^2 \theta_{\mathrm{P}}} \tilde{E}'_{\mathrm{MeV}}. \tag{4}$$

Similarly the recoil energy is according to (1.20)

$$E'_{\mathrm{T,MeV}} = \frac{A_{\mathrm{P}}}{A_{\mathrm{T}}}\left(1 + \frac{A_{\mathrm{P}}}{A_{\mathrm{T}}}\right)^{-2} \frac{\sin^2 \vartheta_{\mathrm{T}}}{\sin^2 \theta_{\mathrm{T}}} \tilde{E}'_{\mathrm{MeV}}. \tag{5}$$

Next we consider the quantities relevant for the Coulomb excitation process. The natural unit of length in the collision is half the distance of closest approach in a head-on collision which is given by

$$a = 0.719990(1 + A_{\mathrm{P}}/A_{\mathrm{T}})Z_1Z_2/E_{\mathrm{P,MeV}}\ \mathrm{fm}. \tag{6}$$

The distance of closest approach in a scattering with centre-of-mass scattering angle ϑ_{P} is given by eq. (I.12)

$$b(\vartheta) = a[1 + 1/\sin(\tfrac{1}{2}\vartheta_{\mathrm{P}})]. \tag{7}$$

For the evaluation of Coulomb excitation cross sections (cf. e.g. eq. (VI.8.3)) one uses the symmetrized expression a_{if} instead of a, i.e. according to (VI.8.2)

$$a_{\mathrm{if}} = 0.719990\left(1 + \frac{A_{\mathrm{P}}}{A_{\mathrm{T}}}\right) \frac{Z_1 Z_2}{\sqrt{E_{\mathrm{P,MeV}} \tilde{E}'_{\mathrm{MeV}}}} \text{ fm}$$
$$\approx a\left[1 + \tfrac{1}{2}\frac{\Delta E}{E_{\mathrm{P}}}\left(1 + \frac{A_{\mathrm{P}}}{A_{\mathrm{T}}}\right) + \cdots\right]. \tag{8}$$

In multiple Coulomb excitation one associates to each nuclear level N a corresponding relative velocity v_N defined by eq. (2) where $\tilde{E}'$ according to (3) is given in terms of E_N and the ground state energy E_0 since

$$\Delta E = \Delta E_N = (E_N - E_0). \tag{9}$$

In terms of the velocities v_N one may instead of eq. (8) use

$$a_{NM} = 1.54590 \times 10^{-3} \frac{A_{\mathrm{P}} + A_{\mathrm{T}}}{A_{\mathrm{P}} A_{\mathrm{T}}} \frac{Z_1 Z_2}{(v_N/c)(v_M/c)} \text{ fm}. \tag{10}$$

The parameter η also depends on the excitation energy. In the initial state it is given by

$$\eta_{\mathrm{i}} = 0.157484 \frac{Z_1 Z_2 A_{\mathrm{P}}^{1/2}}{E_{\mathrm{P,MeV}}^{1/2}}. \tag{11}$$

Similarly one finds

$$\eta_{\mathrm{f}} = 0.157484 \frac{Z_1 Z_2 A_{\mathrm{P}}^{1/2}}{\tilde{E}'^{1/2}_{\mathrm{MeV}}}. \tag{12}$$

In terms of the relative velocities v_N one may instead of eqs. (11) and (12) use

$$\eta_N = 7.29735 \times 10^{-3} \frac{Z_1 Z_2}{(v_N/c)}. \tag{13}$$

From the η values one obtains directly the following expression for the symmetrized adiabaticity parameter ξ_{NM}, i.e.

$$\xi_{NM} = \eta_N - \eta_M = 7.29735 \times 10^{-3} Z_1 Z_2 \left[\frac{1}{(v_N/c)} - \frac{1}{(v_M/c)}\right]. \tag{14}$$

The corresponding expression in terms of the energies $\tilde{E}'$ can conveniently be written as an expansion in powers of $\Delta E/E_{\mathrm{P}}$, i.e.

$$\xi_{NM} \approx \frac{Z_1 Z_2 A_{\mathrm{P}}^{1/2}}{12.70 E_{\mathrm{P,MeV}}^{3/2}}\left(1 + \frac{A_{\mathrm{P}}}{A_{\mathrm{T}}}\right)(E_N - E_M)_{\mathrm{MeV}}$$
$$\times \left\{1 + \tfrac{3}{4}\frac{\Delta E_N + \Delta E_M}{E_{\mathrm{P}}}\left(1 + \frac{A_{\mathrm{P}}}{A_{\mathrm{T}}}\right) + \cdots\right\}. \tag{15}$$

The symmetrized strength parameter χ_{NM} defined e.g. in eq. (VI.8.11) can be written as

$$\chi_{NM}^{(\lambda)} = \frac{1.116537}{(0.719990)^{\lambda}} \frac{(\lambda-1)!}{(2\lambda+1)!!} \frac{A_{\mathrm{P}}^{1/2}(\tilde{E}'_N \tilde{E}'_M)_{\mathrm{MeV}}^{(2\lambda-1)/4}}{Z_1^{\lambda-1} Z_2^{\lambda}(1+A_{\mathrm{P}}/A_{\mathrm{T}})^{\lambda}}$$

$$\times \frac{\langle M \| \mathscr{M}(\mathrm{E}\lambda) \| N \rangle_{e(\mathrm{fm})^{\lambda}}}{\sqrt{2I_M+1}}, \tag{16}$$

where the multipole moments are measured in units of $e(\mathrm{fm})^{\lambda}$. Furthermore, $\tilde{E}_N$ is measured in MeV and is defined by

$$\tilde{E}'_N = E_{\mathrm{P}} - (1 + A_{\mathrm{P}}/A_{\mathrm{T}})(E_N - E_0). \tag{17}$$

Explicit formulae for $\lambda = 1$, 2, 3 and 4 are given in table 3. An alternative expression for χ in terms of the relative velocities v_N/c is given by

$$\chi_{NM}^{(\lambda)} = \frac{0.0517369}{(1.545903 \times 10^{-3})^{\lambda}} \frac{(\lambda-1)!}{(2\lambda+1)!!}$$

$$\times \frac{Z_1\{(v_N/c)(v_M/c)\}^{(2\lambda-1)/2} \langle M \| \mathscr{M}(\mathrm{E}\lambda) \| N \rangle_{e(\mathrm{fm})^{\lambda}}}{Z_1^{\lambda} Z_2^{\lambda}\{(A_{\mathrm{P}}+A_{\mathrm{T}})/A_{\mathrm{P}}A_{\mathrm{T}}\}^{\lambda}\sqrt{2I_M+1}}. \tag{18}$$

TABLE 3

Explicit expressions for the symmetrized strength parameter χ in TORBEN units.

λ	$\chi_{NM}^{(\lambda)} = \chi_{M\to N}^{(\lambda)}$			
1	$\chi_{NM}^{(1)} = 0.516922$	$A_{\mathrm{P}}^{1/2} \dfrac{(\tilde{E}'_N \tilde{E}'_M)_{\mathrm{MeV}}^{1/4}}{Z_2(1+A_{\mathrm{P}}/A_{\mathrm{T}})}$	$(B(\mathrm{E1}, M\to N))^{1/2}$	$e(\mathrm{fm})$
2	$\chi_{NM}^{(2)} = 0.143592$	$A_{\mathrm{P}}^{1/2} \dfrac{(\tilde{E}'_N \tilde{E}'_M)_{\mathrm{MeV}}^{3/4}}{Z_1 Z_2^2(1+A_{\mathrm{P}}/A_{\mathrm{T}})^2}$	$(B(\mathrm{E2}, M\to N))^{1/2}$	$e(\mathrm{fm})^2$
3	$\chi_{NM}^{(3)} = 0.0569816$	$A_{\mathrm{P}}^{1/2} \dfrac{(\tilde{E}'_N \tilde{E}'_M)_{\mathrm{MeV}}^{5/4}}{Z_1^2 Z_2^3(1+A_{\mathrm{P}}/A_{\mathrm{T}})^3}$	$(B(\mathrm{E3}, M\to N))^{1/2}$	$e(\mathrm{fm})^3$
4	$\chi_{NM}^{(4)} = 0.0263807$	$A_{\mathrm{P}}^{1/2} \dfrac{(\tilde{E}'_N \tilde{E}'_M)_{\mathrm{MeV}}^{7/4}}{Z_1^3 Z_2^4(1+A_{\mathrm{P}}/A_{\mathrm{T}})^4}$	$(B(\mathrm{E4}, M\to N))^{1/2}$	$e(\mathrm{fm})^4$
	$\chi_{NM}^{(\mathrm{M}\lambda)} = \chi_{M\to N}^{(\mathrm{M}\lambda)}$			
1	$\chi_{NM}^{(\mathrm{M1})} = 2.51878 \times 10^{-3}$	$\dfrac{(\tilde{E}'_N \tilde{E}'_M)_{\mathrm{MeV}}^{1/2}}{Z_2(1+A_{\mathrm{P}}/A_{\mathrm{T}})}$	$(B(\mathrm{M1}, M\to N))^{1/2}$	μ_n
2	$\chi_{NM}^{(\mathrm{M2})} = 0.699672 \times 10^{-3}$	$\dfrac{(\tilde{E}'_N \tilde{E}'_M)_{\mathrm{MeV}}}{Z_1 Z_2^2(1+A_{\mathrm{P}}/A_{\mathrm{T}})^2}$	$(B(\mathrm{M2}, M\to N))^{1/2}$	$\mu_n(\mathrm{fm})$

The sign of the square root of $B(\lambda)$ should be chosen in accordance with the sign of the reduced matrix element $\langle M \| \mathscr{M}(\mathrm{E}\lambda) \| N\rangle$ or $\langle M \| \mathscr{M}(\mathrm{M}\lambda) \| N\rangle$. The reduced transition probabilities are measured in units of $e^2(\mathrm{fm})^{2\lambda}$ and $\mu_n^2(\mathrm{fm})^{2\lambda-2}$, respectively.

The strength parameters for magnetic excitations, which are defined in eq. (IV.5.5) may similarly be written

$$\chi_{NM}^{(\mathrm{M}\lambda)} = \frac{5.440500 \times 10^{-3}}{(0.719990)^{\lambda}} \frac{(\lambda-1)!}{(2\lambda+1)!!} \frac{(\tilde{E}'_N \tilde{E}'_M)^{\lambda/2}_{\mathrm{MeV}}}{Z_1^{\lambda-1} Z_2^{\lambda}(1 + A_{\mathrm{P}}/A_{\mathrm{T}})^{\lambda}}$$

$$\times \frac{\langle M \| \mathscr{M}(M\lambda) \| N \rangle_{\mu_n(\mathrm{fm})^{\lambda-1}}}{\sqrt{2I_{\mathrm{M}}+1}}, \qquad (19)$$

where the magnetic multipole moment is measured in units of

$$\mu_n(\mathrm{fm})^{\lambda-1} = (e\hbar/2M_{\mathrm{P}}c)(\mathrm{fm})^{\lambda-1},$$

or

$$\chi_{NM}^{(\mathrm{M}\lambda)} = \frac{5.440500 \times 10^{-3}}{(1.545903 \times 10^{-3})^{\lambda}} \frac{(\lambda-1)!}{(2\lambda+1)!!} \frac{\{(v_N/c)(v_M/c)\}^{\lambda}}{Z_1^{\lambda-1} Z_2^{\lambda}\{(A_{\mathrm{P}} + A_{\mathrm{T}})/A_{\mathrm{P}}A_{\mathrm{T}}\}^{\lambda}}$$

$$\times \frac{\langle M \| \mathscr{M}(\mathrm{M}\lambda) \| N \rangle_{\mu_n(\mathrm{fm})^{\lambda-1}}}{\sqrt{2I_M+1}}. \qquad (20)$$

Explicit expressions for $\lambda = 1$ and 2 are given in table 3.

All the formulae given above hold for target excitation as well as for projectile excitation. For target excitation one should identify Z_1 with the projectile charge number and Z_2 with the target charge number, while E_N and E_0 in (17), should be interpreted as the energies of the target nucleus. For projectile excitation Z_1 is the target charge number and Z_2 is the projectile charge number, while E_N and E_0 denote the energies of the nuclear states of the projectile.

From the scattering angle in the centre-of-mass system and the values of ξ_{NM} and χ_{NM} one evaluates the excitation probabilities P_{f} of any given state $|\mathrm{f}\rangle$ and the cross section in the centre-of-mass system is then given by eq. (VI.8.3) which may be written

$$\frac{\mathrm{d}\sigma(\mathrm{i}\to\mathrm{f})}{\mathrm{d}\Omega} = 1.29596 \times 10^{-3} \frac{Z_1^2 Z_2^2}{E^{3/2}_{\mathrm{P,MeV}}(\tilde{E}'_{\mathrm{f,MeV}})^{1/2}} \left(1 + \frac{A_{\mathrm{P}}}{A_{\mathrm{T}}}\right)^2$$

$$\times P_{\mathrm{f}}(\vartheta, \xi_{NM}, \chi_{NM}) \frac{1}{\sin^4(\frac{1}{2}\vartheta)} \times 10^{-24}\ \mathrm{cm}^2. \qquad (21)$$

In the first-order perturbation theory the excitation probability P_{f} is given by

$$P_{\mathrm{f}} = |\chi_{\mathrm{fi}}^{(\lambda)}|^2 R_{\lambda}^2(\vartheta, \xi_{\mathrm{fi}}). \qquad (22)$$

In this case one may write the cross section in the form (see e.g. [ALD 56] p. 458)

$$d\sigma(i \to f) = \frac{Z_1^2 A_P}{4032.04}\left[0.719990\left(1+\frac{A_P}{A_T}\right)Z_1Z_2\right]^{-2\lambda+2}$$

$$\times\ E_{P,MeV}^{\lambda-2}(\tilde{E}'_{f,MeV})^{\lambda-1}B(E\lambda, i \to f)\, df_{E\lambda}(\vartheta, \xi_{fi})b, \qquad (23)$$

where $df_{E\lambda}$ is defined in eq. (IV.4.5) and is illustrated in fig. IV.9. For the total cross section the same formula applies except that $df_{E\lambda}$ should be substituted by $f_{E\lambda}(\xi)$ defined in eq. (IV.4.10). This function is illustrated in fig. IV.11.

The $B(EL)$-values entering in the preceding formulae may be obtained from the γ-decay rates (see eq. (G.5)) between the two levels involved. For the transition amplitude (G.3) from the state $|i\rangle$ to the state $|f\rangle$ one obtains

$$\delta_{\pi L}(i \to f) = i^{s(L)}\cdot 1.6692 \times 10^{10}(0.0050677)^L\sqrt{\frac{L+1}{L}}$$

$$\times\ ((2L+1)!!)^{-1}(\Delta E_{MeV})^{(2L+1)/2}$$

$$\times\ (2I_i+1)^{-1/2}\begin{Bmatrix}\langle I_f\|\mathscr{M}(E\lambda)\|I_i\rangle \\ 0.105157\langle I_f\|\mathscr{M}(M\lambda)\|I_i\rangle\end{Bmatrix}\sec^{-1/2}. \qquad (24)$$

TABLE 4

The connection between the partial decay amplitude $\delta_{\pi L}(i \to f)$, in $\sec^{-1/2}$, and the reduced matrix elements $\langle f\|\mathscr{M}(\pi L)\|i\rangle$ and the connection between the partial decay rate $\gamma_{i\to f}$ and the reduced transition probabilities $B(\pi L)$. (The energy of the emitted γ-quantum in MeV is denoted by ΔE_{MeV}.)

πL	$\delta_{\pi L}(i \to f)$	
E1	$i3.9877(\Delta E_{MeV})^{3/2}\langle I_f\|\mathscr{M}(E1)\|I_i\rangle_{efm}(2I_i+1)^{-1/2}$	$10^7 \sec^{-1/2}$
E2	$-3.5002(\Delta E_{MeV})^{5/2}\langle I_f\|\mathscr{M}(E2)\|I_i\rangle_{efm^2}(2I_i+1)^{-1/2}$	$10^4 \sec^{-1/2}$
E3	$-i2.3890(\Delta E_{MeV})^{7/2}\langle I_f\|\mathscr{M}(E3)\|I_i\rangle_{efm^3}(2I_i+1)^{-1/2}$	$10^1 \sec^{-1/2}$
E4	$1.3025(\Delta E_{MeV})^{9/2}\langle I_f\|\mathscr{M}(E4)\|I_i\rangle_{efm^4}(2I_i+1)^{-1/2}$	$10^{-2} \sec^{-1/2}$
M1	$-4.1933(\Delta E_{MeV})^{3/2}\langle I_f\|\mathscr{M}(M1)\|I_i\rangle_{\mu_n}(2I_i+1)^{-1/2}$	$10^6 \sec^{-1/2}$
M2	$-i3.6807(\Delta E_{MeV})^{5/2}\langle I_f\|\mathscr{M}(M2)\|I_i\rangle_{\mu_n fm}(2I_i+1)^{-1/2}$	$10^3 \sec^{-1/2}$

πL	$\gamma_{i\to f}$	
E1	$1.59016 \times (\Delta E_{MeV})^3 B(E1, i \to f)_{e^2(fm)^2}$	$10^{15} \sec^{-1}$
E2	$1.22513 \times (\Delta E_{MeV})^5 B(E2, i \to f)_{e^2(fm)^4}$	$10^9 \sec^{-1}$
E3	$5.70757 \times (\Delta E_{MeV})^7 B(E3, i \to f)_{e^2(fm)^6}$	$10^2 \sec^{-1}$
E4	$1.69651 \times (\Delta E_{MeV})^9 B(E4, i \to f)_{e^2(fm)^8}$	$10^{-4} \sec^{-1}$
M1	$1.75840 \times (\Delta E_{MeV})^3 B(M1, i \to f)_{\mu_n^2}$	$10^{13} \sec^{-1}$
M2	$1.35475 \times (\Delta E_{MeV})^5 B(M2, i \to f)_{\mu_n^2(fm)^2}$	$10^7 \sec^{-1}$

Explicit expressions for $\delta_{\pi L}$ and $\gamma_{i\to f} = |\delta_{\pi L}(i \to f)|^2$ for E1, E2, E3, E4, M1 and M2 transitions are given in table 4. It is noted that $\gamma_{i\to f}$ is only the inverse of the life time of the nuclear state if no branching occurs including internal conversion. It should also be remembered that the reduced transition probability is not symmetric under interchange of initial and final state, but satisfy

$$(2I_i + 1)\, B(L, i \to f) = (2I_f + 1)\, B(L, f \to i). \tag{25}$$

§ 3. *Projectile and target excitation*

As was mentioned in the previous sections the evaluation of projectile excitation only differs from the evaluation of target excitation through kinematic effects associated with the centre-of-mass motion. In the preceding chapters we have worked in the centre-of-mass system and have not distinguished target and projectile excitation. The two colliding particles were given the index 1 and 2 and we have throughout assumed that particle 2 was excited. In § 1 where the centre-of-mass transformation was discussed we distinguished target and projectile by the indices T and P and the experimental conditions for target excitation is thus achieved by identifying 1 with P and 2 with T, while projectile excitation is obtained by identifying the indices 1 with T and 2 with P. In the semiclassical approximation the simultaneous excitation of projectile and target is obtained by performing two independent calculations with the two above identifications and the resulting simultaneous excitation probabilities are the product of the excitation probabilities thus obtained (see, e.g. § VI.7).

In table 5 we have made a compilation of the rules for evaluating target and projectile excitation in the laboratory system from the knowledge of the kinematics in this coordinate system. We note that the scattering angle ϑ, as used in the theory of Coulomb excitation, is always the projectile scattering angle ϑ_P in the centre-of-mass system. This scattering angle is not always uniquely determined from the laboratory scattering angle θ. Thus if

$$\tau = (A_P/A_T)\sqrt{E_P/\tilde{E}'} > 1, \tag{1}$$

which occurs if $A_P \geqslant A_T$ there will be two solutions for $\vartheta = \vartheta_P$ for a given scattering angle θ_P of the projectile. For the two solutions the energies of the projectile E'_P are different and the two cases may thus be distinguished by detecting the projectile energy after the scattering. If this energy is not measured the cross section for the two centre-of-mass situations should be added.

If the inelastic cross section is measured by detecting the recoil scattering angle instead of the scattering angle of the projectile there will always be

TABLE 5

Rules for connecting centre-of-mass quantities to laboratory quantities for target and projectile excitation.

		Target excitation	Projectile excitation
Z_1 Z_2		Z_P Z_T	Z_T Z_P
scattering angle and laboratory cross section for *detection of projectile*		$\vartheta = \vartheta_P(\theta_P)$ $\left(\frac{d\sigma(\vartheta)}{d\Omega}\right)\frac{d\Omega}{d\omega_P}$	$\vartheta = \vartheta_P(\theta_P)$ $\left(\frac{d\sigma(\vartheta)}{d\Omega}\right)\frac{d\Omega}{d\omega_P}$
scattering angle and laboratory cross section for *detection of recoiling target*		$\vartheta = \vartheta_P(\theta_T)$ $\left(\frac{d\sigma(\vartheta)}{d\Omega}\right)\frac{d\Omega}{d\omega_T}$	$\vartheta = \vartheta_P(\theta_T)$ $\left(\frac{d\sigma(\vartheta)}{d\Omega}\right)\frac{d\Omega}{d\omega_T}$
origin of coordinate systems		origin in target	origin in projectile
axes of coordinate system A	x-axis	bisects $-\boldsymbol{u}_P$ and $\boldsymbol{u}'_P$	bisects $\boldsymbol{u}_P$ and $-\boldsymbol{u}'_P$
	y-axis	y-component of $v_i > 0$	y-component of $v_i < 0$
	z-axis	along $-\boldsymbol{w}_P \times \boldsymbol{w}'_P$	
axes of coordinate system B	x-axis	along $\boldsymbol{w}_P \times \boldsymbol{w}'_P$	
	y-axis	y-component of $v_i > 0$	y-component of $v_i < 0$
	z-axis	bisects $-\boldsymbol{u}_P$ and $\boldsymbol{u}'_P$	bisects $\boldsymbol{u}_P$ and $-\boldsymbol{u}'_P$
axes of coordinate system C	x-axis	x-component of $v_f > 0$	x-component of $v_f < 0$
	y-axis	along $\boldsymbol{w}_P \times \boldsymbol{w}'_P$	
	z-axis	along $\boldsymbol{v}_i$	along $-\boldsymbol{v}_i$

The table contains besides the rules for connecting Z_1 and Z_2 with Z_P and Z_T the formulae for the centre-of-mass scattering angle and the cross section in the laboratory system under the two possible situations, where one observes the reaction through the detection of the scattered projectile or through the recoiling target nucleus. The definitions of the three coordinate systems A, B and C are also given.

two solutions for $\vartheta_T = \pi - \vartheta$, except when θ_T attains its maximum value ($<\frac{1}{2}\pi$) determined by $\sin\theta_T = 1/\tilde{\tau}$. This follows from eq. (1.9) because

$$\tilde{\tau} = \sqrt{E_P/\tilde{E}'} > 1. \tag{2}$$

Since this number is always close to unity, one solution will always correspond to small forward scattering angles, i.e.

$$\vartheta \approx \frac{1}{2}\frac{\Delta E}{E_P}\left(1 + \frac{A_P}{A_T}\right)\tan\theta_T \lessapprox \sqrt{\frac{\Delta E}{E_P}\left(1 + \frac{A_P}{A_T}\right)}. \tag{3}$$

The recoil energy of *this solution* can be found from (2.5) to be given approximately by

$$E'_{\mathrm{T}} \approx \tfrac{1}{4}\Delta E\left(\frac{\Delta E}{E_{\mathrm{P}}}\right)\frac{A_{\mathrm{P}}}{A_{\mathrm{T}}}\frac{1}{\cos^2\theta_{\mathrm{T}}} \leqslant \frac{A_{\mathrm{P}}}{A_{\mathrm{P}}+A_{\mathrm{T}}}\,\Delta E. \tag{4}$$

Actually, the solution (3) for ϑ is of no practical interest in Coulomb excitation since the excitation cross sections will vanish for these forward angles. One may thus verify that the effective ξ value $\xi(\vartheta)$ defined in (I.14) is of the order of magnitude of η.

The other solution for given θ_{T} corresponds to a scattering angle ϑ close to $\pi - 2\theta_{\mathrm{T}}$. In this case the recoil moves almost along the symmetry axis in the hyperbolic centre-of-mass motion, i.e. along the negative z-axis in coordinate system B. Actually, the recoil direction is moved slightly towards the beam direction, i.e.

$$\theta_{\mathrm{T}} \approx \frac{\pi-\vartheta}{2} - \frac{1}{4}\frac{\Delta E}{E_{\mathrm{P}}}\left(1+\frac{A_{\mathrm{P}}}{A_{\mathrm{T}}}\right)\cot\left(\tfrac{1}{2}\vartheta\right). \tag{5}$$

In table 5, formulae are given for the evaluation of the differential cross section in the laboratory system, both for detection of target and projectile. The factor $\mathrm{d}\sigma/\mathrm{d}\Omega$ is the theoretical cross section in the centre-of-mass system, while the second factor $\mathrm{d}\Omega/\mathrm{d}\omega$ is the solid angle ratio given in § 1.

Finally, we have collected the definition of the three coordinate systems used in the theory. These coordinate systems could have been chosen the same for projectile and target excitation. This is due to the fact that the electromagnetic interaction between two colliding particles is symmetric under an interchange of the particles 1 and 2 (cf. § II.1). We obtain the same result if we change the coordinate system in such a way that the target takes the role of the projectile, while the centre of the coordinate system is in the projectile. These coordinate systems are obtained from the previous one by a rotation of 180 degrees around an axis perpendicular to the plane of the orbit. For the measurable quantities which can always be expressed by the statistical tensors $\rho_{k\kappa}$ this rotation is irrelevant. This can be seen explicitly from the symmetry relation of $\rho_{k\kappa}$ as they are given in table III.1.

The angular distributions given in chapter III are evaluated in coordinate systems with axes parallel to those described in table 5, but with the origin in the excited target or projectile nucleus. In evaluating the angular distribution of subsequent γ-quanta in the laboratory system one must therefore usually take into account the distortion due to Doppler shift and aberration (cf. § 6 below).

§ 4. *Range of parameters*

In this section we shall discuss the range of variation which is experimentally possible for those parameters which are important for Coulomb excitation. The most important limitation is due to the fact that no penetration into the range of nuclear forces should occur. Of course, this does not lead to any sharp upper limit of the bombarding energy since the effect of nuclear interaction only gradually sets in, when the surfaces of the two nuclei approach each other. Also, one may, for forward scattering angles, have a rather pure Coulomb excitation even if the bombarding energy is above the Coulomb barrier.

The onset of effects of nuclear forces in heavy ion collisions is expected to depend exponentially on the distance Δ between the nuclear surfaces and on the diffuseness parameter. The bombarding energy at which nuclear effects become important is thus determined by an expression of the type

$$E_{\mathrm{P}}^{\max} = \frac{Z_1 Z_2 e^2}{R_1 + R_2 + \Delta}\left(1 + \frac{A_{\mathrm{P}}}{A_{\mathrm{T}}}\right) \tag{1}$$

where R_1 and R_2 are proportional $A_1^{1/3}$ and $A_2^{1/3}$, respectively. For the estimates in the present section we shall use

$$R = 1.44 \times A^{1/3}\ \mathrm{fm} \tag{2}$$

and

$$\Delta = 2.88\ \mathrm{fm}, \tag{3}$$

which leads to the simple expression

$$E_{\mathrm{P,MeV}}^{\max} = \frac{Z_1 Z_2 (1 + A_{\mathrm{P}}/A_{\mathrm{T}})}{A_1^{1/3} + A_2^{1/3} + 2}. \tag{4}$$

Present experiments indicate, however, that nuclear effects are often important at this bombarding energy. In fig. 3 we have plotted the maximum energy (4) as a function of the target charge number for various projectiles and we have here used the approximate relation

$$Z = 0.487\,\frac{A}{1 + A^{2/3}/166} \tag{5}$$

between charge and mass numbers for stable nuclei. The rule (4) for the maximum bombarding energy is not expected to hold for protons, since the wave length here becomes comparable to the distance Δ.

From the maximum value $E_{\mathrm{P}}^{\max}$ of the bombarding energy one may evaluate the minimum value of the parameter η_{i} for given values of Z_{P} and Z_{T} by means of eq. (2.11). The result is illustrated in fig. 4 as a function of

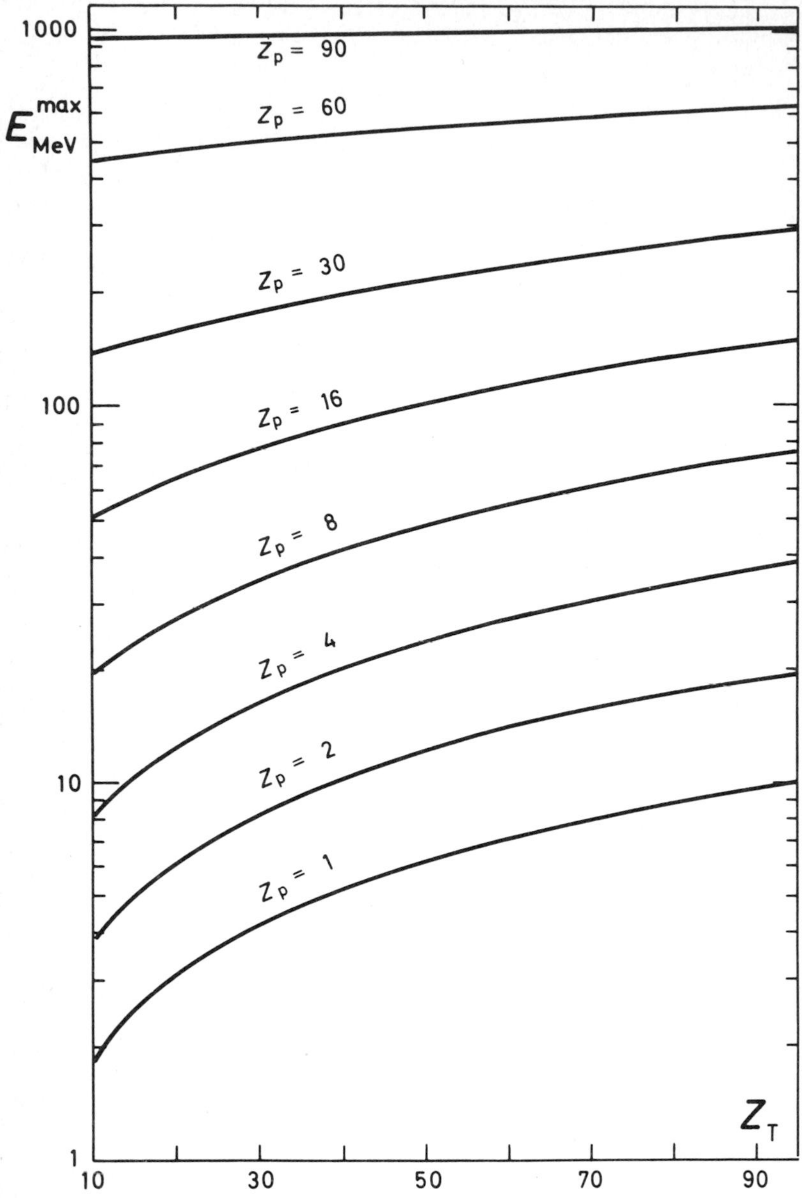

Fig. 3. The maximum bombarding energy according to eq. (4.4) as a function of the target charge for different projectile charges. The relation between charge and mass number is assumed to be given by eq. (4.5).

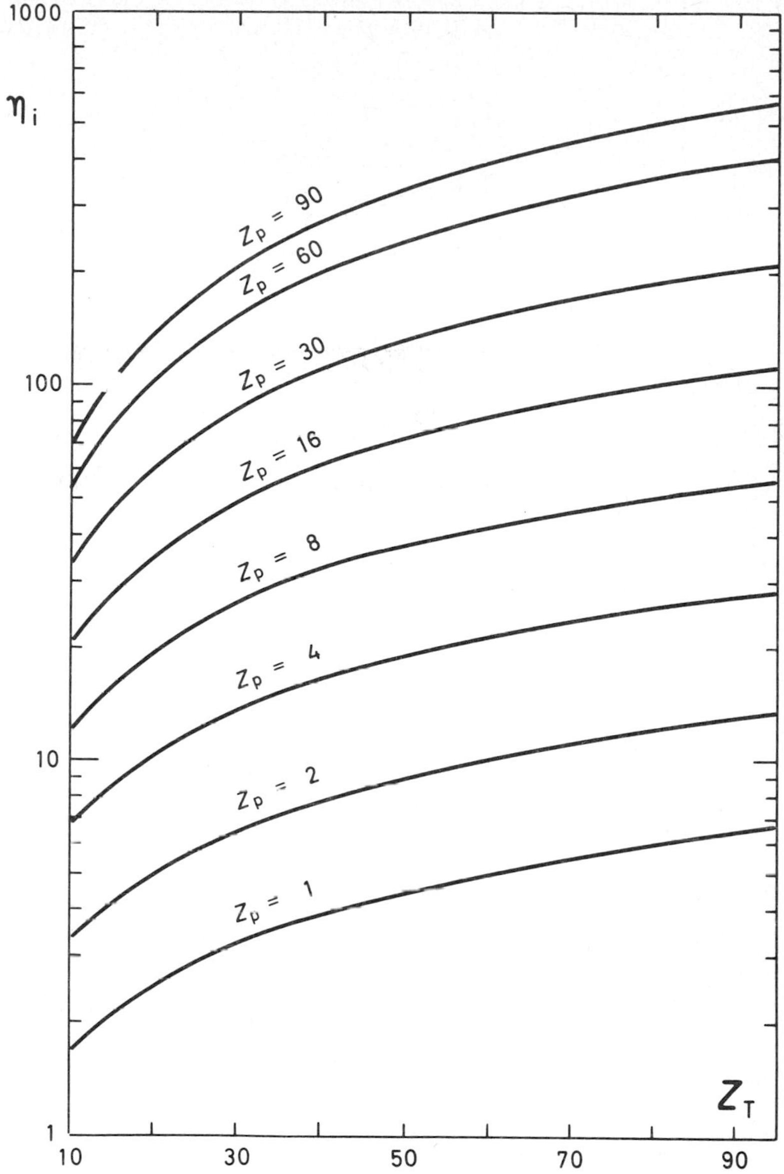

Fig. 4. The parameter η_i given by eq. (2.11) at the bombarding energy (4.4) as a function of the target charge number for various projectiles. The empirical relation (4.5) between charge and mass number has been used.

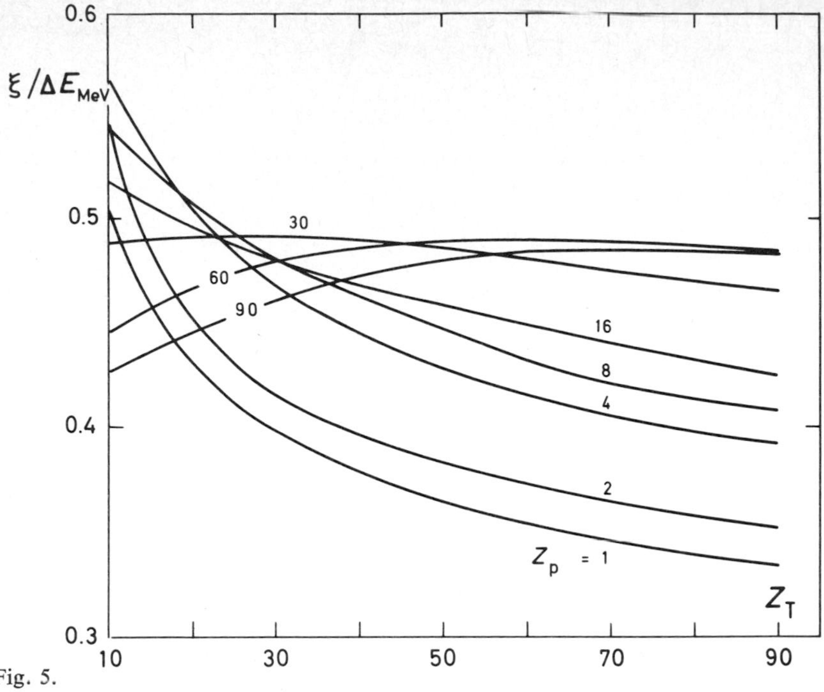
0.6
0.5
0.4
0.3
ξ / ΔE_MeV
30
60
90
16
8
4
2
Z_p = 1
Z_T
10
30
50
70
90

Fig. 5.

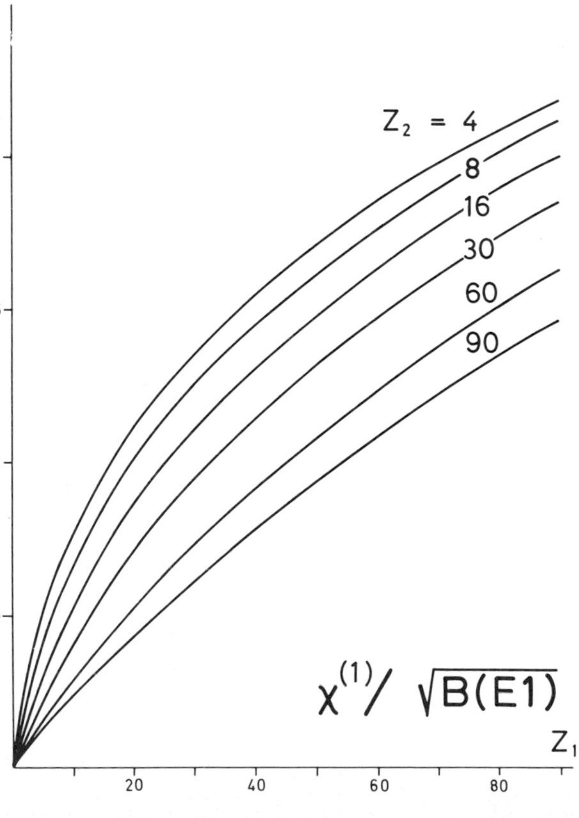
2.0
1.5
1.0
0.5
Z_2 = 4
8
16
30
60
90
χ^(1)/ √B(E1)
Z_1
20
40
60
80

Fig. 6.

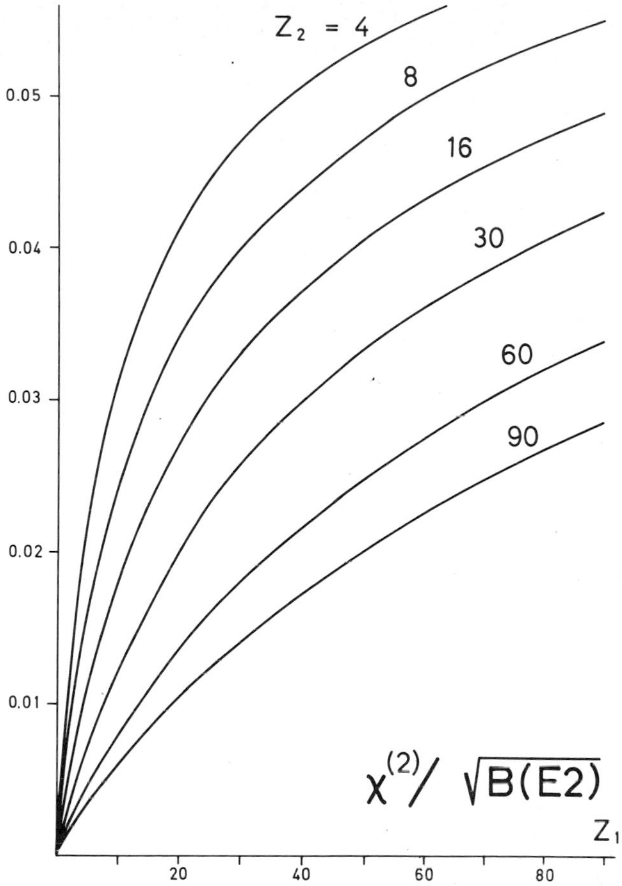

Fig. 7. The ratio $\chi^{(2)}/\sqrt{B(\mathrm{E2})}$ according to table 3 at the bombarding energy (4.4) as a function of Z_1 for different values of Z_2. The quantity $\sqrt{B(\mathrm{E2})}$ should be measured in units of $e(\mathrm{fm})^2$. For target excitation one should use $Z_1 = Z_\mathrm{P}$ and $Z_2 = Z_\mathrm{T}$ while for projectile excitation $Z_1 = Z_\mathrm{T}$ and $Z_2 = Z_\mathrm{P}$.

Fig. 5. The ratio $\xi/\Delta E_{\mathrm{MeV}}$ given by (2.15) at the bombarding energy (4.4) is plotted as a function of target charge number for different projectiles. The excitation energy ΔE is measured in MeV.

Fig. 6. The ratio $\chi^{(1)}/\sqrt{B(\mathrm{E1})}$ according to table 3 at the bombarding energy (4.4) as a function of Z_1 for different values of Z_2. The quantity $\sqrt{B(\mathrm{E1})}$ should be measured in units of $e \times$ fm. For target excitation one should use $Z_1 = Z_\mathrm{P}$ and $Z_2 = Z_\mathrm{T}$ while for projectile excitation $Z_1 = Z_\mathrm{T}$ and $Z_2 = Z_\mathrm{P}$.

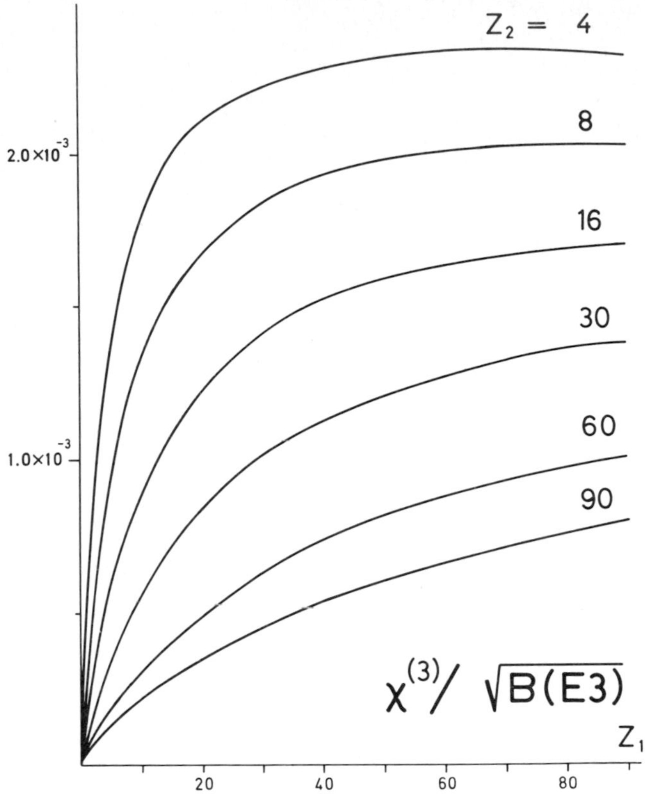

Fig. 8. The ratio $\chi^{(3)}/\sqrt{B}(\text{E3})$ according to table 3 at the bombarding energy (4.4) as a function of Z_1 for different values of Z_2. The quantity $\sqrt{B}(\text{E3})$ should be measured in units of $e(\text{fm})^3$. For target excitation one should use $Z_1 = Z_P$ and $Z_2 = Z_T$ while for projectile excitation $Z_1 = Z_T$ and $Z_2 = Z_P$.

Fig. 9. The ratio $\chi^{(4)}/\sqrt{B}(\text{E4})$ according to table 3 at the bombarding energy (4.4) as a function of Z_1 for different values of Z_2. The quantity $\sqrt{B}(\text{E4})$ should be measured in units of $e(\text{fm})^4$. For target excitation one should use $Z_1 = Z_P$ and $Z_2 = Z_T$ while for projectile excitation $Z_1 = Z_T$ and $Z_2 = Z_P$.

Fig. 10. The ratio $\chi^{(\text{M1})}/\sqrt{B}(\text{M1})$ according to table 3 at the bombarding energy (4.4) as a function of Z_1 for different values of Z_2. The quantity $\sqrt{B}(\text{M1})$ should be measured in units of $(e\hbar/2M_Pc)$. For target excitation one should use $Z_1 = Z_P$ and $Z_2 = Z_T$ while for projectile excitation $Z_1 = Z_T$ and $Z_2 = Z_P$.

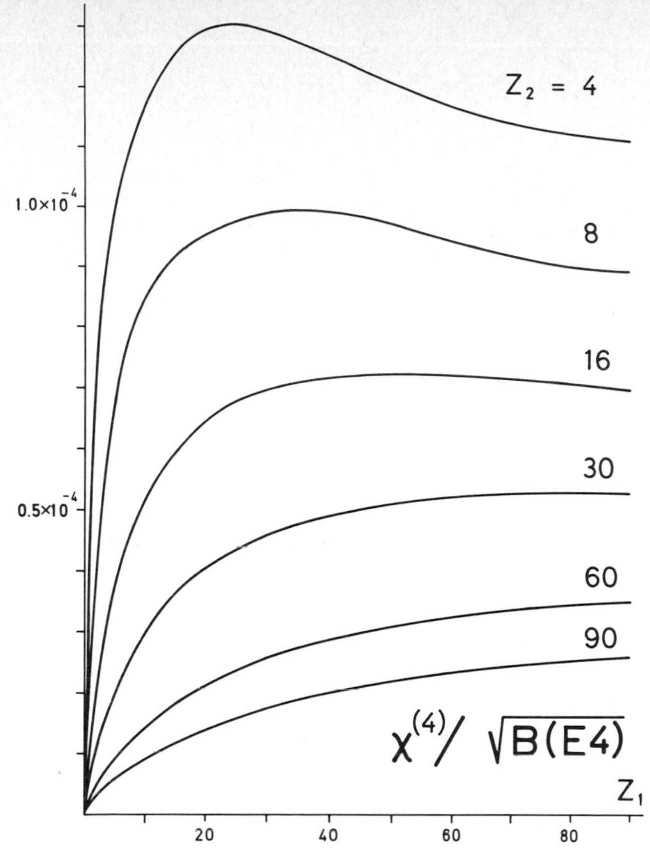

Z₂ = 4
8
16
30
60
90
χ(4)/ √B(E4)
Z₁
1.0×10⁻⁴
0.5×10⁻⁴
20
40
60
80

Fig. 9.

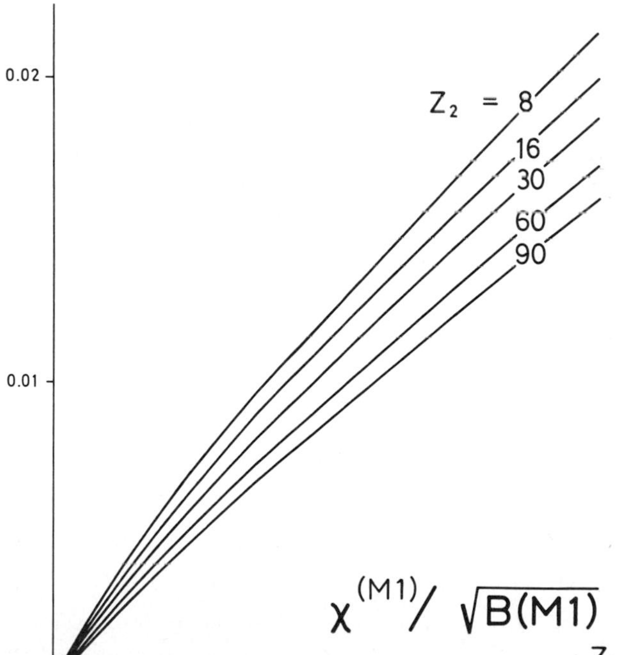

0.02
0.01
Z₂ = 8
16
30
60
90
χ(M1)/ √B(M1)
Z₁
20
40
60
80

Fig. 10.

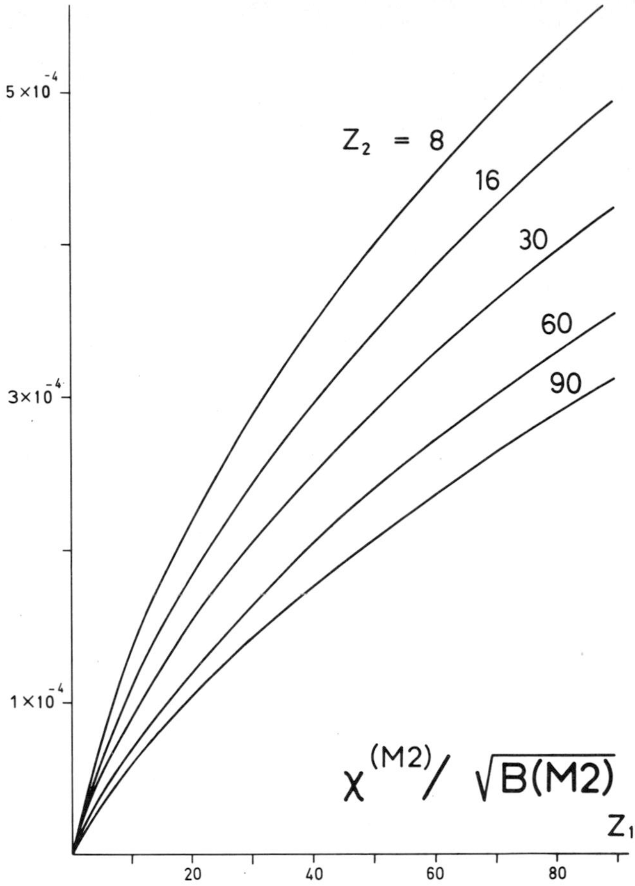

Fig. 11. The ratio $\chi^{(\mathrm{M2})}/\sqrt{B(\mathrm{M2})}$ according to table 3 at the bombarding energy (4.4) as a function of Z_1 for different values of Z_2. The quantity $\sqrt{B(\mathrm{M2})}$ should be measured in units of $(e\hbar/2M_{\mathrm{P}}c)$ fm. For target excitation one should use $Z_1 = Z_{\mathrm{P}}$ and $Z_2 = Z_{\mathrm{T}}$ while for projectile excitation $Z_1 = Z_{\mathrm{T}}$ and $Z_2 = Z_{\mathrm{P}}$.

target charge number for various projectiles. We have again used the empirical relation (5). It is noted that the minimum value of η is closely proportional to Z_{P} if $Z_{\mathrm{P}} < Z_{\mathrm{T}}$.

As was noted in § 2 the adiabaticity parameter ξ (see eq. (2.15)) is nearly proportional to the excitation energy $\Delta E = E_N - E_M$. In fig. 5 we have plotted the quantity $\xi/\Delta E_{\mathrm{MeV}}$ for the maximum value of the bombarding energy as a function of the target charge for different projectiles. In most cases $\xi/\Delta E_{\mathrm{MeV}}$ is of the order of 0.4, i.e.

$$\xi_{\min} \sim 0.4\, \Delta E_{\mathrm{MeV}}. \tag{6}$$

It is noted that the minimal values of η and ξ at the Coulomb barrier are symmetric in Z_P and Z_T.

To estimate the range of variations of the strength parameter χ we consider first the case of target excitation. Inserting the maximum energy (4) into the expressions of table 3, one obtains expressions for the maximum χ-values. The ratios $\chi^{(\lambda)}/\sqrt{B(\mathrm{E}\lambda)}$ and $\chi^{(\mathrm{M}\lambda)}/\sqrt{B(\mathrm{M}\lambda)}$ have been plotted in figs. 6–11 where one should insert $Z_1 = Z_P$ and $Z_2 = Z_T$. It is seen from these figures that for $\lambda = 1$ and 2 the largest χ-values are obtained for the heaviest targets and projectiles, while for higher multipole excitations the χ-value may go through a maximum at some intermediate Z-value. This observation is related to the fact that the higher multipole interactions depend so strongly on the distance of closest approach that the increase in χ due to the increase in charge is compensated by the increase in nuclear radius.

The figs. 6–11 can also be used for determining the maximum χ-value for projectile excitation by interpreting $Z_1 = Z_T$ and $Z_2 = Z_P$.

The order of magnitude of the χ-values can be estimated from the reduced transition probability as given by the so-called single particle or Weisskopf units [BOH 69]:

$$B_{\mathrm{W}}(\mathrm{E}\lambda) = \frac{(1.2)^{2\lambda}}{4\pi}\left(\frac{3}{\lambda+3}\right)^2 A^{2\lambda/3} \tag{7}$$

and

$$B_{\mathrm{W}}(\mathrm{M}\lambda) = \frac{10}{\pi}(1.2)^{2\lambda-2}\left(\frac{3}{\lambda+3}\right)^2 A^{(2\lambda-2)/3}. \tag{8}$$

The quantities (7) and (8) are measured in the units of table 2, i.e. $B(\mathrm{E}\lambda)$ in units of $e^2(\mathrm{fm})^{2\lambda}$ and $B(\mathrm{M}\lambda)$ in units of $(e\hbar/2M_Pc)^2 \times (\mathrm{fm})^{2\lambda-2}$.

§ 5. *Thick targets*

In actual Coulomb excitation experiments one has to work with targets of a finite thickness. During the penetration through the target the projectile loses energy mostly through excitation and ionization of electrons in the target material. The energy loss through collisions with nuclei in the target is of importance only for very low energies where no Coulomb excitation occurs (see e.g. [DAV 72]).

The energy loss can be described to a good approximation as a continuous process, where on the path length $\mathrm{d}x$ the energy is diminished by a definite amount $\mathrm{d}E$. The stopping power $\mathrm{d}E/\mathrm{d}x$ is a function of the projectile velocity v, as well as of target and projectile charge. The stopping power is often expressed in terms of the stopping cross section S defined by

$$\mathrm{d}E/\mathrm{d}x = nS, \tag{1}$$

where n is the number of atoms per unit volume. The stopping cross section S has a dimension of cross section times energy. If the target is a compound or a mixture of various elements, the stopping cross section S may be evaluated from the stopping cross section S_i for pure targets of material i by means of the relation

$$S = \sum_i \nu_i S_i, \tag{2}$$

where ν_i is the relative atomic abundance of material i

$$\nu_i = n_i/n, \tag{3}$$

i.e.

$$n = \sum_i n_i. \tag{4}$$

To describe the counting rate for a process p in the scattering on the nucleus k one uses the yield Y_{pk} which is the number of events per incoming projectile. In the above mentioned approximation of continuous stopping and under the condition that no shadowing effects occur (as with channelling in perfect crystals) one finds

$$Y_{pk} = \int_{x_1}^{x_2} \sigma_{pk}(E) n_k \, \mathrm{d}x, \tag{5}$$

where $\sigma_{pk}(E)$ is the cross section for the process p in the nucleus k.

In terms of the stopping cross section this yield can also be written

$$Y_{pk} = \nu_k \int_{E_\mathrm{i}}^{E_\mathrm{f}} \frac{\sigma_{pk}(E)}{S(E)} \, \mathrm{d}E. \tag{6}$$

The energy at the front surface (position x_1) of the target is denoted by E_i, while E_f is the exit energy at the back surface.

If the total energy loss $E_\mathrm{i} - E_\mathrm{f}$ is so small that the cross section is almost unchanged, i.e.

$$\{\sigma(E_\mathrm{i}) - \sigma(E_\mathrm{f})\}/\sigma(E_\mathrm{i}) \ll 1 \tag{7}$$

we may approximate (5) by

$$Y_{pk} = \sigma_{pk}(E_\mathrm{i}) n_k t, \tag{8}$$

where t is the target thickness. One usually measures this thickness in mg/cm^2 in which case (8) takes the form

$$Y_{pk} = 6.02217 \times 10^{-4} \sigma_{pk}(E_\mathrm{i}) \frac{\nu_k}{\sum_i \nu_i A_i} t_{\mathrm{mg/cm}^2}, \tag{9}$$

where σ is measured in 10^{-24} cm^2 and A_i denotes the mass number of the ith constituent in the target.

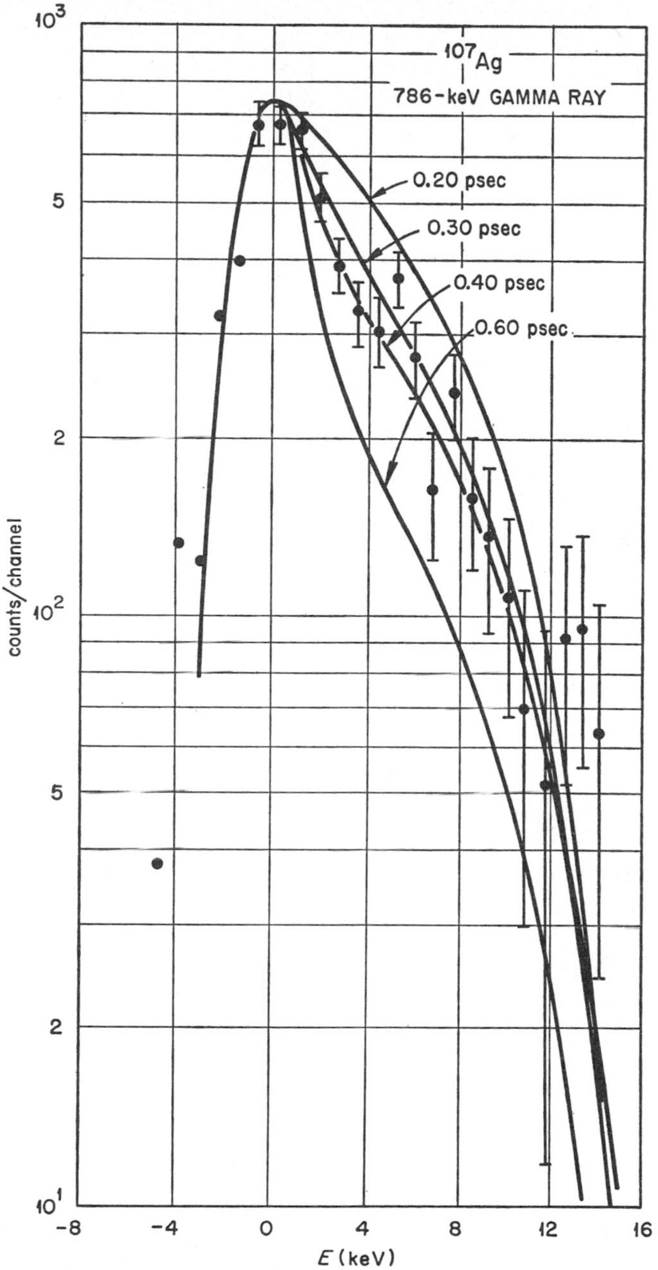

Fig. 12. Comparison of the full-energy peak of the 786 keV gamma line of ^{107}Ag, excited by 45 MeV ^{16}O in a thick target, with line shapes calculated for the half-lives given in the figure. (From Robinson, R. L. et al., 1970, Nucl. Phys. **A150**, 225.)

A convenient measure of the yield is the number of counts N per micro-Coulomb incoming beam. For this counting rate one finds

$$N = \frac{3.75870 \times 10^9}{Z_p^*} \sigma_{pk}(E_i) \frac{\nu_k}{\sum_i \nu_i A_i} t_{\mathrm{mg/cm}^2}, \tag{10}$$

where Z_p^* indicates the charge state of the beam.

Formulae (9) and (10) may be used for somewhat thicker targets if one inserts for the energy E_i the average energy $\frac{1}{2}(E_i + E_f)$. If the lifetime of a nuclear state is of the same order of magnitude as the stopping time of the excited recoil in a thick target, the shape of the gamma line due to Doppler shift may be analyzed to determine the lifetime. The sensitivity of the method for a particular case is illustrated in fig. 12.

§ 6. *Angular distributions of gamma rays*

If one wants to detect the γ-quanta which are emitted from the Coulomb excited nuclear states, a number of practical problems arise which we shall discuss in this section.

The γ-ray angular distribution can be determined from the tensors $\rho_{k\kappa}$ as was shown in chapter III. The most convenient coordinate system for the evaluation of these tensors is, according to § 3, evidently the coordinate system C because only in this system are the directions of the coordinate axes directly related to the laboratory system. Thus the z-axis is parallel to the incoming beam direction, while the x–z plane coincides with the scattering plane.

The angular distribution of the γ-quanta can always be written in the form

$$W(\vartheta_\gamma, \varphi_\gamma) = \sum_{k\kappa} A^*_{k\kappa}(\vartheta)\, Y_{k\kappa}(\vartheta_\gamma, \varphi_\gamma), \tag{1}$$

with

$$A_{k\kappa}(\vartheta) = \sum_{k'\kappa'} \rho^{\mathrm{C}}_{k'\kappa'}(\vartheta)\, K_{k'\kappa',k\kappa} \tag{2}$$

where ϑ_γ, φ_γ are the polar coordinates of the emitted γ-quantum and ϑ is the scattering angle. The coefficients $K_{k'\kappa',k\kappa}$ may partly describe the effects of unobserved γ-quanta and conversion electrons, or the effects of external fields or other attenuation effects which have been discussed in chapter III.

The coordinate system which we have introduced has fixed directions with respect to the laboratory system but is moving with a velocity $\boldsymbol{w}'$ which is the laboratory velocity of the decaying nucleus. In many cases the nucleus will be brought to rest in the target or in a backing before it decays and the angular distribution of the γ-quanta in the laboratory system is then given by eq. (1). If the nucleus decays in flight a distortion of the angular distribution will take place, which may be quite important especially for projectile excitation.

The distortion of the angular distribution is most easily evaluated in a coordinate system where the z-axis is parallel to the velocity $\boldsymbol{w}'$. The probability of detecting a γ-quantum in the direction ϑ'_γ, φ'_γ, within the solid angle $d\Omega'_\gamma$ may then be written

$$W(\vartheta'_\gamma, \varphi'_\gamma)\, d\Omega'_\gamma = \sum_{k\kappa\kappa'} A^*_{k\kappa} D^k_{\kappa\kappa'}(0, \theta, 0)\, Y_{k\kappa'}(\vartheta'_\gamma, \varphi'_\gamma)\, d\Omega'_\gamma. \tag{3}$$

The scattering angle of the decaying nucleus in the laboratory system is denoted by θ, which may either be θ_{T} or θ_{P}. The angles ϑ'_γ and φ'_γ are measured in the system which moves with velocity $\boldsymbol{w}'$. If this γ-quantum is to be observed in the laboratory system the counter must be put at a position with polar angles θ_γ, ϕ_γ which are related to ϑ'_γ, φ'_γ by the relation

$$\tan(\tfrac{1}{2}\vartheta'_\gamma) = \sqrt{\frac{1 + w'/c}{1 - w'/c}}\, \tan(\tfrac{1}{2}\theta_\gamma), \tag{4}$$

and

$$\phi_\gamma = \varphi'_\gamma. \tag{5}$$

The angular distribution in the laboratory system is therefore given by

$$W_{\mathrm{lab}}(\theta_\gamma, \phi_\gamma) = W[\vartheta'_\gamma(\theta_\gamma), \varphi'_\gamma]\, d\Omega'_\gamma/d\omega_\gamma. \tag{6}$$

The solid angle ratio determined from eqs. (4)–(5) is given by

$$\frac{d\Omega'_\gamma}{d\omega_\gamma} = \frac{1 - (w'/c)^2}{(1 - (w'/c)\cos\theta_\gamma)^2}. \tag{7}$$

In Coulomb excitation the velocity w' may reach one tenth of the velocity of light and the distortion effects may thus be quite important. However, quadratic terms in w'/c are of minor importance and we shall therefore only evaluate the explicit expressions for the angular distribution to first order in w'/c. One finds to this accuracy

$$\vartheta'_\gamma = \theta_\gamma + (w'/c)\sin\theta_\gamma \tag{8}$$

and

$$d\Omega'_\gamma/d\omega_\gamma = 1 + 2(w'/c)\cos\theta_\gamma. \tag{9}$$

Expanding the Legendre polynomial in eq. (3) we may write eq. (6) in the form

$$W_{\mathrm{lab}}(\theta_\gamma, \phi_\gamma) = W(\theta_\gamma, \phi_\gamma) + (w'/c)\Delta W(\theta_\gamma, \phi_\gamma). \tag{10}$$

The first-order correction ΔW can again be expressed in terms of spherical harmonics as

$$\Delta W = \sum_{k\kappa\kappa'} A^*_{k\kappa} D^k_{\kappa\kappa'}(0, \theta, 0)(-1)^{k+\kappa'+1}$$
$$\times \left[\begin{pmatrix} k & k-1 & 1 \\ \kappa' & -\kappa' & 0 \end{pmatrix} \sqrt{k(k-1)}\, Y_{k-1,\kappa'}(\theta_\gamma, \phi_\gamma) \right.$$
$$\left. + \begin{pmatrix} k & k+1 & 1 \\ \kappa' & -\kappa' & 0 \end{pmatrix} \sqrt{k+1}(k+2)\, Y_{k+1,\kappa'}(\theta_\gamma, \phi_\gamma)\right]. \quad (11)$$

In this form, the correction can be transformed back to the coordinate system C with z-axis along the beam direction. One finds

$$\Delta W^{\mathrm{C}} = \sum_{k\kappa\kappa'} A^*_{k\kappa}(-1)^{k+\kappa'+1}$$
$$\times \left[\sqrt{\tfrac{4}{3}\pi k(k-1)} \begin{pmatrix} k & k-1 & 1 \\ \kappa & -\kappa' & \mu \end{pmatrix} Y_{1\mu}(\theta, 0)\, Y_{k-1,\kappa'}(\theta^{\mathrm{C}}_\gamma, \phi^{\mathrm{C}}_\gamma) \right.$$
$$\left. + \sqrt{\tfrac{4}{3}\pi(k+1)(k+2)} \begin{pmatrix} k & k+1 & 1 \\ \kappa & -\kappa' & \mu \end{pmatrix} Y_{1\mu}(\theta, 0)\, Y_{k+1,\kappa'}(\theta^{\mathrm{C}}_\gamma, \phi^{\mathrm{C}}_\gamma)\right]. \quad (12)$$

Usually the sum over k contains only even values of k and the distortion (12) is then seen to contain only Legendre polynomials of odd order. It is noted that the expression (12) is explicitly a scalar under rotation.

Besides the distortion (aberration) of the angular distribution the moving source gives rise to a distortion of the frequency of the observed γ-quanta due to a Doppler shift. The relativistic expression for the observed frequency ω' is

$$\omega' = \omega \frac{1 + (w'/c)\cos\theta_\gamma}{\sqrt{1-(w'/c)^2}}, \quad (13)$$

where ω is decay frequency in the rest system, while θ_γ is the angle between the velocity $\boldsymbol{w}'$ of the decaying nucleus and angular observation in the laboratory system. To first order in w'/c the apparent decay energy E'_γ is given by

$$E'_\gamma = E_\gamma(1 + (w'/c)\cos\theta_\gamma) \quad (14)$$

in terms of the true decay energy E_γ.

This Doppler shift makes it possible to detect the recoil angle and through it the scattering angle ϑ by an accurate γ-energy measurement. For excited ions of high velocity (excited projectiles or recoils of mass similar to that of the projectile) this effect may be used to observe the influence of Q_2 on $d\sigma_{2^+}(\vartheta)$. An analysis of an experiment of this type is shown in fig. 13.

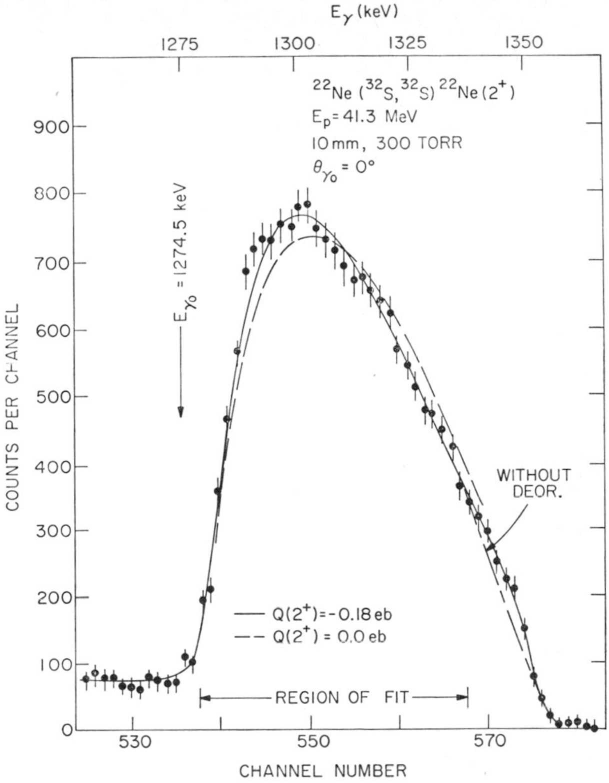

Fig. 13. The 1274 keV line of ^{22}Ne after subtraction of the background spectrum. The solid line represents the theoretical line shape for $Q(2^+) = -0.21$ b (best fit). For comparison, the theoretical curve for $Q(2^+) = 0$ is also given. (From Schwalm, D. and B. Povh, 1969, Phys. Lett. **29B**, 103.)

If the energy shifts are not detected, γ-counters of finite opening will give rise to a smearing of the angular distribution. For γ-counters of finite opening it is convenient to introduce an auxiliary coordinate system where the z-axis points towards the centre of the γ-counter which we assume to have the polar coordinates ϑ_c, φ_c in the laboratory system. In this coordinate system the spherical harmonics in the angular distributions take the form

$$Y_{k\kappa}(\vartheta_\gamma, \varphi_\gamma) = \sum_{\kappa'} D^k_{\kappa\kappa'}(\varphi_c, \vartheta_c, 0)\, Y_{k\kappa'}(\vartheta', \varphi'), \tag{15}$$

where ϑ', φ' defines the direction of the γ-quantum with respect to the counter axis. The integration over a finite γ-detector is then reduced to an

TABLE 6
The attenuation coefficients B_k defined in eq. (6.16) for axially symmetric γ-counters with constant efficiency over their front faces of opening angle ϑ_0.

$B_0 = 1$
$B_1 = \cos^2(\frac{1}{2}\vartheta_0)$
$B_2 = \cos\vartheta_0 \cos^2(\frac{1}{2}\vartheta_0)$
$B_3 = \frac{1}{4}\cos^2(\frac{1}{2}\vartheta_0)(5\cos^2\vartheta_0 - 1)$
$B_4 = \frac{1}{4}\cos^2(\frac{1}{2}\vartheta_0)(7\cos^2\vartheta_0 - 3)\cos\vartheta_0$

integration over ϑ' and φ'. The resulting expressions are especially simple for γ-counters, possessing an axial symmetry. The integration over φ' leads then to the condition $\kappa' = 0$ and the D-function reduces to a spherical harmonics $Y_{k\kappa}(\vartheta_c, \varphi_c)$.

In this simple case the spherical harmonics (15) is replaced by the same spherical harmonics in the angles of the counter axis times a factor B_k defined by

$$B_k = \frac{1}{\Omega_\gamma}\sqrt{\frac{4\pi}{2k+1}}\int_{\Omega_\gamma} d\Omega_\gamma\, Y_{k0}(\vartheta', \varphi') \tag{16}$$

where we have assumed a constant efficiency.

The effect of a finite symmetric γ-counter can thus be incorporated in the angular distributions by the substitution

$$Y_{k\kappa}(\vartheta_\gamma, \varphi_\gamma) \to B_k Y_{k\kappa}(\vartheta_c, \varphi_c). \tag{17}$$

Explicit expressions for B_k are given in table 6.

§ 7. *Corrections to the Rutherford orbit*

In this section we collect a number of effects which, besides the quantum mechanical effects (see chapter IX), may be of interest in high precision experiments. A number of such effects can be described through a slight change of the scattering potential from the pure Coulomb field $Z_1Z_2e^2/r$. To estimate the influence of such deviations we assume that the main effect is associated with the change in the distance of closest approach in the collision. If the additional potential $\Delta V(r)$ does not change too rapidly in the neighbourhood of the distance

$$b(\vartheta) = a(1 + 1/\sin(\tfrac{1}{2}\vartheta)), \tag{1}$$

we may estimate the ratio of the change Δb in b to the distance of closest approach b by the expression

$$\frac{\Delta b}{b} \approx \frac{\Delta V(b(\vartheta))}{E} \tfrac{1}{2}(1 + \sin(\tfrac{1}{2}\vartheta)). \tag{2}$$

The true trajectory may be simulated by a Rutherford orbit with the experimental scattering angle and with the correct distance of closest approach $b + \Delta b$. By adjusting the centre-of-mass energy to be

$$E_{\text{eff}} = E + \delta E \tag{3}$$

the following energy change is found from (2) and (1)

$$\delta E = -\tfrac{1}{2}\Delta V(b(\vartheta))(1 + \sin(\tfrac{1}{2}\vartheta)). \tag{4}$$

The first-order effects of small changes in the Coulomb potential are thus simulated by using the bombarding energy

$$(E_{\text{P}})_{\text{eff}} = E_{\text{P}} - \tfrac{1}{2}(1 + A_{\text{P}}/A_{\text{T}})\, \Delta V(b(\vartheta))(1 + \sin(\tfrac{1}{2}\vartheta)). \tag{5}$$

As possible changes in the Coulomb potential we mention first the *polarization potential* discussed in appendix J. According to eq. (J.16) the polarization potential is given by

$$\Delta V_{\text{pol}}(r) = -1.7 \times 10^{-3} \frac{Z_1^2 A_2^{5/3} + Z_2^2 A_1^{5/3}}{r^4} e^2(\text{fm})^3. \tag{6}$$

From eq. (6) we obtain the corresponding effective change in the centre-of-mass energy

$$\delta E_{\text{MeV}} = 4.5 \times 10^{-3} \frac{Z_1^2 A_2^{5/3} + Z_2^2 A_1^{5/3}}{Z_1^4 Z_2^4} E_{\text{MeV}}^4 \frac{\sin^4(\tfrac{1}{2}\vartheta)}{(1 + \sin(\tfrac{1}{2}\vartheta))^3}. \tag{7}$$

The maximum value of this quantity is obtained for bombardment of uranium on uranium at the Coulomb barrier, where it is of the order of magnitude of 2 MeV. For O-ions on Pb one finds $\delta E \sim 0.2$ MeV.

Next we consider the effect of *screening by the atomic electrons*. We describe the screening in terms of a screening potential ΔV_{screen} which for distances smaller than the Thomas-Fermi radius

$$a_{\text{TF}} = 4.685 \times 10^{-9} Z^{-1/3} \text{ cm} \tag{8}$$

can be approximated by [GOM 49]

$$\Delta V_{\text{screen}}(r) = -4.78 \times 10^{-5}(Z_1 Z_2^{4/3} + Z_1^{4/3} Z_2) \text{ MeV}. \tag{9}$$

The effect of this potential can be simulated by changing the centre-of-mass energy by the amount

$$\delta E_{\mathrm{MeV}} = 2.39 \times 10^{-5}(Z_1 Z_2^{4/3} + Z_1^{4/3} Z_2)(1 + \sin(\tfrac{1}{2}\vartheta)). \tag{10}$$

The maximum value of this quantity obtains for U on U collisions where $\delta E \sim 2$ MeV, while for O on Pb one finds $\delta E \sim 0.2$ MeV.

Finally, we consider the effect of *vacuum polarization* for which the following estimate has been given by [FOL 54]

$$\Delta V_{\mathrm{vac}}(r) = \frac{Z_1 Z_2 e^2}{r} \frac{2\alpha}{3\pi} \int_1^\infty \exp\left\{-\frac{2\xi r}{\lambda\!\!\!^{-}_{\mathrm{c}}}\right\}\left(1 + \frac{1}{2\xi^2}\right) \frac{(\xi^2 - 1)^{1/2}}{\xi^2}\, \mathrm{d}\xi, \tag{11}$$

where α is the fine-structure constant and $\lambda\!\!\!^{-}_{\mathrm{c}}$ the Compton wave length of the electron

$$\lambda\!\!\!^{-}_{\mathrm{c}} = \hbar/mc = 386.17\ \mathrm{fm}. \tag{12}$$

Since the distance r in Coulomb excitation is considerably smaller than $\lambda\!\!\!^{-}_{\mathrm{c}}$ one may expand the potential (11) in powers of r. One finds

$$\Delta V_{\mathrm{vac}}(r) = \frac{Z_1 Z_2 e^2}{r} \frac{2\alpha}{3\pi}\left\{\log \frac{\gamma \lambda\!\!\!^{-}_{\mathrm{c}}}{r} - \frac{5}{6} + \frac{3\pi}{4} \frac{r}{\lambda\!\!\!^{-}_{\mathrm{c}}} + \cdots\right\}, \tag{13}$$

where γ is the Eulerian constant

$$\gamma = 0.5772\ldots. \tag{14}$$

For the effective change δE in the centre-of-mass energy one finds

$$\delta E_{\mathrm{MeV}} = 1.55 \times 10^{-3} \log \frac{96.86}{(b(\vartheta))_{\mathrm{fm}}} E_{\mathrm{MeV}} \sin(\tfrac{1}{2}\vartheta). \tag{15}$$

This shows that vacuum polarization effects are of the same order of magnitude as polarization and screening.

For collisions close to the Coulomb barrier the *nuclear forces* may give rise to a change to the Rutherford trajectory. It seems that the following ion-ion potential can be used for order of magnitude estimates

$$\Delta V_{\mathrm{nuc}}(r) = -Sr^2 \exp\{(R - r)/a\}, \tag{16}$$

where the parameters have the following values

$$\begin{aligned} S &\approx 2\ \mathrm{MeV\,fm^{-2}}, \\ a &\approx 0.59\ \mathrm{fm}, \\ R &\sim 1.14(A_1^{1/3} + A_2^{1/3})\ \mathrm{fm}. \end{aligned} \tag{17}$$

It should be noted, however, that in many cases the main effect of the nuclear forces is associated with the non-diagonal effects.

Other deviations from the pure Coulomb trajectory are due to deviations of the field from spherical symmetry. Thus a static quadrupole moment of the nucleus gives rise to orbits which do not lie in a plane. On the average, for unpolarized targets these effects vanish but may be taken into account for polarized targets [SMI 68, ROS 73].

In the present review we have completely neglected *relativistic effects* which are expected to be of the order of magnitude $(v/c)^2$, i.e.

$$(v/c)^2 = 0.00214711 E_{\mathrm{P,MeV}}/A_{\mathrm{P}}. \tag{18}$$

The maximum value of E_{P} should be taken from fig. 4 and it is thus seen that corrections of the order of magnitude up to 1% could be expected.

In principle it should be possible, neglecting radiation loss, to formulate a Coulomb excitation theory which includes terms of order $(v/c)^2$ [LAN 62]. Some relativistic effects are due to the change in the centre-of-mass transformation as compared to the expressions in § 1. Others may be included by using retarded expressions for the multipole operators (cf. eq. (G.4)). We shall here only estimate a third type of relativistic corrections associated with the change of the orbit. The relativistic treatment has been solved in first-order perturbation theory and is used in inelastic electron scattering.

The relativistic motion of a particle of charge $Z_1 e$ and rest mass m_o in the point Coulomb field of a charge $Z_2 e$ is determined by the equation of motion

$$m_o \frac{\mathrm{d}^2 \boldsymbol{r}}{\mathrm{d}\tau^2} = \frac{Z_1 Z_2 e^2}{r^3} \boldsymbol{r} \left[\frac{W}{m_o c^2} - \frac{Z_1 Z_2 e^2}{m_o c^2 r} \right], \tag{19}$$

where τ is the proper time and W the total energy

$$W = m_o c^2 + E. \tag{20}$$

If the second term in (19) would not be present, the solution would lead to Rutherford trajectories but with slightly changed initial conditions. Thus, the scattering angle would be determined by the relation

$$\tan(\tfrac{1}{2}\vartheta) = \frac{Z_1 Z_2 e^2}{(m_o^2 c^2/W)(\mathrm{d}r/\mathrm{d}\tau)^2_{r=\infty}} \frac{1}{\rho}, \tag{21}$$

where ρ is the impact parameter. Expressing the denominator in terms of the relativistic kinetic energy E one finds

$$\tan(\tfrac{1}{2}\vartheta) = \frac{Z_1 Z_2 e^2}{2 E_{\mathrm{eff}} \rho} \tag{22}$$

where

$$E_{\text{eff}} = E(1 - \tfrac{1}{2}E/m_o c^2). \tag{23}$$

The second term in (19) gives rise to a modification of the trajectory which can be estimated from the potential

$$\Delta V(r) = -\frac{(Z_1 Z_2 e^2)^2}{2m_o c^2}\frac{1}{r^2} \tag{24}$$

by means of eq. (4). One finds

$$\delta E = \frac{E^2}{m_o c^2}\frac{\sin^2(\frac{1}{2}\vartheta)}{1 + \sin(\frac{1}{2}\vartheta)}. \tag{25}$$

The total change in the trajectory can therefore be estimated by changing the bombarding energy by the amount

$$\delta E = -\frac{E^2}{2m_o c^2}\frac{1 - \sin(\frac{1}{2}\vartheta)}{1 + \sin(\frac{1}{2}\vartheta)}(1 + 2\sin(\tfrac{1}{2}\vartheta)). \tag{26}$$

This correction is somewhat smaller than the estimate expected from (18).

In the estimates in this section we have only considered the corrections due to the change in the orbit. Additional corrections arise from the difference in the time development of the collision process. This change in the effective collision time could be especially important for excitations with large values of ξ where the excitation probability depends exponentially on ξ.

Finally, it should be remembered that in violent ion-ion collisions, that may lead to Coulomb excitation, energy may be lost partly due to ejection of electrons in the two colliding atoms and partly due to bremsstrahlung. The ejection of K and L electrons has been studied recently [HAN 72]. The average energy loss is typically of the order of magnitude 10 keV. The cross section for bremsstrahlung emitted in heavy ion collision has been discussed in [ALD 56]. Due to the fact that the charge-to-mass ratio is similar for all heavy ions, the dipole bremsstrahlung is strongly reduced. The average energy loss to bremsstrahlung is typically of the order of magnitude 1 keV.

APPENDIX A

Multipole-Multipole Interaction

A problem of rather general interest in atomic and nuclear physics is that of the electromagnetic interaction between two extended charge-current distributions (cf. [HIR 54] p. 835 ff.). A convenient form of this interaction can be obtained when the two charge-current distributions do not overlap, since the distributions can then be described by means of the electric and magnetic multipole moments.

Let us first consider the case where the two charge-current distributions are stationary. As is illustrated in fig. 1 we describe the charge-current distribution of system 1 in a coordinate system S_1 by $\rho_1(\boldsymbol{r}_1')$ and $\boldsymbol{j}_1(\boldsymbol{r}_1')$ while the corresponding quantities of system 2 are given by $\rho_2(\boldsymbol{r}_2'), \boldsymbol{j}_2(\boldsymbol{r}_2')$ in a coordinate system S_2 with axes parallel to those of S_1. The position vector of the origin of 1 with respect to the origin of 2 is denoted by $\boldsymbol{r}$, which we first consider time-independent. Neglecting retardation the interaction Hamiltonian between the systems 1 and 2 is given by (cf. [LAN 62] p. 193)

$$W(1, 2) = \int\int \mathrm{d}\tau_1\, \mathrm{d}\tau_2\, \frac{\rho_1(\boldsymbol{r}_1')\rho_2(\boldsymbol{r}_2') - (1/c^2)\boldsymbol{j}_1(\boldsymbol{r}_1')\,\boldsymbol{j}_2(\boldsymbol{r}_2')}{|\boldsymbol{r} + \boldsymbol{r}_1' - \boldsymbol{r}_2'|}. \tag{1}$$

The expansion of $1/|\boldsymbol{r} + \boldsymbol{r}_1' - \boldsymbol{r}_2'|$ into powers of either r_1' or r_2' can be made by means of the formula

$$\frac{1}{|\boldsymbol{r} + \boldsymbol{r}_1' - \boldsymbol{r}_2'|} = \sum_{\lambda_2\mu_2} \frac{4\pi}{(2\lambda_2 + 1)} \frac{(r_2')^{\lambda_2}}{|\boldsymbol{r} + \boldsymbol{r}_1'|^{\lambda_2+1}}\, Y_{\lambda_2\mu_2}(\boldsymbol{r}_2')\, Y^*_{\lambda_2\mu_2}(\boldsymbol{r} + \boldsymbol{r}_1'), \tag{2}$$

where we have assumed that $r_2' < |\boldsymbol{r} + \boldsymbol{r}_1'|$. A similar expansion can be made in powers of r_1' under the condition $r_1' < |\boldsymbol{r} - \boldsymbol{r}_2'|$. From these two expansions one may conclude

$$\frac{1}{|\boldsymbol{r} + \boldsymbol{r}_1' - \boldsymbol{r}_2'|} = \sum_{\substack{\lambda_1\lambda_2\lambda \\ \mu_1\mu_2\mu}} a(\lambda_1, \lambda_2, \lambda, \mu_1, \mu_2, \mu) \frac{(r_1')^{\lambda_1}(r_2')^{\lambda_2}}{r^{\lambda+1}}\, Y_{\lambda_1\mu_1}(\boldsymbol{r}_1')\, Y_{\lambda_2\mu_2}(\boldsymbol{r}_2')\, Y_{\lambda\mu}(\boldsymbol{r}) \tag{3}$$

provided that $r > r_1' + r_2'$. The dependence of the coefficients on the indices μ_1, μ_2 and μ can be determined from the condition that the expression (3) is a

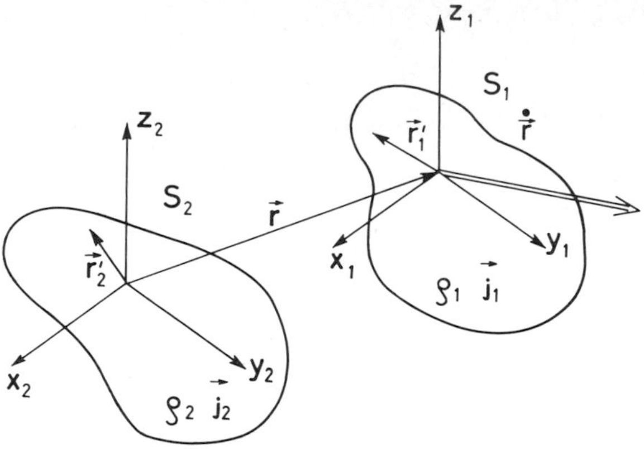

Fig. 1. Illustration of the coordinate systems S_1 and S_2 used in the evaluation of the interaction between two extended charge-current distributions ρ_1, $\boldsymbol{j}_1$ and ρ_2, $\boldsymbol{j}_2$. The system S_1 is produced by a parallel displacement of the system S_2 through the vector $\boldsymbol{r}$. The relative velocity $\dot{\boldsymbol{r}}$ of the origin of S_1 is indicated as well as position vectors $\boldsymbol{r}_1'$ and $\boldsymbol{r}_2'$ within the two frames of reference.

scalar under rotations of the coordinate system. It must thus be of the form

$$a(\lambda_1, \lambda_2, \lambda, \mu_1, \mu_2, \mu) = \begin{pmatrix} \lambda_1 & \lambda_2 & \lambda \\ \mu_1 & \mu_2 & \mu \end{pmatrix} c(\lambda_1, \lambda_2, \lambda), \tag{4}$$

where the first factor is the 3-j symbol which couples the three spherical harmonics in (3) to a scalar, while the second factor depends only on λ_1, λ_2 and λ. We note furthermore that from dimensional arguments

$$\lambda = \lambda_1 + \lambda_2. \tag{5}$$

The dependence of the coefficient c on λ_1 and λ_2 can be found by considering a special case e.g. where the three vectors $\boldsymbol{r}$, $\boldsymbol{r}_1'$ and $\boldsymbol{r}_2'$ are parallel. One finds

$$c(\lambda_1, \lambda_2) = (4\pi)^{3/2}(-1)^{\lambda_2}\sqrt{\frac{(2\lambda_1 + 2\lambda_2)!}{(2\lambda_1 + 1)!\,(2\lambda_2 + 1)!}}. \tag{6}$$

The result is thus

$$\frac{1}{|\boldsymbol{r} + \boldsymbol{r}_1' - \boldsymbol{r}_2'|} = \sum_{\lambda_1\lambda_2\mu_1\mu_2} c(\lambda_1\lambda_2)\begin{pmatrix} \lambda_1 & \lambda_2 & \lambda_1 + \lambda_2 \\ \mu_1 & \mu_2 & -(\mu_1 + \mu_2) \end{pmatrix} \frac{(r_1')^{\lambda_1}(r_2')^{\lambda_2}}{r^{\lambda_1+\lambda_2+1}}$$
$$\times\; Y_{\lambda_1\mu_1}(\boldsymbol{r}_1')\; Y_{\lambda_2\mu_2}(\boldsymbol{r}_2')\; Y_{\lambda_1+\lambda_2,-(\mu_1+\mu_2)}(\boldsymbol{r}). \tag{7}$$

In the stationary case we may write (1) in the form

$$W_{\text{stat}}(1, 2) = W_{\text{E}}(1, 2) + W_{\text{M}}(1, 2), \tag{8}$$

where the electric-electric multipole interaction $W_E(1, 2)$ originates from the first term in eq. (1). One finds from eq. (7)

$$W_E(1, 2) = \sum_{\lambda_1\lambda_2\mu_1\mu_2} c(\lambda_1\lambda_2)\begin{pmatrix}\lambda_1 & \lambda_2 & \lambda_1 + \lambda_2 \\ \mu_1 & \mu_2 & -(\mu_1 + \mu_2)\end{pmatrix}$$
$$\times \mathcal{M}_1(E\lambda_1, \mu_1)\mathcal{M}_2(E\lambda_2, \mu_2)\frac{1}{r^{\lambda_1+\lambda_2+1}} Y_{\lambda_1+\lambda_2, -(\mu_1+\mu_2)}(\boldsymbol{r}), \quad (9)$$

where the electric multipole moments are defined in eq. (II.1.3).

The magnetic-magnetic interaction which arises from the second term in (1) can be evaluated by inserting (7) in (1) and expanding the current vectors $\boldsymbol{j}$ on the spherical unit vectors

$$\boldsymbol{e}_{\pm 1} = \mp \frac{1}{\sqrt{2}}(\boldsymbol{e}_x \pm i\boldsymbol{e}_y),$$
$$\boldsymbol{e}_0 = \boldsymbol{e}_z. \quad (10)$$

Rewriting the product of the spherical harmonics and the unit vectors in terms of vector spherical harmonics defined by [EDM 57] p. 83

$$\boldsymbol{\Phi}_{\lambda JM}(\boldsymbol{r}) = \sum_{\mu q} Y_{\lambda\mu}(\boldsymbol{r}) \langle\lambda\mu 1 q\,|JM\rangle\, \boldsymbol{e}_q, \quad (11)$$

and utilizing that

$$\operatorname{div} \boldsymbol{j} = 0, \quad (12)$$

one obtains

$$W_M(1, 2) = \sum_{\lambda_1\lambda_2\mu_1\mu_2} c(\lambda_1, \lambda_2)\begin{pmatrix}\lambda_1 & \lambda_2 & \lambda_1 + \lambda_2 \\ \mu_1 & \mu_2 & -(\mu_1 + \mu_2)\end{pmatrix}$$
$$\times \mathcal{M}_1(M\lambda_1, \mu_1)\, \mathcal{M}_2(M\lambda_2, \mu_2)\frac{1}{r^{\lambda_1+\lambda_2+1}} Y_{\lambda_1+\lambda_2, -(\mu_1+\mu_2)}(\boldsymbol{r}), \quad (13)$$

where the magnetic multipole moments are defined by eq. (II.1.4).

It is noted that the expressions (9) and (13) are identical except for the fact that the summation in (9) includes the terms with $\lambda_1 = 0$ and $\lambda_2 = 0$.

The formulae (9) and (13) can also be applied for time dependent charge-current distributions as long as the retardation effects can be neglected. In this approximation the expressions for the interaction energy can be generalized to the case where the systems 1 and 2 have a relative velocity i.e. where the vector $\boldsymbol{r}$ is a function of time, and where

$$|\dot{\boldsymbol{r}}|/c \ll 1, \quad (14)$$

so that quadratic terms in this quantity can be neglected.

For the electromagnetic interaction of the two moving systems we may again use expression (1) with modified charge-current densities which take into account the translatory motion. If we evaluate the interaction in a rest system with respect to 2 the charge-current density of system 1 is given by

$$\boldsymbol{j}_1' = \boldsymbol{j}_1 + \dot{\boldsymbol{r}}\rho_1,$$

$$\rho_1' = \rho_1 + \frac{1}{c^2}\dot{\boldsymbol{r}}\boldsymbol{j}_1 \tag{15}$$

to lowest order in $\dot{\boldsymbol{r}}/c$.

Inserting (15) in (1) we obtain the following interaction energy

$$W(1, 2) = \int\int \frac{\rho_1\rho_2 - (1/c^2)\boldsymbol{j}_1\boldsymbol{j}_2}{|\boldsymbol{r} + \boldsymbol{r}_1' - \boldsymbol{r}_2'|}\,\mathrm{d}\tau_1\,\mathrm{d}\tau_2 - \frac{\dot{\boldsymbol{r}}}{c^2}\int\int \frac{(\rho_1\boldsymbol{j}_2 - \rho_2\boldsymbol{j}_1)}{|\boldsymbol{r} + \boldsymbol{r}_1' - \boldsymbol{r}_2'|}\,\mathrm{d}\tau_1\,\mathrm{d}\tau_2. \tag{16}$$

Neglecting retardation effects, the first integral leads again to the results (8), (9) and (13). The second integral, however, gives rise to an interaction between the electric multipole moments of one system with the magnetic multipole moments of the other system. We may thus write

$$W(1, 2) = W_{\mathrm{E}}(1, 2) + W_{\mathrm{M}}(1, 2) + W_{\mathrm{EM}}(1, 2). \tag{17}$$

The electric-magnetic multipole-multipole expansion can be obtained by inserting (7) into (16) and expanding $\boldsymbol{j}_1$ and $\boldsymbol{j}_2$ on the unit vectors (10). By a similar technique to that applied above for the magnetic-magnetic interaction one obtains the following result:*

$$W_{\mathrm{EM}}(1, 2)$$

$$= \sum_{\lambda_1\lambda_2\mu_1\mu_2} ic(\lambda_1\lambda_2)\{\mathscr{M}_1(\mathrm{E}\lambda_1, \mu_1)\,\mathscr{M}_2(\mathrm{M}\lambda_2, \mu_2) - \mathscr{M}_1(\mathrm{M}\lambda_1, \mu_1)\,\mathscr{M}_2(\mathrm{E}\lambda_2, \mu_2)\}$$

$$\times \frac{\dot{\boldsymbol{r}}}{c}\frac{1}{r^{\lambda_1+\lambda_2+1}}\left\{\begin{pmatrix}\lambda_1 & \lambda_2 & \lambda_1+\lambda_2\\ \mu_1 & \mu_2 & -(\mu_1+\mu_2)\end{pmatrix}\sqrt{\frac{\lambda_1+\lambda_2+1}{\lambda_1+\lambda_2}}\,\boldsymbol{\Phi}_{\lambda_1+\lambda_2,\lambda_1+\lambda_2,-(\mu_1+\mu_2)}(\boldsymbol{r})\right.$$

$$\left. - \begin{pmatrix}\lambda_1 & \lambda_2 & \lambda_1+\lambda_2-1\\ \mu_1 & \mu_2 & -(\mu_1+\mu_2)\end{pmatrix}\sqrt{\frac{\lambda_1}{\lambda_2(\lambda_1+\lambda_2)}}\,\boldsymbol{\Phi}_{\lambda_1+\lambda_2,\lambda_1+\lambda_2-1,-(\mu_1+\mu_2)}(\boldsymbol{r})\right\}. \tag{18}$$

It is noted that for $\lambda_1 = 0$ or $\lambda_2 = 0$ the second term vanishes due to a vanishing 3-j symbol. The interaction energy is then proportional to

$$\boldsymbol{\Phi}_{\lambda\lambda\mu}(\boldsymbol{r}) = \frac{1}{\sqrt{\lambda(\lambda+1)}}\,\boldsymbol{L}Y_{\lambda\mu}(\boldsymbol{r}). \tag{19}$$

* The formula given in [ALD 69a] contains some misprints in the second term.

APPENDIX B

Condition for the Semiclassical Description

The condition for the applicability of classical physics for the scattering in a Coulomb field was studied by N. Bohr [BOH 48]. We shall first reproduce and slightly generalize this discussion for the case of elastic scattering in an arbitrary central potential.

We thus consider the scattering of a particle on a target nucleus where the particle experiences a deflection ϑ. In classical physics the deflection angle is completely specified by the impact parameter ρ and the bombarding energy E, i.e.

$$\vartheta = \vartheta(\rho, E). \tag{1}$$

In a quantum mechanical treatment we may try to reproduce the classical situation by setting up a diaphragm with a small hole so as to create a wavepacket which moves along the classical orbit (see fig. 1). The finite length of the wavepacket introduces an uncertainty ΔE in the bombarding energy.

The hole of diameter d gives rise to an uncertainty in the deflection angle ϑ for two reasons. Firstly, the quantum mechanical diffraction results in an angular spread which is given by

$$\delta\vartheta_{\mathrm{diff}} = \lambda\!\!\!^{-}/d, \tag{2}$$

where $\lambda\!\!\!^{-}$ is the wave length $\lambda\!\!\!^{-} = \hbar/mv$ of the scattered particle. Secondly, the uncertainty in the definition of the impact parameter gives rise to an angular spread

$$\delta\vartheta_{\rho} = \tfrac{1}{2}(\partial\vartheta/\partial\rho)d \tag{3}$$

due to the relation (1).

Finally, the uncertainty in the bombarding energy gives rise to an uncertainty in the deflection angle which is given by

$$\delta\vartheta_{E} = (\partial\vartheta/\partial E)\Delta E. \tag{4}$$

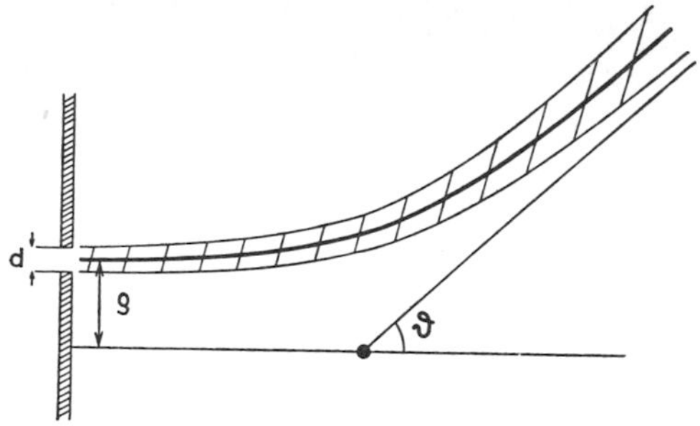

Fig. 1. The scattering of a wavepacket through the scattering angle ϑ. The impact parameter ρ of the wavepacket has been determined by a diaphragm with a circular hole of diameter d.

The total spread in the deflection angle can be estimated as the root of the sum of the squares of (2), (3) and (4). By choosing the radius in the hole of the diaphragm to be

$$d = \left(\frac{2\lambda\!\!\!^{-}}{\partial\vartheta/\partial\rho}\right)^{1/2}, \tag{5}$$

one minimizes the contribution from (2) and (3) and one obtains the following optimal angular spread

$$(\delta\vartheta_{\min})^2 = \lambda\!\!\!^{-}\, \partial\vartheta/\partial\rho + (\partial\vartheta/\partial E)^2(\Delta E)^2. \tag{6}$$

It is seen that classical physics can be applied if the quantity $\delta\vartheta_{\min}$ is small compared to the deflection angle ϑ itself, i.e.

$$\lambda\!\!\!^{-}(\partial\vartheta/\partial\rho) + (\partial\vartheta/\partial E)^2(\Delta E)^2 \ll \vartheta^2. \tag{7}$$

For a Coulomb excitation where the classical orbit is determined by elastic scattering in the monopole Coulomb field the expression (1) for the classical deflection angle is given by

$$\vartheta = 2 \arctan(Z_1 Z_2 e^2/2\rho E). \tag{8}$$

We thus find from (6) and (7)

$$(\delta\vartheta_{\min})^2 = \frac{2}{\eta}\sin^2(\tfrac{1}{2}\vartheta) + \left(\frac{\Delta E}{E}\right)^2 \sin^2\vartheta, \tag{9}$$

where the parameter η is given by eq. (II.2.1).

This expression shows that the classical treatment is applicable for all scattering angles if

$$2\eta \gg 1$$

and (10)

$$\Delta E/E \ll 1.$$

In practice one rarely* has available a diaphragm which can define the classical trajectory on an atomic scale in the way indicated on fig. 1. The incoming beam can, however, always be thought of as an incoherent superposition of wavepackets. To illustrate this we consider the beam as an ensemble of wavepackets $|\boldsymbol{R}\rangle$ which all have the average momentum $\boldsymbol{p}_o$, i.e. have the wave functions

$$\langle \boldsymbol{r}|\boldsymbol{R}\rangle = \exp\left(\frac{\mathrm{i}}{\hbar}\boldsymbol{p}_o\boldsymbol{r}\right)\exp\left(-\frac{(x-X)^2}{4\sigma_x^2}\right)\exp\left(-\frac{(y-Y)^2}{4\sigma_y^2}\right)\exp\left(-\frac{(z-Z)^2}{4\sigma_z^2}\right) \times ((2\pi)^{3/2}\sigma_x\sigma_y\sigma_z)^{-1/2}. \tag{11}$$

We assume that all wavepackets in the ensemble have the uncertainty σ_x, σ_y and σ_z in the x, y and z coordinates, respectively, while the average position $\langle \boldsymbol{r}\rangle = \boldsymbol{R} = (X, Y, Z)$ is assumed to be random within the (large) volume V of the beam. The density operator for the beam is then given by

$$\rho = \int |\boldsymbol{R}\rangle \frac{\mathrm{d}X\,\mathrm{d}Y\,\mathrm{d}Z}{V}\langle \boldsymbol{R}|. \tag{12}$$

In the corresponding density matrix $\langle \boldsymbol{r}'|\,\rho\,|\boldsymbol{r}\rangle$ the integration over X, Y and Z can be carried out leading to the result

$$\langle \boldsymbol{r}'|\,\rho\,|\boldsymbol{r}\rangle = \exp\left(\frac{\mathrm{i}}{\hbar}\boldsymbol{p}_o(\boldsymbol{r}'-\boldsymbol{r})\right)\exp\left(-\frac{(x-x')^2}{8\sigma_x^2}\right) \times \exp\left(-\frac{(y-y')^2}{8\sigma_y^2}\right)\exp\left(-\frac{(z-z')^2}{8\sigma_z^2}\right), \tag{13}$$

where $\boldsymbol{r}' = (x', y', z')$.

Since it is not customary to think of the beam as an incoherent superposition of wavepackets, we also write down a more intuitively correct density operator. We assume that in the beam each particle is described by a plane wave $|\boldsymbol{p}\rangle$ with momentum $\boldsymbol{p}$ and that the ensemble consists of an incoherent superposition of these states with a probability distribution $P(\boldsymbol{p})$ which describes the fact that the accelerator voltage and beam alignment is not perfectly accurate. Assuming a Gaussian probability distribution

$$P(\boldsymbol{p}) = ((2\pi)^{3/2}\sigma_{p_x}\sigma_{p_y}\sigma_{p_z})^{-1/2}\exp\left(-\frac{(p_x-p_{ox})^2}{2\sigma_{p_x}^2}\right) \times \exp\left(-\frac{(p_y-p_{oy})^2}{2\sigma_{p_y}^2}\right)\exp\left(-\frac{(p_z-p_{oz})^2}{2\sigma_{p_z}^2}\right), \tag{14}$$

* An approximation to the situation is achieved through the string effect in perfect crystals.

with uncertainties $\Delta \boldsymbol{p} = (\sigma_{p_x}, \sigma_{p_y}, \sigma_{p_z})$ around an average $\langle \boldsymbol{p} \rangle = \boldsymbol{p}_0$ the density matrix is

$$\langle \boldsymbol{r}' | \rho | \boldsymbol{r} \rangle = \int \exp\left(\frac{\mathrm{i}}{\hbar} \boldsymbol{p}\boldsymbol{r}'\right) P(\boldsymbol{p}) \exp\left(-\frac{\mathrm{i}}{\hbar} \boldsymbol{p}\boldsymbol{r}\right) \mathrm{d}^3 p. \tag{15}$$

The integral can be evaluated, and the result is identical to (13) if

$$\sigma_x = \hbar/2\sigma_{p_x}, \qquad \sigma_y = \hbar/2\sigma_{p_y}, \qquad \sigma_z = \hbar/2\sigma_{p_z}. \tag{16}$$

There is thus no way in which one can distinguish between an incoherent superposition of plane waves and an incoherent superposition of wavepackets. Whether one or the other description is to be used is rather a matter of convenience, which depends on the type of experiment which is performed. Thus an experiment where coincidence between the scattered projectile and target is detected accurately, calls for a description in terms of plane waves, while a detection of the time of the impact is most easily understood by a representation in terms of wavepackets. The symmetry between the bombarding condition and the detection can be further elucidated by noting that the detection of the scattering of a well-aligned beam into a large counter is equivalent to the scattering of a beam with bad angular resolution into a small counter. The latter situation can be described in terms of small wavepackets.

APPENDIX C

Wavepacket Description of Scattering

In this appendix we discuss the inelastic scattering of heavy ions in terms of wavepackets following [BRO 72].

We aim at solving the Schrödinger equation for Coulomb excitation

$$i\hbar \frac{\partial}{\partial t} \Psi(t) = H\Psi(t), \tag{1}$$

by expanding $\Psi(t)$ on wavepackets. The Hamiltonian H in (II.1.1) will be written in the form

$$H = H_o(1) + H_o(2) + W(1, 2; \boldsymbol{r}) - Z_1Z_2e^2/r + T_{\text{rel}} + Z_1Z_2e^2/r, \tag{2}$$

in the centre-of-mass system. The quantity T_{rel} is the relative kinetic energy of the centres of mass of the colliding ions while $H_o(1)$ and $H_o(2)$ are the intrinsic free Hamiltonians of the two nuclei.

In the absence of the interaction $W(1, 2; \boldsymbol{r}) - Z_1Z_2e^2/r$ the solution of (1) is approximately given by the wavepacket

$$\Psi_s(t) = \psi_n^{(1)}(\zeta_1)\, \psi_m^{(2)}(\zeta_2)\, X_s(\boldsymbol{r}, t) \exp\left(-\frac{i}{\hbar} (E_n^{(1)} + E_m^{(2)})t \right), \tag{3}$$

where $\psi_n^{(1)}(\zeta_1)$ and $\psi_m^{(2)}(\zeta_2)$ are the intrinsic eigenstates of the nuclei 1 and 2 with eigenvalues $E_n^{(1)}$ and $E_m^{(2)}$, respectively, i.e.

$$\begin{aligned} H_o(1)\, \psi_n^{(1)}(\zeta_1) &= E_n^{(1)} \psi_n^{(1)}(\zeta_1), \\ H_o(2)\, \psi_m^{(2)}(\zeta_2) &= E_m^{(2)} \psi_m^{(2)}(\zeta_2). \end{aligned} \tag{4}$$

The channel index s indicates both quantum numbers n and m. The wavepackets X_s in the relative motion is given by

$$X_s(\boldsymbol{r}, t) = \exp\left\{ \frac{i}{\hbar} m_o \boldsymbol{v}_s(t)(\boldsymbol{r} - \boldsymbol{R}_s(t)) \right\}$$

$$\times \exp\left\{ -\frac{i}{\hbar} \int_0^t [Z_1Z_2e^2/R_s(t') - \tfrac{1}{2}m_o(v_s(t'))^2]\, dt' \right\} \chi_s(\boldsymbol{r} - \boldsymbol{R}_s(t)), \tag{5}$$

where $\boldsymbol{R}_s(t)$ is the solution of the classical equations of motion in the Coulomb field, i.e.

$$m_o \ddot{\boldsymbol{R}}_s(t) = -\frac{Z_1 Z_2 e^2}{(R_s(t))^3} \boldsymbol{R}_s(t), \tag{6}$$

while

$$\boldsymbol{v}_s(t) = \dot{\boldsymbol{R}}_s(t). \tag{7}$$

The shape of the wavepacket is given by the normalized function $\chi_s(\boldsymbol{x})$ which is assumed to be non-vanishing only in a small region around $\boldsymbol{R}_s(t)$.

It is seen that (3) and (5) satisfy the wave equation

$$i\hbar \frac{\partial}{\partial t} \Psi_s(t) = (H_o(1) + H_o(2) + T_{\mathrm{rel}} + Z_1 Z_2 e^2/r)\Psi_s(t), \tag{8}$$

provided that the width σ of the function χ (e.g. a Gaussian) satisfies the inequalities

$$\sigma^2 \gg \hbar^2/m_o^2 v^2 = \lambda\!\!\!^{-2} \tag{9}$$

and

$$\sigma^2 \ll (2Z_1 Z_2 e^2/m_o v^2)^2 = (2a)^2. \tag{10}$$

The first inequality ensures that the wavepacket contains sufficiently many wave lengths that the momentum is relatively well defined, while the second inequality (10) ensures that the wavepacket is narrow enough that it does not break up in the Coulomb field. The two inequalities (9) and (10) can be satisfied simultaneously provided

$$\lambda\!\!\!^{-2} \ll (2a)^2 \tag{11}$$

or

$$(2a/\lambda\!\!\!^{-})^2 = 4\eta^2 \gg 1. \tag{12}$$

For heavy-ion collision the left-hand side of this inequality is of the order of magnitude of 10^5 (cf. fig. X.4).

The wave equation (1) should be solved with the initial condition

$$\Psi(t = -\infty) = \Psi_0, \tag{13}$$

where the entrance channel, in which both nuclei are usually in their ground state, is indicated by the index 0. The final state $\Psi(t = +\infty)$ will consist of a superposition of different channels, where the nuclei are in excited states. Due to energy and angular momentum transfer each channel will be associated with a different wavepacket describing the relative motion.

In order to describe this situation we introduce the wave function

$$\Psi(t) = \sum_s a_s(t)\, \Psi_s(t). \tag{14}$$

Introducing this into the wave equation (1) it is found that the amplitudes a_s must satisfy the coupled equations

$$i\hbar\dot{a}_s(t) = \sum_{s'} \int d^3r\, X_s^*(\boldsymbol{r}, t) \langle s|\, W(1, 2; \boldsymbol{r}) - \frac{Z_1Z_2e^2}{r} |s'\rangle\, X_{s'}(\boldsymbol{r}, t) a_{s'}(t), \quad (15)$$

where

$$\langle s|\, W(1, 2; \boldsymbol{r}) - \frac{Z_1Z_2e^2}{r} |s'\rangle$$

$$= \int d\zeta_1\, d\zeta_2\, \psi_n^{(1)*}(\zeta_1)\, \psi_m^{(2)*}(\zeta_2) \left(W(1, 2; \boldsymbol{r}) - \frac{Z_1Z_2e^2}{r} \right) \psi_{n'}^{(1)}(\zeta_1)\, \psi_{m'}^{(2)}(\zeta_2). \quad (16)$$

We shall now make the assumption that among the many possible wavepackets in a given channel s' only one contributes. The trajectory $\boldsymbol{R}_{s'}(t)$ is defined such that the wavepacket $\chi_{s'}(\boldsymbol{r} - \boldsymbol{R}_{s'}(t))$ has the maximum overlap (≈ 1) with the wavepacket $\chi_s(\boldsymbol{r} - \boldsymbol{R}_s(t))$ in all the channels that take part in the reaction. This assumption is clearly only reasonable if the energy and angular momentum transfer is small.

Assuming furthermore that the variation of $W(1, 2; \boldsymbol{r})$ over the size of the wavepackets is small, the integration over the relative coordinate $\boldsymbol{r}$ may be performed leading to the result

$$i\hbar\dot{a}_s = \sum_{s'} \langle s|\, W(1, 2, \boldsymbol{R}_{ss'}(t)) - \frac{Z_1Z_2e^2}{R_{ss'}(t)} |s'\rangle \exp\left\{\frac{i}{\hbar}(E_s - E_{s'})t\right\} a_{s'}(t), \quad (17)$$

where $E_s = E_n^{(1)} + E_m^{(2)}$, and

$$\boldsymbol{R}_{ss'}(t) = \tfrac{1}{2}(\boldsymbol{R}_s(t) + \boldsymbol{R}_{s'}(t)) \quad (18)$$

indicate the average trajectory.

APPENDIX D

Rotation Matrices

We define the physical (active) rotation through an angle ω around an axis $\hat{\boldsymbol{\omega}}$ by the operator $R_{\hat{\omega}}(\omega)$, such that the rotated state vector $|\psi\rangle$ is given by $R_{\hat{\omega}}(\omega)\,|\psi\rangle$. In order that the expectation values of the dynamical variables transform according to this physical rotation, the coordinate $\boldsymbol{r}_i$, momentum $\boldsymbol{p}_i$ and spin $\boldsymbol{s}_i$ of particle i must fulfill the following equations

$$\begin{aligned} R_{\hat{\omega}}^{-1}(\delta\omega)\boldsymbol{r}_i R_{\hat{\omega}}(\delta\omega) &= \boldsymbol{r}_i + \delta\boldsymbol{\omega} \times \boldsymbol{r}_i, \\ R_{\hat{\omega}}^{-1}(\delta\omega)\boldsymbol{p}_i R_{\hat{\omega}}(\delta\omega) &= \boldsymbol{p}_i + \delta\boldsymbol{\omega} \times \boldsymbol{p}_i, \\ R_{\hat{\omega}}^{-1}(\delta\omega)\boldsymbol{s}_i R_{\hat{\omega}}(\delta\omega) &= \boldsymbol{s}_i + \delta\boldsymbol{\omega} \times \boldsymbol{s}_i, \end{aligned} \tag{1}$$

where we have assumed the angle $\delta\omega$ to the infinitesimal. From these equations it follows that the transformation operator $R_{\hat{\omega}}(\delta\omega)$ is given by

$$R_{\hat{\omega}}(\delta\omega) = 1 - \frac{\mathrm{i}}{\hbar} I_{\hat{\omega}}\delta\omega, \tag{2}$$

where $I_{\hat{\omega}}$ is the total angular momentum around the $\hat{\boldsymbol{\omega}}$-axis, i.e.

$$I_{\hat{\omega}} = \sum_i (\boldsymbol{r}_i \times \boldsymbol{p}_i + \boldsymbol{s}_i)\hat{\boldsymbol{\omega}}. \tag{3}$$

The transformation operator $R_{\hat{\omega}}(\omega)$ for a finite rotation ω around the $\hat{\boldsymbol{\omega}}$-axis is consequently given by

$$R_{\hat{\omega}}(\omega) = \exp\left\{-\frac{\mathrm{i}}{\hbar} I_{\hat{\omega}}\omega\right\}. \tag{4}$$

The rotation operators associated with two different rotation axes $\hat{\boldsymbol{\omega}}$ and $\hat{\boldsymbol{\omega}}'$ do not commute, and similarly R does not commute with the translation operator.

In order to specify the rotation one must besides ω know e.g. the two polar coordinates of the axis $\hat{\boldsymbol{\omega}}$. A more convenient way of specifying the rotation is obtained by decomposing the transformation into consecutive rotations around coordinate axes. We thus define the rotation of the coordinate system S into the coordinate system S′ through the three Eulerian angles α, β and γ by the following process

1. rotation around z-axis by angle α
2. rotation around new y-axis by angle β (5)
3. rotation around new z-axis by angle γ.

By the new y-axis we mean here the y-axis after the rotation 1 has been performed, i.e. an axis which lies along the line of intersection between the xy and the $x'y'$ planes.

We note that the angles representing the inverse transformation $S' \to S$ are $(\alpha, \beta, \gamma)^{-1} = (-\gamma, -\beta, -\alpha)$.

While in the above definition the transformation $S \to S'$ is specified by rotations around the axes of the rotating system, the same transformation may also be expressed by consecutive rotations around the axes of the fixed system S. By considering the transformation (5) from the point of view of the rotating system one sees that the coordinate transformation $S \to S'$ may also be obtained by the following process

1′. rotating around z-axis by angle γ
2′. rotating around old y-axis by angle β (6)
3′. rotating around old z-axis by angle α.

In terms of the Eulerian angles the transformation operator R takes the form

$$R(\alpha, \beta, \gamma) = R_z(\alpha)R_y(\beta)R_z(\gamma)$$
$$= \exp\left(-\frac{\mathrm{i}}{\hbar} I_z\alpha\right) \exp\left(-\frac{\mathrm{i}}{\hbar} I_y\beta\right) \exp\left(-\frac{\mathrm{i}}{\hbar} I_z\gamma\right). \tag{7}$$

We have here generated the rotation by the definition (6) as referred to a fixed frame of reference, because the operator R describes the rotation of the physical state $|A\rangle$ into the state $R\,|A\rangle$ in the same reference system.

We shall especially be interested in the relation between an eigenstate $|jm\rangle$ of total angular momentum j and z-component m and the state which results from a rotation of this state through the Eulerian angles α, β, γ. We find from (7)

$$R\,|jm'\rangle = \sum_m \langle jm|\, R\,|jm'\rangle\,|jm\rangle$$
$$= \sum_m (D^j_{mm'}(\alpha, \beta, \gamma))^*\,|jm\rangle, \tag{8}$$

where we have utilized that R is diagonal in j. The rotation matrix $D^j_{mm'}$ is the angular momentum representation of R^{-1}, and is defined by

$$D^j_{mm'}(\alpha, \beta, \gamma) = \langle jm'|\exp\left(\frac{\mathrm{i}}{\hbar} I_z\gamma\right)\exp\left(\frac{\mathrm{i}}{\hbar} I_y\beta\right)\exp\left(\frac{\mathrm{i}}{\hbar} I_z\alpha\right)|jm\rangle. \tag{9}$$

The fundamental transformation property (8) can also be written in terms of wave functions. Thus we may express the wave function $\psi_{jm}(\boldsymbol{r}_i, s_{i,z})$ corresponding to the unrotated state $|j, m\rangle$ in terms of the wave function ψ^{R} for the rotated state. We find

$$\begin{aligned}\psi_{jm}(\boldsymbol{r}_i, s_{i,z}) &= \langle \boldsymbol{r}_i, s_{i,z}|jm\rangle \\ &= \sum_{m'} D^j_{mm'}(\alpha, \beta, \gamma)\langle \boldsymbol{r}_i, s_{i,z}|R|jm'\rangle \\ &= \sum_{m'} D^j_{mm'}(\alpha, \beta, \gamma)\psi^{\mathrm{R}}_{jm'}(\boldsymbol{r}_i, s_{i,z}).\end{aligned} \tag{10}$$

The wave functions $\psi^{\mathrm{R}}_{jm'}(\boldsymbol{r}_i, s_{i,z})$ describe the rotated state vector in the same coordinate system as the original state vector. The quantum number m' is not the eigenvalue of I_z, but of the z-component of the rotated angular momentum.

The second equation in (10) can be interpreted in a different way by acting with R on the state vector $\langle \boldsymbol{r}_i, s_{i,z}|$. One thereby obtains

$$\begin{aligned}\psi_{jm}(\boldsymbol{r}_i, s_{i,z}) &= \sum_{m'} D^j_{mm'}(\alpha, \beta, \gamma)\langle R^{-1}(\boldsymbol{r}_i, s_{i,z})|jm'\rangle \\ &= \sum_{m'} D^j_{mm'}(\alpha, \beta, \gamma)\psi_{jm'}(\boldsymbol{r}'_i, s_{i,z'}).\end{aligned} \tag{11}$$

The coordinates $\boldsymbol{r}'$ indicate the position of the particle as seen from a coordinate system S′ which is related to the original system S through the rotation R. Similarly the degree of freedom $s_{i,z'}$ indicates the z'-component of the spin in the new coordinate system.

The explicit evaluation of the rotation matrix as a function of α, β and γ can be found in textbooks on angular momentum.* Since in many texts the notation differs from the one used in the present work, we shall in the following, for purpose of reference, give a list of the relevant properties of the rotation matrices.

* Our definition of the D-function is the same as that of A. Bohr and B. Mottelson [BOH 69], but differs from the definition of Edmonds [EDM 57], i.e.

$$D^j_{mm'}(\alpha, \beta, \gamma) = (-1)^{m-m'}[D^j_{mm'}(\alpha, \beta, \gamma)]_{\mathrm{Edmonds}}.$$

It should be noted furthermore that the definition of Edmonds contains some inconsistencies in the text.

From the definition of $D^j_{mm'}(\alpha, \beta, \gamma)$ it follows that

$$D^j_{mm'}(\alpha, \beta, \gamma) = e^{im\alpha} d^j_{mm'}(\beta)\, e^{im'\gamma}. \tag{12}$$

Explicit expressions for $d^j_{mm'}(\beta)$ for $j = \frac{1}{2}, 1, \frac{3}{2}, 2, \frac{5}{2}, 3, \frac{7}{2}$ and 4 are given in table 1.

For higher values of j, one may calculate $d^j_{mm'}$, by means of the recursion relations

$$d^j_{mm'}(\beta) = \sqrt{\frac{j-m}{j-m'}}\, d^{j-1/2}_{m+1/2,m'+1/2}(\beta)\cos(\tfrac{1}{2}\beta) - \sqrt{\frac{j+m}{j-m'}}\, d^{j-1/2}_{m-1/2,m'+1/2}(\beta)\sin(\tfrac{1}{2}\beta), \tag{13}$$

and

$$d^j_{mm'}(\beta) = \frac{1}{2}\sqrt{\frac{(j-m-1)(j-m)}{(j+m')(j-m')}}\, d^{j-1}_{m+1,m'}(\beta)\sin\beta + \sqrt{\frac{(j+m)(j-m)}{(j+m')(j-m')}}\, d^{j-1}_{mm'}(\beta)\cos\beta - \frac{1}{2}\sqrt{\frac{(j+m-1)(j+m)}{(j+m')(j-m')}}\, d^{j-1}_{m-1,m'}(\beta)\sin\beta, \tag{14}$$

or from the explicit expression

$$d^j_{mm'}(\beta) = \left[\frac{(j+m')!\,(j-m')!}{(j+m)!\,(j-m)!}\right]^{1/2} \sum_\sigma \binom{j+m}{j-m'-\sigma}\binom{j-m}{\sigma}(-1)^{j-m'-\sigma} \times (\cos(\tfrac{1}{2}\beta))^{2\sigma+m+m'}(\sin(\tfrac{1}{2}\beta))^{2j-2\sigma-m-m'}, \tag{15}$$

which for $m = m' = j$ gives the simple result

$$d^j_{jj}(\beta) = \cos^{2j}(\tfrac{1}{2}\beta). \tag{16}$$

We quote the following symmetry relations for $d^j_{mm'}(\beta)$

$$d^j_{mm'}(-\beta) = d^j_{m'm}(\beta), \tag{17}$$

$$d^j_{mm'}(\pi + \beta) = (-1)^{j+m} d^j_{-mm'}(\beta), \tag{18}$$

$$d^j_{mm'}(\beta) = (-1)^{m-m'} d^j_{-m-m'}(\beta), \tag{19}$$

and

$$d^j_{mm'}(\beta) = (-1)^{m-m'} d^j_{m'm}(\beta). \tag{20}$$

TABLE 1

The explicit expressions for the rotation matrices $d^j_{mm'}(\beta)$ for $j = 0, \frac{1}{2}, 1, \frac{3}{2}, 2, \frac{5}{2}, 3, \frac{7}{2}$ and 4 and $m > 0$ and $m \geqslant m'$. [For other m values, the expressions can be obtained by the symmetry relations $d^j_{mm'}(\beta) = d_{-m'-m}(\beta) = (-1)^{m-m'} d^j_{m'm}(\beta) = (-1)^{m-m'} d^j_{-m-m'}(\beta)$.]

j	m	m'	$d^j_{mm'}(\beta)$
0	0	0	1
$\frac{1}{2}$	$\frac{1}{2}$	$\frac{1}{2}$	$\cos(\frac{1}{2}\beta)$
	$\frac{1}{2}$	$-\frac{1}{2}$	$-\sin(\frac{1}{2}\beta)$
1	1	1	$\frac{1}{2}(1 + \cos\beta) = \cos^2(\frac{1}{2}\beta)$
	1	0	$-\sqrt{\frac{1}{2}}\sin\beta = -\sqrt{2}\sin(\frac{1}{2}\beta)\cos(\frac{1}{2}\beta)$
	1	-1	$\frac{1}{2}(1 - \cos\beta) = \sin^2(\frac{1}{2}\beta)$
	0	0	$\cos\beta = \cos^2(\frac{1}{2}\beta) - \sin^2(\frac{1}{2}\beta)$
$\frac{3}{2}$	$\frac{3}{2}$	$\frac{3}{2}$	$\cos^3(\frac{1}{2}\beta)$
	$\frac{3}{2}$	$\frac{1}{2}$	$-\sqrt{3}\cos^2(\frac{1}{2}\beta)\sin(\frac{1}{2}\beta)$
	$\frac{3}{2}$	$-\frac{1}{2}$	$\sqrt{3}\cos(\frac{1}{2}\beta)\sin^2(\frac{1}{2}\beta)$
	$\frac{3}{2}$	$-\frac{3}{2}$	$-\sin^3(\frac{1}{2}\beta)$
	$\frac{1}{2}$	$\frac{1}{2}$	$\cos^3(\frac{1}{2}\beta) - 2\cos(\frac{1}{2}\beta)\sin^2(\frac{1}{2}\beta)$
	$\frac{1}{2}$	$-\frac{1}{2}$	$\sin^3(\frac{1}{2}\beta) - 2\sin(\frac{1}{2}\beta)\cos^2(\frac{1}{2}\beta)$
2	2	2	$\frac{1}{4}(1 + \cos\beta)^2 = \cos^4(\frac{1}{2}\beta)$
	2	1	$-\frac{1}{2}\sin\beta(1 + \cos\beta) = -2\cos^3(\frac{1}{2}\beta)\sin(\frac{1}{2}\beta)$
	2	0	$\frac{1}{2}\sqrt{\frac{3}{2}}\sin^2\beta = \sqrt{6}\sin^2(\frac{1}{2}\beta)\cos^2(\frac{1}{2}\beta)$
	2	-1	$-\frac{1}{2}\sin\beta(1 - \cos\beta) = -2\cos(\frac{1}{2}\beta)\sin^3(\frac{1}{2}\beta)$
	2	-2	$\frac{1}{4}(1 - \cos\beta)^2 = \sin^4(\frac{1}{2}\beta)$
	1	1	$\cos^2\beta + \frac{1}{2}\cos\beta - \frac{1}{2} = \cos^4(\frac{1}{2}\beta) - 3\sin^2(\frac{1}{2}\beta)\cos^2(\frac{1}{2}\beta)$
	1	0	$-\sqrt{\frac{3}{2}}\sin\beta\cos\beta = -\sqrt{6}\cos^3(\frac{1}{2}\beta)\sin(\frac{1}{2}\beta) + \sqrt{6}\cos(\frac{1}{2}\beta)\sin^3(\frac{1}{2}\beta)$
	1	-1	$-\cos^2\beta + \frac{1}{2}\cos\beta + \frac{1}{2} = -\sin^4(\frac{1}{2}\beta) + 3\sin^2(\frac{1}{2}\beta)\cos^2(\frac{1}{2}\beta)$
	0	0	$\cos^2\beta - \frac{1}{2}\sin^2\beta = \cos^4(\frac{1}{2}\beta) + \sin^4(\frac{1}{2}\beta) - 4\cos^2(\frac{1}{2}\beta)\sin^2(\frac{1}{2}\beta)$
$\frac{5}{2}$	$\frac{5}{2}$	$\frac{5}{2}$	$\cos^5(\frac{1}{2}\beta)$
	$\frac{5}{2}$	$\frac{3}{2}$	$-\sqrt{5}\cos^4(\frac{1}{2}\beta)\sin(\frac{1}{2}\beta)$
	$\frac{5}{2}$	$\frac{1}{2}$	$\sqrt{10}\cos^3(\frac{1}{2}\beta)\sin^2(\frac{1}{2}\beta)$
	$\frac{5}{2}$	$-\frac{1}{2}$	$-\sqrt{10}\cos^2(\frac{1}{2}\beta)\sin^3(\frac{1}{2}\beta)$
	$\frac{5}{2}$	$-\frac{3}{2}$	$\sqrt{5}\cos(\frac{1}{2}\beta)\sin^4(\frac{1}{2}\beta)$
	$\frac{5}{2}$	$-\frac{5}{2}$	$-\sin^5(\frac{1}{2}\beta)$
	$\frac{3}{2}$	$\frac{3}{2}$	$\cos^5(\frac{1}{2}\beta) - 4\cos^3(\frac{1}{2}\beta)\sin^2(\frac{1}{2}\beta)$
	$\frac{3}{2}$	$\frac{1}{2}$	$-2\sqrt{2}\cos^4(\frac{1}{2}\beta)\sin(\frac{1}{2}\beta) + 3\sqrt{2}\cos^2(\frac{1}{2}\beta)\sin^3(\frac{1}{2}\beta)$
	$\frac{3}{2}$	$-\frac{1}{2}$	$-2\sqrt{2}\sin^4(\frac{1}{2}\beta)\cos(\frac{1}{2}\beta) + 3\sqrt{2}\sin^2(\frac{1}{2}\beta)\cos^3(\frac{1}{2}\beta)$
	$\frac{3}{2}$	$-\frac{3}{2}$	$\sin^5(\frac{1}{2}\beta) - 4\sin^3(\frac{1}{2}\beta)\cos^2(\frac{1}{2}\beta)$
	$\frac{1}{2}$	$\frac{1}{2}$	$\cos^5(\frac{1}{2}\beta) - 6\cos^3(\frac{1}{2}\beta)\sin^2(\frac{1}{2}\beta) + 3\cos(\frac{1}{2}\beta)\sin^4(\frac{1}{2}\beta)$
	$\frac{1}{2}$	$-\frac{1}{2}$	$-\sin^5(\frac{1}{2}\beta) + 6\sin^3(\frac{1}{2}\beta)\cos^2(\frac{1}{2}\beta) - 3\sin(\frac{1}{2}\beta)\cos^4(\frac{1}{2}\beta)$

(continued)

TABLE 1—*continued*

j	m	m'	$d^j_{mm'}(\beta)$
3	3	3	$\frac{1}{8}(1 + \cos\beta)^3$
	3	2	$-\frac{1}{4}\sqrt{\frac{3}{2}}\sin\beta(1 + \cos\beta)^2$
	3	1	$\frac{1}{8}\sqrt{15}\sin^2\beta(1 + \cos\beta)$
	3	0	$-\frac{1}{4}\sqrt{5}\sin^3\beta$
	3	-1	$\frac{1}{8}\sqrt{15}\sin^2\beta(1 - \cos\beta)$
	3	-2	$-\frac{1}{4}\sqrt{\frac{3}{2}}\sin\beta(1 - \cos\beta)^2$
	3	-3	$\frac{1}{8}(1 - \cos\beta)^3$
	2	2	$\frac{1}{4}(1 + \cos\beta)^2(3\cos\beta - 2)$
	2	1	$-\frac{1}{4}\sqrt{\frac{5}{2}}\sin\beta(3\cos^2\beta + 2\cos\beta - 1) = -\frac{1}{4}\sqrt{\frac{5}{2}}\sin\beta(1 + \cos\beta)(3\cos\beta - 1)$
	2	0	$\frac{1}{2}\sqrt{\frac{15}{2}}\cos\beta\sin^2\beta$
	2	-1	$\frac{1}{4}\sqrt{\frac{5}{2}}\sin\beta(3\cos^2\beta - 2\cos\beta - 1) = -\frac{1}{4}\sqrt{\frac{5}{2}}\sin\beta(1 - \cos\beta)(3\cos\beta + 1)$
	2	-2	$\frac{1}{4}(1 - \cos\beta)^2(3\cos\beta + 2)$
	1	1	$\frac{1}{8}(1 + \cos\beta)(15\cos^2\beta - 10\cos\beta - 1)$
	1	0	$-\frac{1}{4}\sqrt{3}\sin\beta(5\cos^2\beta - 1)$
	1	-1	$\frac{1}{8}(1 - \cos\beta)(15\cos^2\beta + 10\cos\beta - 1)$
	0	0	$\frac{1}{2}\cos\beta(5\cos^2\beta - 3)$
$\frac{7}{2}$	$\frac{7}{2}$	$\frac{7}{2}$	$\cos^7(\frac{1}{2}\beta)$
	$\frac{7}{2}$	$\frac{5}{2}$	$-\sqrt{7}\cos^6(\frac{1}{2}\beta)\sin(\frac{1}{2}\beta)$
	$\frac{7}{2}$	$\frac{3}{2}$	$\sqrt{21}\cos^5(\frac{1}{2}\beta)\sin^2(\frac{1}{2}\beta)$
	$\frac{7}{2}$	$\frac{1}{2}$	$-\sqrt{35}\cos^4(\frac{1}{2}\beta)\sin^3(\frac{1}{2}\beta)$
	$\frac{7}{2}$	$-\frac{1}{2}$	$\sqrt{35}\cos^3(\frac{1}{2}\beta)\sin^4(\frac{1}{2}\beta)$
	$\frac{7}{2}$	$-\frac{3}{2}$	$-\sqrt{21}\cos^2(\frac{1}{2}\beta)\sin^5(\frac{1}{2}\beta)$
	$\frac{7}{2}$	$-\frac{5}{2}$	$\sqrt{7}\cos(\frac{1}{2}\beta)\sin^6(\frac{1}{2}\beta)$
	$\frac{7}{2}$	$-\frac{7}{2}$	$-\sin^7(\frac{1}{2}\beta)$
	$\frac{5}{2}$	$\frac{5}{2}$	$7\cos^7(\frac{1}{2}\beta) - 6\cos^5(\frac{1}{2}\beta)$
	$\frac{5}{2}$	$\frac{3}{2}$	$-\sqrt{3}[7\cos^6(\frac{1}{2}\beta) - 5\cos^4(\frac{1}{2}\beta)]\sin(\frac{1}{2}\beta)$
	$\frac{5}{2}$	$\frac{1}{2}$	$\sqrt{5}[7\cos^5(\frac{1}{2}\beta) - 4\cos^3(\frac{1}{2}\beta)]\sin^2(\frac{1}{2}\beta)$
	$\frac{5}{2}$	$-\frac{1}{2}$	$\sqrt{5}[7\sin^5(\frac{1}{2}\beta) - 4\sin^3(\frac{1}{2}\beta)]\cos^2(\frac{1}{2}\beta)$
	$\frac{5}{2}$	$-\frac{3}{2}$	$-\sqrt{3}[7\sin^6(\frac{1}{2}\beta) - 5\sin^4(\frac{1}{2}\beta)]\cos(\frac{1}{2}\beta)$
	$\frac{5}{2}$	$-\frac{5}{2}$	$7\sin^7(\frac{1}{2}\beta) - 6\sin^5(\frac{1}{2}\beta)$
	$\frac{3}{2}$	$\frac{3}{2}$	$21\cos^7(\frac{1}{2}\beta) - 30\cos^5(\frac{1}{2}\beta) + 10\cos^3(\frac{1}{2}\beta)$
	$\frac{3}{2}$	$\frac{1}{2}$	$-\sqrt{15}[7\cos^6(\frac{1}{2}\beta) - 8\cos^4(\frac{1}{2}\beta) + 2\cos^2(\frac{1}{2}\beta)]\sin(\frac{1}{2}\beta)$
	$\frac{3}{2}$	$-\frac{1}{2}$	$\sqrt{15}[7\sin^6(\frac{1}{2}\beta) - 8\sin^4(\frac{1}{2}\beta) + 2\sin^2(\frac{1}{2}\beta)]\cos(\frac{1}{2}\beta)$
	$\frac{3}{2}$	$-\frac{3}{2}$	$-21\sin^7(\frac{1}{2}\beta) + 30\sin^5(\frac{1}{2}\beta) - 10\sin^3(\frac{1}{2}\beta)$
	$\frac{1}{2}$	$\frac{1}{2}$	$35\cos^7(\frac{1}{2}\beta) - 60\cos^5(\frac{1}{2}\beta) + 30\cos^3(\frac{1}{2}\beta) - 4\cos(\frac{1}{2}\beta)$
	$\frac{1}{2}$	$-\frac{1}{2}$	$35\sin^7(\frac{1}{2}\beta) - 60\sin^5(\frac{1}{2}\beta) + 30\sin^3(\frac{1}{2}\beta) - 4\sin(\frac{1}{2}\beta)$

(continued)

TABLE 1—*continued*

j	m	m'	$d^j_{mm'}(\beta)$
4	4	4	$\frac{1}{16}(1+\cos\beta)^4$
	4	3	$-\frac{1}{8}\sqrt{2}\sin\beta(1+\cos\beta)^3$
	4	2	$\frac{1}{8}\sqrt{7}\sin^2\beta(1+\cos\beta)^2$
	4	1	$-\frac{1}{4}\sqrt{\frac{7}{2}}\sin^3\beta(1+\cos\beta)$
	4	0	$\frac{1}{8}\sqrt{\frac{35}{2}}\sin^4\beta$
	4	−1	$-\frac{1}{4}\sqrt{\frac{7}{2}}\sin^3\beta(1-\cos\beta)$
	4	−2	$\frac{1}{8}\sqrt{7}\sin^2\beta(1-\cos\beta)^2$
	4	−3	$-\frac{1}{8}\sqrt{2}\sin\beta(1-\cos\beta)^3$
	4	−4	$\frac{1}{16}(1-\cos\beta)^4$
	3	3	$\frac{1}{8}(1+\cos\beta)^3(4\cos\beta-3)$
	3	2	$-\frac{1}{4}\sqrt{\frac{7}{2}}\sin\beta(1+\cos\beta)^2(2\cos\beta-1)$
	3	1	$\frac{1}{8}\sqrt{7}\sin^2\beta(1+\cos\beta)(4\cos\beta-1)$
	3	0	$-\frac{1}{4}\sqrt{35}\sin^3\beta\cos\beta$
	3	−1	$\frac{1}{8}\sqrt{7}\sin^2\beta(1-\cos\beta)(4\cos\beta+1)$
	3	−2	$-\frac{1}{4}\sqrt{\frac{7}{2}}\sin\beta(1-\cos\beta)^2(2\cos\beta+1)$
	3	−3	$\frac{1}{8}(1-\cos\beta)^3(4\cos\beta+3)$
	2	2	$\frac{1}{4}(1+\cos\beta)^2(7\cos^2\beta-7\cos\beta+1)$
	2	1	$-\frac{1}{8}\sqrt{2}\sin\beta(1+\cos\beta)(14\cos^2\beta-7\cos\beta-1)$
	2	0	$\frac{1}{8}\sqrt{10}\sin^2\beta(7\cos^2\beta-1)$
	2	−1	$-\frac{1}{8}\sqrt{2}\sin\beta(1-\cos\beta)(14\cos^2\beta+7\cos\beta-1)$
	2	−2	$\frac{1}{4}(1-\cos\beta)^2(7\cos^2\beta+7\cos\beta+1)$
	1	1	$\frac{1}{8}(1+\cos\beta)(28\cos^3\beta-21\cos^2\beta-6\cos\beta+3)$
	1	0	$-\frac{1}{4}\sqrt{5}\sin\beta\cos\beta(7\cos^2\beta-3)$
	1	−1	$\frac{1}{8}(1-\cos\beta)(28\cos^3\beta+21\cos^2\beta-6\cos\beta-3)$
	0	0	$\frac{1}{8}(35\cos^4\beta-30\cos^2\beta+3)$

From these relations follow a number of properties of the rotation matrix:

$$(D^j_{mm'}(\alpha,\beta,\gamma))^* = D^j_{m'm}(-\gamma,-\beta,-\alpha) = (-1)^{m-m'}D^j_{-m-m'}(\alpha,\beta,\gamma), \tag{21}$$

$$D^j_{mm'}(\alpha,\beta,\gamma) = (-1)^{m-m'}D^j_{m'm}(\gamma,\beta,\alpha). \tag{22}$$

For special values of angles and indices the rotation matrices reduce in the following manner:

$$d^j_{mm'}(0) = \delta_{mm'}, \tag{23}$$

$$d^j_{mm'}(\pi) = (-1)^{j+m}\delta_{m-m'}, \tag{24}$$

$$D^l_{m0}(\alpha,\beta,\gamma) = \sqrt{\frac{4\pi}{2l+1}}\,Y_{lm}(\beta,\alpha), \tag{25}$$

and

$$D^l_{0m}(\alpha, \beta, \gamma) = (-1)^m \sqrt{\frac{4\pi}{2l+1}}\, Y_{lm}(\beta, \gamma). \tag{26}$$

Furthermore we quote the formulae for the reduction of products of rotation matrices.

For two successive rotations $S \to S''$ and $S'' \to S'$ one finds

$$D^j_{mm'}(S \to S') = \sum_{m''} D^j_{mm''}(S'' \to S') D^j_{m''m'}(S \to S''), \tag{27}$$

which for $S' \equiv S$ leads to the unitarity relation

$$\sum_{m''} (D^j_{m''m}(\alpha, \beta, \gamma))^*\, D^j_{m''m'}(\alpha, \beta, \gamma) = \delta_{mm'}$$

and (28)

$$\sum_{m''} D^j_{mm''}(\alpha, \beta, \gamma)(D^j_{m'm''}(\alpha, \beta, \gamma))^* = \delta_{mm'}.$$

For the product of two rotation matrices of the same Eulerian angles ϑ_i one finds furthermore:

$$D^{j_1}_{m_1m_1'}(\vartheta_i)\, D^{j_2}_{m_2m_2'}(\vartheta_i) = \sum_{jmm'} (2j+1)\begin{pmatrix} j_1 & j & j_2 \\ m_1 & m & m_2 \end{pmatrix}(D^j_{mm'}(\vartheta_i))^*\begin{pmatrix} j_1 & j & j_2 \\ m_1' & m' & m_2' \end{pmatrix}, \tag{29}$$

where m, m' are seen to be fixed by the 3-j symbols to be $-(m_1 + m_2)$, $-(m_1' + m_2')$. The formula (29) may also be written

$$\sum_{m_1m_2} \begin{pmatrix} j_1 & j_2 & j \\ m_1 & m_2 & m \end{pmatrix} D^{j_1}_{m_1m_1'}(\vartheta_i) D^{j_2}_{m_2m_2'}(\vartheta_i) = \begin{pmatrix} j_1 & j_2 & j \\ m_1' & m_2' & m' \end{pmatrix}(D^j_{mm'}(\vartheta_i))^*. \tag{30}$$

The integral properties of the D-functions are expressed by the orthogonality relation

$$\frac{1}{8\pi^2}\int_0^{2\pi} \mathrm{d}\alpha \int_0^{\pi} \mathrm{d}\beta \sin\beta \int_0^{2\pi} \mathrm{d}\gamma\, (D^{j_1}_{m_1m_1'}(\alpha, \beta, \gamma))^*\, D^{j_2}_{m_2m_2'}(\alpha, \beta, \gamma) = \delta_{m_1m_2}\delta_{m_1'm_2'}\delta_{j_1j_2}(2j_1+1)^{-1}. \tag{31}$$

From this relation one may by means of (29) derive the following formula

$$\frac{1}{8\pi^2}\int_0^{2\pi} \mathrm{d}\alpha \int_0^{\pi} \mathrm{d}\beta \sin\beta \int_0^{2\pi} \mathrm{d}\gamma\, D^{j_1}_{m_1m_1'}(\alpha, \beta, \gamma)\, D^{j_2}_{m_2m_2'}(\alpha, \beta, \gamma)\, D^{j_3}_{m_3m_3'}(\alpha, \beta, \gamma) = \begin{pmatrix} j_1 & j_2 & j_3 \\ m_1 & m_2 & m_3 \end{pmatrix}\begin{pmatrix} j_1 & j_2 & j_3 \\ m_1' & m_2' & m_3' \end{pmatrix}. \tag{32}$$

APPENDIX E

Phase Conventions of Nuclear Wave Functions

The fundamental set of common eigenstates of the Hamiltonian H_0 and the total angular momentum operators $\boldsymbol{I}^2$ and I_z, which we denote by $|IM\rangle$, are only defined up to an arbitrary phase. It is convenient to introduce phase conventions in order that some physical quantities like vector addition coefficients and various other matrix elements become real. Thus one has for many years used the so-called Condon and Shortley [CON 51] convention for the relative phases of the state vectors $|IM\rangle$ for a fixed value of I. This convention can be expressed by means of the relations

$$(I_x \pm iI_y)\,|IM\rangle = \sqrt{(I \mp M)(I \pm M + 1)}\,|I, M \pm 1\rangle \tag{1}$$

connecting $|IM\rangle$ with $|IM'\rangle$.

In later years it has become customary to introduce a convention for the relative phase of states with different I, in such a way that the matrix elements of the electromagnetic multipole moments become real. This is achieved by demanding that the phase of the state $|IM\rangle_0$ be chosen such that under time reversal T the state transforms according to the formula

$$T\,|IM\rangle_0 = (-1)^{I+M}\,|I - M\rangle_0. \tag{2}$$

This condition is equivalent to the rule that the state $|IM\rangle_0$ is invariant under the combined operation RT, where R is the rotation $R_y(\pi)$ of 180 degrees around the y-axis in the positive sense of rotation, i.e.

$$RT\,|IM\rangle_0 = |IM\rangle_0. \tag{3}$$

Orbital angular momentum eigenstates of the form $Y_{lm}(\vartheta, \varphi)$ do not satisfy the relation (3) and should therefore be multiplied with the factor i^l. Furthermore the matrix elements of the electric multipole operators $\mathscr{M}(E\lambda, \mu)$ are only real if multiplied by a factor i^λ, while the magnetic multipole moments $\mathscr{M}(M\lambda, \mu)$ should be multiplied by $i^{\lambda-1}$.

These inconveniences can be avoided by using instead the phase convention

$$RPT\,|IM\rangle = |IM\rangle, \tag{4}$$

where P is the parity operator. With this definition the orbital angular momentum eigenfunction $Y_{lm}(\vartheta, \varphi)$ has the correct transformation property without the additional factor i^l. The usual spin eigenfunctions satisfy (4) as well as (3), and the property (4) is conserved under addition of angular momenta just like (3). Actually, a state vector $|IM\rangle$ which satisfies (4) can be obtained from a state vector $|IM\rangle_o$ satisfying (3) by means of the rule

$$|IM\rangle = \mathrm{i}^{\pi}\,|IM\rangle_o, \tag{5}$$

where π is the (even or odd) parity of the state $|IM\rangle$. It follows e.g. from (5) that the states $|IM\rangle$ transform as follows

$$T\,|IM\rangle = (-1)^{I+M+\pi}\,|I-M\rangle \tag{6}$$

under time reversal.

Since the transformation properties of the electromagnetic multipole operators (II.1.3) and (II.1.4) are seen to be

$$RPT\,\mathcal{M}(\lambda,\mu)(RPT)^{-1} = \mathcal{M}(\lambda,\mu), \tag{7}$$

it follows from (4) that all matrix elements of the electromagnetic multipole operators are real. Thus especially the reduced matrix elements defined by

$$\langle I_1\,\|\mathcal{M}(\lambda)\,\|I_2\rangle = \frac{(-1)^{I_1-M_1}}{\begin{pmatrix} I_1 & \lambda & I_2 \\ -M_1 & \mu & M_2 \end{pmatrix}}\,\langle I_1M_1|\,\mathcal{M}(\lambda,\mu)\,|I_2M_2\rangle \tag{8}$$

are real numbers. Since the multipole operators satisfy the relation*

$$\mathcal{M}(\lambda,\mu)^{\dagger} = (-1)^{\mu}\,\mathcal{M}(\lambda,-\mu) \tag{9}$$

characteristic for Hermitian tensor operators the reduced matrix element (8) satisfies the symmetry relation

$$\langle I_1\|\,\mathcal{M}(\lambda)\,\|I_2\rangle = (-1)^{I_1-I_2}\,\langle I_2\|\,\mathcal{M}(\lambda)\,\|I_1\rangle. \tag{10}$$

The phase convention (4) is used throughout the present work. The convenience of using the time reversal operator in conjunction with the parity operator is well known in field theory, where it is often combined furthermore with the rotation of 180 degrees around the y-axis in the isospin space.

* If the electric multiple operator is defined in terms of the nuclear current density, this relation is not fulfilled (cf. appendix G and K. Alder and R. M. Steffen [HAM 74]).

With the phase convention (4) (or (3)) the sign of the nuclear wave function is still left undetermined. This means that the sign of the non-diagonal reduced matrix elements of the multipole operators is undetermined. We shall sometimes use the convention that the matrix element of the lowest electric multipole matrix element connecting a given state with the ground state is positive.

APPENDIX F

Density Matrices

The theory of density matrices as it applies to the evaluation of angular distributions and angular correlations of nuclear reaction products is given in several text books and review articles (e.g. [FRA 65, DEV 57, ROS 67, GAB 70]). We shall reproduce here a few relevant features of the theory in order to define the notation used in the present review. The application of the theory to the case of γ-decay is given in appendix G.

A system about which we have an incomplete information must be described in terms of a statistical operator ρ. This operator is defined by means of the probabilities P_s with which the system is known to be in the states $|s\rangle$. Thus the positive numbers P_s fulfill the condition

$$\sum_s P_s = 1 . \tag{1}$$

We assume furthermore that the states $|s\rangle$ are normalized, but they are not necessarily orthogonal. The density operator is then defined by

$$\rho = \sum_s |s\rangle P_s \langle s| . \tag{2}$$

The expectation value of any physical quantity A for the ensemble described by the density operator ρ is given by

$$\langle \bar{A} \rangle = \mathrm{Tr}(\rho A) = \mathrm{Tr}(A\rho) = \sum_s P_s \langle s|\, A\, |s\rangle . \tag{3}$$

The density operator ρ is according to its definition (2) a Hermitian operator with trace unity, i.e.

$$\rho^\dagger = \rho$$

and

$$\mathrm{Tr}\, \rho = 1 . \tag{4}$$

The eigenvalues of ρ are positive numbers P_n and we may thus always write a density operator in the form

$$\rho = \sum_n |n\rangle P_n \langle n| ,$$

where $|n\rangle$ are mutually orthogonal eigenstates. The probabilities P_n are given in terms of the probabilities P_s in (1) by the relation

$$P_n = \sum_s P_s |\langle n|s\rangle|^2 . \tag{5}$$

A pure state $|r\rangle$ can be described by the density operator

$$\rho_r = |r\rangle \langle r| \tag{6}$$

which is a projection operator, i.e.

$$\rho_r^2 = \rho_r . \tag{7}$$

This relation characterizes a pure state. We shall use the interaction representation for the density operator and the time-dependence of ρ is then determined by

$$\dot{\rho} = \frac{1}{i\hbar} [\tilde{H}_{\text{int}}(t), \rho], \tag{8}$$

where $\tilde{H}_{\text{int}}$ is defined similarly to $\tilde{V}(t)$ in (II.3.3). Introducing the time development operator $U(t, t_o)$ defined in (II.3.6) i.e.

$$U(t, t_o) = \mathscr{T} \exp\left\{ -\frac{i}{\hbar} \int_{t_o}^{t} \tilde{H}_{\text{int}}(t') \, dt' \right\} \tag{9}$$

we may write ρ at time t in terms of ρ at time t_o as

$$\rho(t) = U(t, t_o)\rho(t_o)U^{\dagger}(t, t_o). \tag{10}$$

The density operator changes in time partly due to the equation of motion (8) and partly due to additional information about the system obtained by measurements. The observations which are made in angular correlations experiments are always of the simple type where a decay or reaction product is detected at large distances from the residual system where no direct interaction with this system takes place. We ask for the density matrix of the residual nucleus after a successful detection of the decay products and have no interest in the future of this product.

A rather common type of measurement is a counter experiment where one asks for the probability that a particle is detected with momentum $\boldsymbol{p}$. This probability P_p is according to (3) given by the expectation value of the projection operator

$$\varepsilon_p = |p\rangle \langle p|, \tag{11}$$

i.e.

$$P_p = \mathrm{Tr}(\rho\varepsilon_p). \tag{12}$$

By the measurement of a decay product we do not disturb the residual system directly and the density operator of the final system ρ_f is obtained from the initial density operator ρ_i by leaving out such components which by the measurement were ensured not to be there. Thus, we have

$$\rho_f = \varepsilon_p \rho_i \varepsilon_p / P_p \tag{13}$$

where ε_p is the projection operator (11) for the measurement which was successfully performed.

The same expression can be used if the counter makes a registration (a click) for particles with momentum within a finite solid angle and a finite energy interval. The projection operator in (11), (12) and (13) should then be substituted by

$$\varepsilon(\Delta\Omega, \Delta E) = \int_V \mathrm{d}^3p\, |p\rangle \langle p|, \tag{14}$$

where V is the domain in p-space defined by the solid angle $\Delta\Omega$ and the energy interval ΔE.

We shall only be concerned with the case where we are not interested any more in the state of the decay product observed and the density operator for the residual system ρ (res) is, therefore, given by contraction over the degrees of freedom of the decay product as given e.g. by the momentum vector $\boldsymbol{p}$. Thus we find

$$\begin{aligned}\rho(\text{res}) &= \sum_{p'} \langle p'|\, \rho_f(\text{total})\, |p'\rangle \\ &= \sum_{p'} \langle p'|\, \varepsilon\rho_i(\text{total})\varepsilon\, |p'\rangle \\ &= \int_V \mathrm{d}^3p\, \langle p|\, \rho_i(\text{total})\, |p\rangle / P,\end{aligned} \tag{15}$$

where P is the detection probability

$$P = \mathrm{Tr} \int_V \mathrm{d}^3p\, \langle p|\, \rho_i(\text{total})\, |p\rangle. \tag{16}$$

The trace is taken over all degrees of freedom of the residual nucleus (including e.g. summation over magnetic quantum numbers).

These results are, of course, only applicable for the ideal situation where the counter has the perfect efficiency unity within the solid angle $\Delta\Omega$ and the energy interval ΔE. In practice there will be a finite detection efficiency $\varepsilon(\boldsymbol{p}) < 1$, which e.g. reaches a maximum when $\hat{\boldsymbol{p}}$ is in the direction of the axis of the counter and has a magnitude which corresponds to the centre of the energy interval ΔE.

The generalization of formula (13) to the case where the detection is non-ideal and cannot be described by a simple projection operator has been treated by W. Franz [FRA 65a]. The expressions (15) and (16) can, however, easily be generalized to this case by including the factor $\varepsilon(\boldsymbol{p})$ inside the integral over $\boldsymbol{p}$.

If the particle which is detected has a spin, one must include a projection operator which corresponds to the type of spin measurement performed. Thus, if the particle has spin of magnitude $s\hbar$, one may detect the component of the spin along the momentum $\boldsymbol{p}$ i.e. the helicity

$$h = \boldsymbol{s}\boldsymbol{p}/p, \tag{17}$$

which has the eigenvalues

$$h = -s, -s+1, \ldots, s \tag{18}$$

in units of $\hbar$.

The projection operator for the measurement, which determines the helicity to have the value h, is given by

$$\varepsilon = |h\rangle\langle h| \tag{19}$$

or

$$\langle h'|\,\varepsilon\,|h''\rangle = \delta_{hh'}\delta_{hh''}. \tag{20}$$

It may well be, however, that one performs instead a measurement which determines e.g. the z-component of $\boldsymbol{s}$. In this case the projection operator has non-diagonal matrix elements in the h representation i.e. the spin measurement of the particle will in general be described by the non-diagonal matrix

$$\varepsilon_{hh'} = \langle h|\,\varepsilon\,|h'\rangle. \tag{21}$$

For non-ideal spin measurements one introduces the efficiency matrix

$$\varepsilon_{hh'} = \langle h|\sum_s |s\rangle\,\varepsilon_s\,\langle s|\,|h'\rangle, \tag{22}$$

where ε_s is the efficiency with which the spinstate $|s\rangle$ is detected.

The expressions (15) and (16) can thus be generalized to include the non-ideal detection of momentum as well as spin. One finds

$$\rho(\text{res}) = \sum_{hh'} \int \varepsilon(\boldsymbol{p})\varepsilon_{hh'} \langle \boldsymbol{p}h'|\ \rho_{\text{i}}(\text{total})\ |\boldsymbol{p}h\rangle\ \mathrm{d}^3p/P, \tag{23}$$

where

$$P = \mathrm{Tr}\Big(\sum_{hh'} \int \varepsilon(\boldsymbol{p})\varepsilon_{hh'} \langle \boldsymbol{p}h'|\ \rho_{\text{i}}(\text{total})\ |\boldsymbol{p}h\rangle\ \mathrm{d}^3p\Big) \tag{24}$$

is the probability of detecting the particle with momentum determined by the detection efficiency $\varepsilon(\boldsymbol{p})$ and with a spin determined by the efficiency $\varepsilon_{hh'}$ in the helicity representation. The state $|\boldsymbol{p}h\rangle$ is a free state of the particle with momentum $\boldsymbol{p}$ and helicity quantum number h.

We have assumed in eqs. (23) and (24) that the efficiency matrix for the measurement is diagonal in $|\boldsymbol{p}|$. In actual experiments where the arrival time of the decay product in a counter is registered, the corresponding efficiency matrix must be non-diagonal in energy within an energy interval $\Delta E \gtrsim \hbar/\Delta t$, where Δt is the time resolution. This complication is not important for most experiments, since the nuclear energy levels are so far apart that they do not violate the above condition. On the other hand, the energy splitting of a nuclear level in external fields, i.e. hyperfine structure splittings, are so small ($\sim 10^{-6}$ eV) that non-diagonal matrix elements are of importance if the time resolution is better than $\hbar/\Delta E \sim 10^{-9}$ sec. Under normal conditions, the results (23) and (24) then describe the density matrix $\rho(\text{res})$ at the time t_0 where it was established that the reaction product was emitted from the nucleus, while P gives the total probability of detection up to this time. The probability of detection ΔP within the time interval $t_0 - \Delta t$ to t_0 is obtained as

$$\Delta P = P(t_0) - P(t_0 - \Delta t). \tag{25}$$

Not only are the energy differences between the nuclear levels much larger than $\hbar/\Delta t$ where Δt is the resolution time, but they are also in most cases much larger than the natural line width $\hbar/\tau$ due to the lifetime τ. Therefore, the density matrices will never contain non-diagonal matrix elements except between different magnetic substates of the same spin. The corresponding density matrix $\langle IM|\ \rho\ |IM'\rangle$ describes the state of polarization of the energy level of spin I. The above facts give rise to the important simplification that the time development of the density matrix can be decomposed into independent stages which can be treated separately and can be combined in different ways according to the actual experiment performed.

Below we will give the relevant formulae for the change in the density matrix which takes place in a nuclear reaction and in a decay. For a nuclear reaction one finds according to (10) that the density matrix ρ_f after the reaction is given in terms of the density matrix ρ_i before the reaction by

$$\langle I_f M_f \boldsymbol{p}_f | \rho_f | I_f M_f' \boldsymbol{p}_f \rangle = \sum_{M_i M_i' \boldsymbol{p}_i} \langle I_f M_f \boldsymbol{p}_f | S | I_i M_i \boldsymbol{p}_i \rangle \times \langle I_i M_i \boldsymbol{p}_i | \rho_i | I_i M_i' \boldsymbol{p}_i \rangle \langle I_i M_i' \boldsymbol{p}_i | S^\dagger | I_f M_f' \boldsymbol{p}_f \rangle, \tag{26}$$

where $\langle I_f M_f \boldsymbol{p}_f | S | I_i M_i \boldsymbol{p}_i \rangle$ is the S-matrix element connecting the initial nuclear state $|I_i M_i\rangle$ and the state of the incoming projectile $|\boldsymbol{p}_i\rangle$ with the final nuclear state $|I_f M_f\rangle$ and the reaction product in the state $|\boldsymbol{p}_f\rangle$. The initial density matrix specifies the orientation of the target nucleus as well as the bombarding conditions.

Limiting the ensemble of nuclei to those for which a nuclear reaction has taken place one should substitute the S-matrix by the reaction matrix T. Furthermore, it is convenient to leave out the density matrix describing the bombarding condition, thereby normalizing the final density matrix such that the trace is equal to the total cross section, i.e.

$$\langle I_f M_f \boldsymbol{p}_f | \rho_\sigma | I_f M_f' \boldsymbol{p}_f \rangle = \frac{m_0^2 p_f}{4\pi^2 \hbar^4 p_i} \sum_{M_i, M_i'} \langle I_f M_f \boldsymbol{p}_f | T | I_i M_i \boldsymbol{p}_i \rangle \times \langle I_i M_i | \rho_i | I_i M_i' \rangle \langle I_i M_i' \boldsymbol{p}_i | T^\dagger | I_f M_f' \boldsymbol{p}_f \rangle. \tag{27}$$

The diagonal matrix elements $M_f = M_f'$ are equal to the differential cross section for exciting a state $|I_f M_f\rangle$. It is easy to generalize the expression (27) to the case where the projectile and the reaction product have spins. One should in this case include the density matrix which describes the state of polarization of the incoming particle.

If the nucleus undergoes a process which is caused by a time-independent interaction as e.g. a γ-decay, the time dependence of ρ is described by exponential laws. If the nucleus is initially at time $t = 0$ in a state described by the density matrix $\langle I_i M_i | \rho_i | I_i M_i' \rangle$, the density matrix at time t is given by

$$\langle I_i M_i | \rho(t) | I_i M_i' \rangle = \langle I_i M_i | \rho_i | I_i M_i' \rangle \exp(-\gamma_i t) \tag{28}$$

and for other states by

$$\langle \boldsymbol{ph} I_f M_f | \rho(t) | \boldsymbol{ph}' I_f M_f' \rangle = \frac{2\pi}{\hbar} \frac{1}{\gamma_i} \frac{\mathrm{d}N}{\mathrm{d}E} (1 - \exp(-\gamma_i t)) \times \sum_{M_i M_i'} \langle I_f M_f \boldsymbol{ph} | T | I_i M_i \rangle \langle I_i M_i | \rho_i | I_i M_i' \rangle \times \langle I_i M_i' | T^\dagger | I_f M_f' \boldsymbol{ph}' \rangle. \tag{29}$$

The quantity γ_i is the decay rate of the state i while dN/dE denotes the density of final states f. The T-matrix describes the transition amplitude from the initial state of spin I_i to a final state of spin I_f where the decay product is emitted in direction $\boldsymbol{p}$ with helicity h. The density matrix contains the degrees of freedom of the decay product and we consider only diagonal matrix elements in $\boldsymbol{p}$. It is normalized in such a way that the trace over I, M including an integration over $\boldsymbol{p}$ is unity. This follows from the fact that

$$\gamma_{i\to f} = \sum_{M_f} \int d\Omega_p \frac{2\pi}{\hbar} |\langle \boldsymbol{p} I_f M_f | T | I_i M_i \rangle|^2 \, dN/dE \tag{30}$$

denotes the partial decay constant for the transition $I_i \to I_f$ and

$$\gamma_i = \sum_f \gamma_{i\to f}. \tag{31}$$

Fig. 1 illustrates the complete density matrix, where the submatrix $\langle I_i M_i | \rho | I_i M_i' \rangle$ which originally has trace one is changed and the trace is gradually distributed over the complete matrix.

In the above discussion we have assumed that the states $|f\rangle$ do not decay again. If the state $|f\rangle$ decays eq. (29) should be modified and more complicated exponential time-factors would appear which are similar to the populations in radioactive decay sequences.

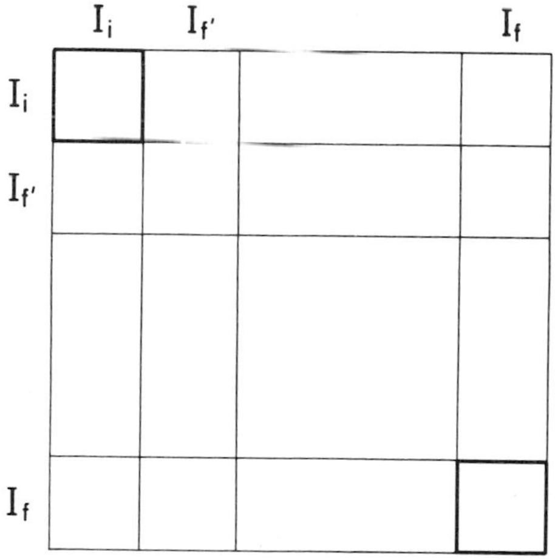

Fig. 1. The figure illustrates the density matrix for a radioactive decay. Initially the density matrix has matrix elements only within the submatrix (I_i, I_i) corresponding to the fact that the nucleus is known to be in this state. Gradually these matrix elements decrease and new matrix elements appear within the submatrices $(I_f', I_f'), \ldots, (I_f, I_f)$ according to eqs. (28) and (29). If the γ-quantum corresponding to the transition to the state I_f is observed the density matrix disappears except for the submatrix (I_f, I_f).

As an example we may consider a Coulomb excitation where the density matrix (27) for the nucleus after the process has nonvanishing matrix elements for a large number of different nuclear levels. If a measurement is performed e.g. of the energy and direction of the scattered projectile determining the final state to be $|\mathrm{f}\rangle$, the density matrix after the measurement is given by (23), i.e.

$$\langle I_\mathrm{f} M_\mathrm{f}|\, \rho\, |I_\mathrm{f} M_\mathrm{f}'\rangle = \langle \boldsymbol{p}_\mathrm{f} I_\mathrm{f} M_\mathrm{f}|\, \rho_\sigma\, |\boldsymbol{p}_\mathrm{f} I_\mathrm{f} M_\mathrm{f}'\rangle / (\mathrm{d}\sigma/\mathrm{d}\Omega)_\mathrm{f}, \tag{32}$$

where

$$(\mathrm{d}\sigma/\mathrm{d}\Omega)_\mathrm{f} = \sum_{M_\mathrm{f}} \langle \boldsymbol{p}_\mathrm{f} I_\mathrm{f} M_\mathrm{f}|\, \rho_\sigma\, |\boldsymbol{p}_\mathrm{f} I_\mathrm{f} M_\mathrm{f}\rangle \tag{33}$$

is the cross section for exciting the state f.

As a second example we may consider the observation of a γ-quantum where the measurement consists of the determination of its direction of emission and of its polarization. The efficiency matrix $\varepsilon_{hh'}$ (22) is often written in the form

$$\varepsilon_{hh'} = \tfrac{1}{2}\varepsilon \langle h|\, (1 + \boldsymbol{Q}\boldsymbol{\sigma})\, |h'\rangle, \tag{34}$$

where $\boldsymbol{\sigma}$ are the Pauli spin matrices while $\boldsymbol{Q}$ describes the type of polarization measurement. Thus for an ideal polarization measurement $|\boldsymbol{Q}| = 1$ while no measurement of polarization corresponds to $\boldsymbol{Q} = 0$. If $\boldsymbol{Q}$ is in the direction or opposite $\boldsymbol{p}$ a measurement of the circular polarization is performed, while $\boldsymbol{Q}$ perpendicular to $\boldsymbol{p}$ describes a measurement of linear polarization [FRA 65]. The quantity ε is the total efficiency $\sum_s \varepsilon_s$ where ε_s is defined in eq. (22).

The probability of observing a gamma quantum in the time interval 0 to t_2 with the specification (34) is according to (24) given by

$$P = \sum_{M_\mathrm{f} hh'} \tfrac{1}{2}\varepsilon \langle h|\, (1 + \boldsymbol{Q}\boldsymbol{\sigma})\, |h'\rangle \langle \boldsymbol{p}h' I_\mathrm{f} M_\mathrm{f}|\, \rho(t_2)\, |\boldsymbol{p}h I_\mathrm{f} M_\mathrm{f}\rangle, \tag{35}$$

where $\rho(t_2)$ is given by (29). If the counter is open only in the time interval t_1 to t_2, one must subtract from (35) the corresponding expression at the time t_1. Once a gamma quantum has been recorded the nucleus is known to be in the state of spin I_f and the density matrix is then given by

$$\langle I_\mathrm{f} M_\mathrm{f}|\, \rho\, |I_\mathrm{f} M_\mathrm{f}'\rangle = \sum_{hh'} \tfrac{1}{2}\varepsilon \langle h'|\, (1 + \boldsymbol{Q}\boldsymbol{\sigma})\, |h\rangle \langle \boldsymbol{p}h I_\mathrm{f} M_\mathrm{f}|\, \rho(t_2)\, |\boldsymbol{p}h' I_\mathrm{f} M_\mathrm{f}'\rangle P^{-1}, \tag{36}$$

where the density matrix on the right-hand side is given by (29).

The above procedure can be used also for the observation of successive γ-quanta in a cascade. Here one has, however, to take into account that the time sequence of the γ-quanta is known and that the density matrix after each γ-quantum is changed in this time sequence. In appendix G explicit expressions for the density matrix (29) as well as for (36) are given.

For the evaluation of angular distributions it is often convenient to use the statistical tensors instead of density matrices $\langle IM|\,\rho\,|I'M'\rangle$. We thus define

$$\rho_{k\kappa}(f_1I_1, f_2I_2) = [(2I_1+1)(2I_2+1)]^{1/4} \sum_{M_1M_2} (-1)^{I_1-M_1} \begin{pmatrix} I_1 & k & I_2 \\ -M_1 & \kappa & M_2 \end{pmatrix} \times \langle I_2M_2f_2|\,\rho\,|I_1M_1f_1\rangle. \tag{37}$$

It follows from the fact that ρ is Hermitian that

$$\rho^*_{k\kappa}(f_1I_1, f_2I_2) = (-1)^{-I_1+I_2+\kappa}\rho_{k-\kappa}(f_2I_2, f_1I_1). \tag{38}$$

Furthermore, it is seen that

$$\sum_{fI} \rho_{00}(fI, fI) = 1, \tag{39}$$

since the trace of the density matrix is unity. The summation in (39) should be extended over all degrees of freedom f. If by measurement with efficiency ε_f it has been confirmed that the nucleus is in the state of spin I the probability for this measurement is given by (24), i.e.

$$P = \sum_f \rho_{00}(fI, fI)\varepsilon_f. \tag{40}$$

The statistical tensor after the measurement is a convenient measure of the state of polarizations of the level of spin I and we shall often denote it by the polarization tensor and use the notation $P_{k\kappa}(I)$. According to (23) one finds

$$P_{k\kappa}(I) = \rho_{k\kappa}(\text{res}) = \frac{\sum_f \rho_{k\kappa}(fI, fI)\varepsilon_f}{\sum_f \rho_{00}(fI, fI)\varepsilon_f}. \tag{41}$$

The polarization tensor has the properties

$$P^*_{k\kappa}(I) = (-1)^{\kappa}P_{k-\kappa}(I) \tag{42}$$

and

$$P_{00}(I) = 1. \tag{43}$$

APPENDIX G

Gamma-Ray Angular Distribution

In many Coulomb excitation experiments it is important to know the angular distribution of the γ-quanta following the Coulomb excitation process. Although the subject of gamma angular distributions from states which have been polarized in a nuclear reaction has been treated many times (see e.g. [DEV 57, FRA 65, ROS 67]), we shall rederive the relevant formulae in a notation consistent with the present work.

1. *Statistical tensors for gamma decay*

We imagine that the initial decaying state of spin I has been polarized by some reaction or through the detection of some preceding radiation. The information about the state of polarization is represented by the density matrix (see appendix F). After the gamma decay the final state will be in a state of polarization which depends on the initial density matrix, as well as on the detection of the gamma quantum, i.e, its direction and polarization (see fig. 1).

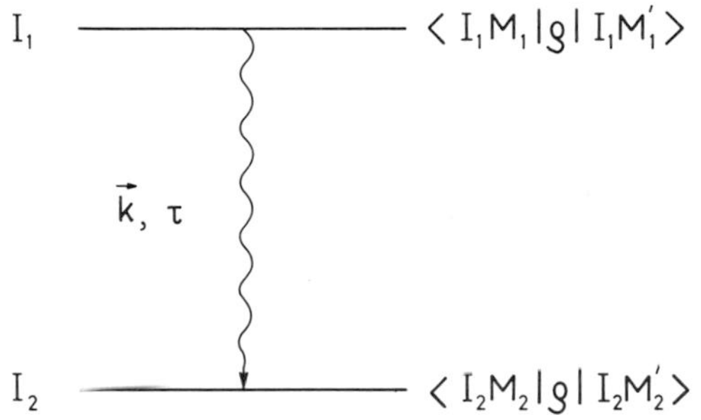

Fig. 1. From a state with spin I_1 whose polarization is described by the density matrix $\langle I_1 M_1 | \rho | I_1 M_1' \rangle$ a gamma quantum is emitted whose direction $\boldsymbol{k}$ and state of polarization τ is detected. The density matrix of the final state of spin I_2 can then be determined.

The connection between the density matrix for the initial state and the final state is given by eqs. (F.28) and (F.29). In the first-order perturbation theory one finds

$$\langle \boldsymbol{k}\tau I_2 M_2 | \, \rho(t) \, | I_2 M_2' \boldsymbol{k}\tau' \rangle$$
$$= \frac{2\pi}{\hbar}\frac{1}{\gamma_1}\frac{\mathrm{d}N}{\mathrm{d}E}(1 - \exp(-\gamma_1 t)) \sum_{M_1 M_1'} \langle I_2 M_2 \boldsymbol{k}\tau | \, H_\gamma \, | I_1 M_1 \rangle$$
$$\times \langle I_1 M_1 | \, \rho \, | I_1 M_1' \rangle \langle I_1 M_1' | \, H_\gamma \, | I_2 M_2' \boldsymbol{k}\tau' \rangle, \tag{1}$$

where we have specified the final state by the direction of emission $\boldsymbol{k}$ of the gamma quantum and its helicity τ, H_γ being the interaction between the nucleus and the electromagnetic radiation field. The total decay probability per unit time of the state of spin I_1 is denoted by γ_1, and $\mathrm{d}N/\mathrm{d}E$ is the density of final states.

The matrix element of the interaction may be written

$$\langle I_2 M_2 \boldsymbol{k}\tau | \, H_\gamma \, | I_1 M_1 \rangle = -\sqrt{\frac{4\pi^2\hbar^2 c}{k^2 V}} \sum_{LM} \sqrt{\frac{2L+1}{8\pi}} \langle I_2 M_2 L M | I_1 M_1 \rangle$$
$$\times (-1)^{I_2 - I_1} D^L_{M\tau}(\boldsymbol{z} \to \boldsymbol{k})[\delta_{\mathrm{E}L}(1 \to 2) + \tau \delta_{\mathrm{M}L}(1 \to 2)]. \tag{2}$$

In eq. (2) V denotes the volume of normalization, ω is the frequency and $k = \omega/c$ the wave number of the gamma quantum. The multipole transition amplitudes $\delta_{\pi L}(1 \to 2)$ are defined through the matrix elements of the electromagnetic multipole operators by

$$\delta_{\pi L} = \mathrm{i}^{s(L)} \sqrt{\frac{8\pi(L+1)}{L[(2L+1)!!]^2 \hbar}} \left(\frac{\omega}{c}\right)^{2L+1} \frac{\langle I_2 \| \mathscr{M}(\pi L) \| I_1 \rangle}{\sqrt{2I_1 + 1}}, \tag{3}$$

where $s(L)$ is given by

$$s(L) = \begin{cases} L, & \text{for E}L \text{ radiation} \\ L+1, & \text{for M}L \text{ radiation}. \end{cases}$$

The multipole operators appearing in (3) contain the nuclear current $\boldsymbol{j}(\boldsymbol{r})$ and the multipole components of the vector potential. They are defined by

$$\mathscr{M}(\mathrm{E}L, \mathrm{M}) = \frac{(2L+1)!!}{k^{L+1} c (L+1)} \int \boldsymbol{j}(\boldsymbol{r}) \nabla \times \boldsymbol{L} j_L(kr) Y_{LM}(\boldsymbol{r}) \, \mathrm{d}^3 r,$$
$$\mathscr{M}(\mathrm{M}L, \mathrm{M}) = -\mathrm{i} \frac{(2L+1)!!}{k^L c (L+1)} \int \boldsymbol{j}(\boldsymbol{r}) \boldsymbol{L} j_L(kr) Y_{LM}(\boldsymbol{r}) \, \mathrm{d}^3 r, \tag{4}$$

where the operator $\boldsymbol{L}$ is given in eq. (II.1.5). The multipole operators (4) depend on k in contrast to the multipole operators defined in § II.1. It can be shown that the expression (4) reduces to the expressions (II.1.3–4) for small values of k. In the derivation of this result, one utilizes the continuity equation of the nuclear current. This introduces an ambiguity in the relative sign of the two expressions for the electric multipole moment and the above relation only holds if the initial state has a higher energy than the final state (see e.g. [HAM 74]). The absolute square of $\delta_{\pi L}(1 \to 2)$ is equal to the partial transition probability per unit time. If the state can only decay by γ-emission the lifetime τ of the state 1 is thus given by

$$\gamma_1 = \frac{1}{\tau} = \sum_{L\pi n} |\delta_{\pi L}(1 \to n)|^2, \tag{5}$$

where we have summed over all final states $|n\rangle$.

If time reversal invariance is valid one may introduce the phase convention of appendix E and the transition amplitudes $\delta_{\pi L}(1 \to 2)$ have the phase $\mathrm{i}^{s(L)}$. One may also introduce the phase $\mathrm{i}^{s(L)}$ in the reduced matrix element and redefine the phases of the nuclear wave functions in such a way that δ_L is real [BOH 69]. With this choice of phases, factors of the type i^{λ} would be introduced in the theory of Coulomb excitation. The mixing ratios which are measured experimentally are ratios of the δ_L's and are obviously real quantities whichever choice is made on the phases of the nuclear states.

Introducing eq. (2) into eq. (1) one finds that the statistical tensor $\rho_{k_2\kappa_2}(I_2)$ for the state of spin I_2 can be expressed in terms of the normalized statistical tensor of the initial state $P_{k_1\kappa_1}(I_1)$ by the following formula (see eq. (F.37))

$$\begin{aligned}\rho_{k_2\kappa_2}(I_2\boldsymbol{k}\tau, I_2\boldsymbol{k}\tau') &= \frac{\sqrt{(2I_1+1)(2I_2+1)}}{8\pi\gamma_1}\,\mathrm{d}\Omega \\ &\times \sum_{\substack{k_1kL\\ L'\kappa_1\kappa}} (2k+1)(2k_1+1)(-1)^{L+k_2+\kappa_1+\tau'}\sqrt{(2L+1)(2L'+1)} \\ &\times [\delta_{\mathrm{E}L} + \tau\delta_{\mathrm{M}L}][\delta^*_{\mathrm{E}L'} + \tau'\delta^*_{\mathrm{M}L'}]P_{k_1\kappa_1}(I_1)[D^k_{\kappa q}(z \to \boldsymbol{k})]^* \\ &\times \begin{pmatrix} L & k & L' \\ \tau & q & -\tau' \end{pmatrix}\begin{pmatrix} k_1 & k & k_2 \\ -\kappa_1 & \kappa & \kappa_2 \end{pmatrix}\begin{Bmatrix} I_1 & I_1 & k_1 \\ L' & L & k \\ I_2 & I_2 & k_2 \end{Bmatrix},\end{aligned} \tag{6}$$

where the last factor is the 9-j symbol defined e.g. in [EDM 57].

The statistical tensor (6) is in fact a density matrix in the two states of polarization ($\tau = \pm 1$) of the gamma quantum. The angular distribution of the γ-quanta is obtained directly from eq. (6) by means of eq. (F.35). The experimental set-up is here described in terms of the efficiency matrix (F.34). Formula (6) may be used to evaluate any type of angular correlation with gamma rays and we shall in the following write down the explicit results for a number of cases.

In the application of eq. (6) one may usually apply the approximate conservation of parity and the time reversal invariance. If parity is conserved we have the relations

$$(-1)^{\pi_1+\pi_2} = \begin{cases} (-1)^L, & \text{for E}L\text{ transition} \\ (-1)^{L+1}, & \text{for M}L\text{ transition,} \end{cases} \tag{7}$$

where π_1 and π_2 are the (even or odd) parities of the nuclear states 1 and 2, respectively.

In the following we shall use eq. (6) to evaluate first the angular distribution of γ-quanta from a state whose polarization is given by the statistical tensor $P_{k\kappa}$. Secondly, we shall evaluate the density matrix of the nuclear states after the detection of the gamma ray, and, finally, we shall give the application to the case of γ–γ correlation. For the detection of gamma quanta we shall consider three types of measurements, viz. detection of circular polarization, detection of linear polarization and detection of direction of emission only.

2. *Angular distribution of gamma rays*

We first consider the situation in which the polarization tensor $P_{k_1\kappa_1}(I_1)$ is known and ask for the probability of observing the gamma quantum. This probability is given by eq. (F.35) where the efficiency matrix is given by eq. (F.34) and where the trace should be taken also over the magnetic quantum numbers of the final nuclear states. The detection probability is thus given by

$$P(\boldsymbol{k}, \varepsilon) = \sum_{\tau\tau'} \rho_{00}(I_2\boldsymbol{k}\tau, I_2\boldsymbol{k}\tau')\varepsilon_{\tau'\tau}, \tag{8}$$

where

$$\rho_{00}(I_2\boldsymbol{k}\tau, I_2\boldsymbol{k}\tau') = \sum_{M_2} \langle I_2M_2\boldsymbol{k}\tau|\, \rho\, |I_2M_2\boldsymbol{k}\tau'\rangle. \tag{9}$$

Setting $k_2 = 0$ in (6) one finds

$$\rho_{00}(I_2\boldsymbol{k}\tau, I_2\boldsymbol{k}\tau') = \frac{\mathrm{d}\Omega}{8\pi\gamma_1} \sum_{kLL'\kappa} (-1)^k\sqrt{2k+1}\, \frac{\begin{pmatrix} L & L' & k \\ \tau & -\tau' & q \end{pmatrix}}{\begin{pmatrix} L & L' & k \\ 1 & -1 & 0 \end{pmatrix}} F_k(LL'I_2I_1)$$

$$\times\, (D^k_{\kappa q}(z \to \boldsymbol{k}))^* P_{k\kappa}(I_1)[\delta_{\mathrm{E}L} + \tau\delta_{\mathrm{M}L}][\delta^*_{\mathrm{E}L'} + \tau'\delta^*_{\mathrm{M}L'}], \tag{10}$$

where $F_k(LL'I_2I_1)$ is the usual γ–γ correlation coefficient defined by

$$F_k(LL'I_2I_1) = (-1)^{I_1+I_2-1}\sqrt{(2k+1)(2I_1+1)(2L+1)(2L'+1)}$$

$$\times \begin{pmatrix} L & L' & k \\ 1 & -1 & 0 \end{pmatrix} \begin{Bmatrix} L & L' & k \\ I_1 & I_1 & I_2 \end{Bmatrix}. \tag{11}$$

The coefficient is symmetric in L and L' and normalized in such a way that for $k = 0$, F_0 is unity. It is tabulated e.g. in [FRA 65].

Circular polarization detected

The angular distribution of γ-quanta with circular polarization τ is obtained by setting $\tau = \tau'$ or by using the efficiency matrix

$$\begin{pmatrix} \varepsilon_{11} & \varepsilon_{1-1} \\ \varepsilon_{-11} & \varepsilon_{-1-1} \end{pmatrix} = \tfrac{1}{2}\begin{pmatrix} 1+\tau & 0 \\ 0 & 1-\tau \end{pmatrix}. \tag{12}$$

One finds

$$P(\boldsymbol{k}, \tau) = \frac{\mathrm{d}\Omega}{8\pi\gamma_1} \sum_{LL'k\kappa} (-\tau)^k \delta_L \delta_{L'}^* F_k(LL'I_2I_1)\sqrt{2k+1} \times (D_{\kappa 0}^k(z \to \boldsymbol{k}))^* P_{k\kappa}(I_1), \tag{13}$$

where δ_L stands for δ_{EL} or δ_{ML} according to the parity selection rule.

Linear polarization detected

The angular distribution of γ-quanta which are linearly polarized along the x-axis can be computed by means of the efficiency matrix

$$\begin{pmatrix} \varepsilon_{11} & \varepsilon_{1-1} \\ \varepsilon_{-11} & \varepsilon_{-1-1} \end{pmatrix} = \tfrac{1}{2}\begin{pmatrix} 1 & 1 \\ 1 & 1 \end{pmatrix}. \tag{14}$$

The resulting angular distribution can be written in the form

$$P(\boldsymbol{k}, \varepsilon_x) = \frac{\mathrm{d}\Omega}{8\pi\gamma_1} \sum_{\substack{k\,\text{even} \\ LL'\kappa}} \sqrt{2k+1}\,\delta_L \delta_{L'}^* F_k(LL'I_2I_1) P_{k\kappa}(I_1) \times \left\{ D_{\kappa 0}^{k*}(\varphi, \vartheta, \psi) + \frac{1}{2} \frac{\begin{pmatrix} L & L' & k \\ 1 & 1 & -2 \end{pmatrix}}{\begin{pmatrix} L & L' & k \\ 1 & -1 & 0 \end{pmatrix}} (-1)^{L'+\pi_1-\pi_2} \times [D_{\kappa -2}^{k*}(\varphi, \vartheta, \psi) + D_{\kappa 2}^{k*}(\varphi, \vartheta, \psi)] \right\}, \tag{15}$$

where $(\varphi, \vartheta, \psi)$ are Eulerian angles describing the rotation from the laboratory system to a system where the z-axis is in the direction $\boldsymbol{k}$ and the x-axis is in the direction of the linear polarization. The quantities π_1 and π_2 are the even and odd parities of the nuclear states.

No polarization detected

If the state of polarization is not detected, the efficiency matrix is

$$\begin{pmatrix} \varepsilon_{11} & \varepsilon_{1-1} \\ \varepsilon_{-11} & \varepsilon_{-1-1} \end{pmatrix} = \begin{pmatrix} 1 & 0 \\ 0 & 1 \end{pmatrix} \tag{16}$$

and the angular distribution takes the form

$$P(\boldsymbol{k}) = \frac{\mathrm{d}\Omega}{4\pi\gamma_1} \sum_{\substack{k\,\text{even} \\ LL'\kappa}} \sqrt{2k+1}\,\delta_L \delta_{L'}^* F_k(LL'I_2I_1) P_{k\kappa}(I_1) D_{\kappa 0}^{k*}(z \to \boldsymbol{k}). \tag{17}$$

3. *Statistical tensor after gamma decay*

In this section we consider the situation where the statistical tensor for the initial state is given, and where we ask for the statistical tensor after a γ-quantum has been detected with a certain polarization. This situation is described in principle by formula (6) and we shall here only specialize this formula to the case where no polarization is measured or where not even the direction of the γ-quantum is detected.

No polarization detected

After the detection of the γ-quantum in the direction $\boldsymbol{k}$ the statistical tensor is given by (F.41), where the efficiency matrix is given by eq. (16). One thus finds according to (6)

$$P_{k_2\kappa_2}(I_2) = \frac{\rho_{k_2\kappa_2}(I_2)}{\rho_{00}(I_2)}, \tag{18}$$

where

$$\rho_{k_2\kappa_2}(I_2) = \frac{\mathrm{d}\Omega}{4\pi\gamma_1} \sum_{\substack{k\,\text{even} \\ LL'k_1\kappa_1}} (-1)^{k_1+\kappa_1} \sqrt{\frac{(2k+1)(2k_1+1)}{2k_2+1}} \begin{pmatrix} k_1 & k & k_2 \\ -\kappa_1 & \kappa & \kappa_2 \end{pmatrix}$$

$$\times\ \delta_L \delta_{L'}^* F_k^{k_2 k_1}(LL'I_2I_1) P_{k_1\kappa_1}(I_1)\ D_{\kappa 0}^{k*}(z \to \boldsymbol{k}) \tag{19}$$

and

$$\rho_{00}(I_2) = \frac{\mathrm{d}\Omega}{4\pi\gamma_1} \sum_{\substack{k\,\text{even} \\ LL'\kappa}} \sqrt{2k+1}\,\delta_L \delta_{L'}^* F_k(LL'I_2I_1) P_{k\kappa}(I_1)\ D_{\kappa 0}^{k*}(z \to \boldsymbol{k}). \tag{20}$$

The geometrical coefficient in eq. (19) is defined by [HAM 74]

$$F_k^{k_2k_1}(LL'I_2I_1) = (-1)^{L'+k_2+k_1+1}\sqrt{(2I_1+1)(2I_2+1)} \times \sqrt{(2L+1)(2L'+1)(2k+1)(2k_1+1)(2k_2+1)} \times \begin{pmatrix} L & L' & k \\ 1 & -1 & 0 \end{pmatrix} \begin{Bmatrix} I_2 & L & I_1 \\ I_2 & L' & I_1 \\ k_2 & k & k_1 \end{Bmatrix}. \quad (21)$$

It is noted that

$$\begin{aligned} F_k^{0k_1}(LL'I_2I_1) &= \delta_{k_1k}F_k(LL'I_2I_1), \\ F_k^{k_20}(LL'I_2I_1) &= \delta_{kk_2}(-1)^{L+L'}F_k(LL'I_1I_2), \\ F_0^{k_2k_1}(LL'I_2I_1) &= \delta_{LL'}\delta_{k_1k_2}(-1)^{I_1+I_2+k_1+L} \\ &\quad \times \sqrt{(2I_1+1)(2I_2+1)(2k_1+1)} \begin{Bmatrix} I_1 & I_1 & k_1 \\ I_2 & I_2 & L \end{Bmatrix} \end{aligned} \quad (22)$$

and

$$F_0^{00}(LL'I_2I_1) = \delta_{LL'}.$$

No gamma quantum observed

If the gamma quantum is not detected, but at some later time it has been verified that the decay has taken place, e.g. by the detection of the decay products of this state, the statistical tensor is obtained by integrating (6) over the direction $\boldsymbol{k}$ and taking the trace over the polarization τ. The result can be written

$$P_{k\kappa}(I_2) = P_{k\kappa}(I_1)\, H_k(I_2I_1), \quad (23)$$

where

$$H_k(I_2I_1) = \frac{1}{\gamma_1}(-1)^{I_1+I_2+k}\sqrt{(2I_1+1)(2I_2+1)}\sum_L(-1)^L|\delta_L|^2 \begin{Bmatrix} I_1 & I_1 & k \\ I_2 & I_2 & L \end{Bmatrix}. \quad (24)$$

4. *Gamma-gamma correlation*

As an illustration of the application of the formulae given above we consider the familiar case of gamma-gamma correlation. An unpolarized state of spin I_1 decays by γ-emission of multipolarity L_1 and L_1' to a state of spin I_2. The direction of emission $\boldsymbol{k}_1$ is detected, but no measurement of the state of polarization is performed. The statistical tensor for the state with spin I_2 is then given by (19) with $k_1 = \kappa_1 = 0$. The probability of detecting the γ-decay of this state into a state with spin I_3 in a direction $\boldsymbol{k}_2$ is then given by eq. (17). The correlated probability of detecting both γ-quanta is thus given by

$$P(\boldsymbol{k}_1, \boldsymbol{k}_2) = \frac{d\Omega_1}{4\pi\gamma_1}\frac{d\Omega_2}{4\pi\gamma_2} \sum_{k \text{ even}} \left(\sum_{L_1 L_1'} \delta_{L_1}\delta^*_{L_1'} F_k(L_1 L_1' I_1 I_2)(-1)^{L_1+L_1'} \right)$$

$$\times \left(\sum_{L_2 L_2'} \delta_{L_2}\delta^*_{L_2'} F_k(L_2 L_2' I_3 I_2) \right) P_k(\cos(\boldsymbol{k}_1 \boldsymbol{k}_2)) . \qquad (25)$$

If the γ-counters are only open within a finite time-interval this formula should be supplemented by the appropriate exponential function in time, as discussed in appendix F.

APPENDIX H

The Coulomb Excitation Functions

The Coulomb excitation functions $I_{\lambda\mu}(\vartheta, \xi)$ play an important role in the theory of Coulomb excitation, not only in the first-order perturbation theory of electric excitations, but also for magnetic excitations and multipole-multipole excitations. Furthermore, the second-order perturbation theory can also be expressed in terms of these integrals and it is noteworthy that the polarization effect is related to those Coulomb excitation functions, where the parameters λ and μ do not satisfy the usual condition of even $\lambda + \mu$.

The Coulomb excitation functions are defined by the relation

$$I_{\lambda\mu}(\vartheta, \xi) = \int_{-\infty}^{+\infty} \frac{[\varepsilon + \cosh w + i\sqrt{\varepsilon^2 - 1}\sinh w]^{\mu}}{(1 + \varepsilon \cosh w)^{\lambda+\mu}} \exp\{i\xi[\varepsilon \sinh w + w]\}\, dw, \tag{1}$$

where λ and μ are integers and $\lambda \geqslant 0$. The quantity ε is connected with the scattering angle ϑ, which is assumed to be positive and less than 180 degrees, i.e.

$$\varepsilon = 1/\sin(\tfrac{1}{2}\vartheta) \geqslant 1. \tag{2}$$

From the definition (1) follows immediately a number of relations. Thus, it follows from the substitution $w \to -w$ that

$$I^{*}_{\lambda\mu}(\vartheta, \xi) = I_{\lambda\mu}(\vartheta, \xi), \tag{3}$$

which means that $I_{\lambda\mu}(\vartheta, \xi)$ is real. Furthermore, we note that a change of sign in ξ leads to the following relation

$$I_{\lambda\mu}(\vartheta, -\xi) = I_{\lambda-\mu}(\vartheta, \xi). \tag{4}$$

For $\vartheta = \pi$, i.e. $\varepsilon = 1$ the integrals (1) are independent of μ, i.e.

$$I_{\lambda\mu}(\pi, \xi) = \int_{-\infty}^{+\infty} \frac{\exp\{i\xi[\sinh w + w]\}\, dw}{(1 + \cosh w)^{\lambda}} = I_{\lambda 0}(\pi, \xi). \tag{5}$$

The Coulomb excitation functions can only be expressed in terms of elementary functions for $\xi = 0$, where one finds

$$\begin{aligned} I_{\lambda\mu}(\vartheta, 0) &= (\varepsilon^2 - 1)^{-\lambda+1/2} \int_{-\phi_0}^{\phi_0} \exp(i\mu\phi)(\varepsilon \cos \phi - 1)^{\lambda-1} \, d\phi \\ &= (2\pi)^{1/2}(\lambda - 1)! \, (\cos(\tfrac{1}{2}\vartheta))^{1/2}(\tan(\tfrac{1}{2}\vartheta))^{\lambda} P_{\mu-1/2}^{-\lambda+1/2}(\sin(\tfrac{1}{2}\vartheta)) \\ &= 2 \frac{(\lambda - 1)!}{(\lambda + \mu - 1)!} (-1)^{\mu}(i \tan(\tfrac{1}{2}\vartheta))^{\lambda} Q_{\lambda-1}^{\mu}(i \tan(\tfrac{1}{2}\vartheta)), \end{aligned} \tag{6}$$

where we have used the substitution

$$\tan \phi = \frac{\sqrt{\varepsilon^2 - 1} \sinh w}{\cosh w + \varepsilon} \tag{7}$$

and where

$$\phi_0 = \tfrac{1}{2}(\pi - \vartheta). \tag{8}$$

The functions P and Q in eq. (6) are the Legendre functions of first and second kind, which can be expressed in terms of elementary functions. The simplest result is obtained for $\mu = \pm\lambda$, where

$$I_{\lambda\pm\lambda}(\vartheta, 0) = 2 \frac{(\lambda - 1)!}{(2\lambda - 1)!!} (\sin(\tfrac{1}{2}\vartheta))^{\lambda}. \tag{9}$$

For other values of λ and μ, the Coulomb excitation functions are most easily obtained by the recursion relation for the Legendre functions Q. For $\lambda = 1, 2, 3$ and 4 explicit results are given in table 1 (see also [ALD 56]).

From eq. (9) follows the important result for backwards scattering and $\xi = 0$, i.e.

$$I_{\lambda\mu}(\pi, 0) = 2(\lambda - 1)!/(2\lambda - 1)!!. \tag{10}$$

For ξ different from zero, the orbital integrals can be expressed in terms of cylinder functions of imaginary arguments if $\lambda = 0$ and $\lambda = 1$. Thus one finds

$$I_{00}(\vartheta, \xi) = 2 \exp(-\tfrac{1}{2}\pi\xi) K_{i\xi}(\varepsilon\xi) \tag{11}$$

and

$$I_{1\pm1}(\vartheta, \xi) = -2\xi \exp(-\tfrac{1}{2}\pi\xi)\left[K'_{i\xi}(\varepsilon\xi) \pm \frac{(\varepsilon^2 - 1)^{1/2}}{\varepsilon} K_{i\xi}(\varepsilon\xi)\right]. \tag{12}$$

For higher values of λ, the Coulomb excitation functions can be evaluated in terms of confluent hypergeometric functions of two variables [ALD 56, ALD 56a]. These expressions offer a series expansion of $I_{\lambda\mu}(\vartheta, \xi)$ for small values of ξ. It is noted that $\xi = 0$ is an essential singularity since the series expansion contains a term $\xi^{\lambda} \log \xi$.

TABLE 1

Explicit expressions for the Coulomb excitation functions $I_{\lambda\mu}(\vartheta, 0)$ for $|\mu| \leqslant \lambda$ and $\lambda = 1, 2, 3$ and 4.

$$I_{10} = 2\frac{\pi - \vartheta}{2}\tan(\tfrac{1}{2}\vartheta)$$

$$I_{1\pm1} = 2\sin(\tfrac{1}{2}\vartheta)$$

$$I_{20} = 2\tan^2(\tfrac{1}{2}\vartheta)[1 - \frac{\pi - \vartheta}{2}\tan(\tfrac{1}{2}\vartheta)]$$

$$I_{2\pm1} = \frac{\tan^2(\frac{1}{2}\vartheta)}{\cos(\frac{1}{2}\vartheta)}\left[\tan(\tfrac{1}{2}\vartheta)\cos^2(\tfrac{1}{2}\vartheta) + \frac{\pi - \vartheta}{2}\right]$$

$$I_{2\pm2} = \tfrac{2}{3}\sin^2(\tfrac{1}{2}\vartheta)$$

$$I_{30} = \tan^3(\tfrac{1}{2}\vartheta)[-3\tan(\tfrac{1}{2}\vartheta) + \frac{\pi - \vartheta}{2}(3\tan^3(\tfrac{1}{2}\vartheta) + 1)]$$

$$I_{3\pm1} = \frac{2}{3}\frac{\tan^3(\frac{1}{2}\vartheta)}{\cos(\frac{1}{2}\vartheta)}\left[(3 - \cos^2(\tfrac{1}{2}\vartheta)) - 3\frac{\pi - \vartheta}{2}\tan(\tfrac{1}{2}\vartheta)\right]$$

$$I_{3\pm2} = \frac{1}{6}\frac{\tan^3(\frac{1}{2}\vartheta)}{\cos^2(\frac{1}{2}\vartheta)}\left[2\tan(\tfrac{1}{2}\vartheta)\cos^4(\tfrac{1}{2}\vartheta) + 3\tan(\tfrac{1}{2}\vartheta)\cos^2(\tfrac{1}{2}\vartheta) - 3\frac{\pi - \vartheta}{2}\right]$$

$$I_{3\pm3} = \tfrac{4}{15}\sin^3(\tfrac{1}{2}\vartheta)$$

$$I_{40} = \tfrac{1}{3}\tan^4(\tfrac{1}{2}\vartheta)\left[4 + 15\tan^2(\tfrac{1}{2}\vartheta) - 3\frac{\pi - \vartheta}{2}(5\tan^3(\tfrac{1}{2}\vartheta) + 3\tan(\tfrac{1}{2}\vartheta))\right]$$

$$I_{4\pm1} = \frac{1}{4}\frac{\tan^4(\frac{1}{2}\vartheta)}{\cos(\frac{1}{2}\vartheta)}\left[-(5\tan^3(\tfrac{1}{2}\vartheta) + 3\tan(\tfrac{1}{2}\vartheta))\cos^2(\tfrac{1}{2}\vartheta) - 10\tan(\tfrac{1}{2}\vartheta) + 3\frac{\pi - \vartheta}{2}(1 + 5\tan^2(\tfrac{1}{2}\vartheta))\right]$$

$$I_{4\pm2} = \frac{1}{10}\frac{\tan^4(\frac{1}{2}\vartheta)}{\cos^2(\frac{1}{2}\vartheta)}\left[-(5\tan^3(\tfrac{1}{2}\vartheta) + 3\tan(\tfrac{1}{2}\vartheta))\cos^4(\tfrac{1}{2}\vartheta) + 3(5\tan^2(\tfrac{1}{2}\vartheta) + 1)\cos^2(\tfrac{1}{2}\vartheta) + 5 - 15\frac{\pi - \vartheta}{2}\tan(\tfrac{1}{2}\vartheta)\right]$$

$$I_{4\pm3} = \frac{1}{60}\frac{\tan^4(\frac{1}{2}\vartheta)}{\cos^3(\frac{1}{2}\vartheta)}\left[-4(5\tan^5(\tfrac{1}{2}\vartheta) + 3\tan^3(\tfrac{1}{2}\vartheta))\cos^6(\tfrac{1}{2}\vartheta) + (50\tan^3(\tfrac{1}{2}\vartheta) + 12\tan(\tfrac{1}{2}\vartheta))\cos^4(\tfrac{1}{2}\vartheta) - 45\tan(\tfrac{1}{2}\vartheta)\cos^2(\tfrac{1}{2}\vartheta) + 15\frac{\pi - \vartheta}{2}\right]$$

$$I_{4\pm4} = \tfrac{4}{35}\sin^4(\tfrac{1}{2}\vartheta)$$

For the numerical evaluation of the Coulomb excitation functions it is most convenient to use the definition (1). In order to improve the convergence for large values of ξ one may translate the path of integration in (1) by an amount $+\frac{1}{2}i\pi$ whereby one obtains

$$I_{\lambda\mu}(\vartheta, \xi) = \exp(-\tfrac{1}{2}\pi\xi)\int_{-\infty}^{+\infty} dw \exp\{-\xi\varepsilon\cosh w + i\xi w\} \times \frac{[i\sinh w + \varepsilon - \sqrt{\varepsilon^2 - 1}\cosh w]^\mu}{(i\varepsilon\sinh w + 1)^{\lambda+\mu}}. \tag{13}$$

Numerical values of $I_{\lambda\mu}$ calculated in this way are given for $\lambda = 1, 2, 3$ and 4 in [ALD 56a].

The behaviour of the Coulomb excitation integrals for large values of ξ and arbitrary scattering angle has been studied by [BRU 62, BRU 64]. For small values of ϑ (i.e. $\sqrt{\varepsilon^2 - 1} \gg \xi^{-1/3}$) the Coulomb excitation functions according to eq. (13) may be written (see [ALD 56])

$$
\begin{aligned}
I_{\lambda\mu}(\vartheta, \xi) &= \exp(-\tfrac{1}{2}\pi\xi)\varepsilon^{-\lambda} \int_{-\infty}^{+\infty} \exp(\mathrm{i}\xi\varepsilon \sinh w) \frac{[1 + \mathrm{i} \sinh w]^{\mu}}{(\cosh w)^{\lambda+\mu}} \mathrm{d}w \\
&= (-1)^{(\lambda-\mu)/2} \varepsilon^{-1} \left(\frac{\xi}{2\varepsilon}\right)^{(\lambda-1)/2} \Gamma\left(\frac{-\lambda + \mu + 1}{2}\right) \\
&\quad \times \exp(-\tfrac{1}{2}\pi\xi) W_{-\mu/2, -\lambda/2}(2\xi\varepsilon), \qquad (14)
\end{aligned}
$$

where W is the Whittaker function. While eq. (14) holds for arbitrary values of ξ, one may in the limit $\xi\varepsilon \gg 1$ use the asymptotic expansion of the Whittaker function to obtain the simple result

$$
\begin{aligned}
&I_{\lambda\mu}(\vartheta, \xi) \\
&= \frac{2\pi}{\Gamma(\frac{1}{2}(\lambda - \mu + 1))} \exp\left\{ -\xi\left(\frac{1}{\sin(\frac{1}{2}\vartheta)} + \tfrac{1}{2}\pi\right)\right\} \xi^{(\lambda-\mu-1)/2} (\tfrac{1}{2} \sin(\tfrac{1}{2}\vartheta))^{(\lambda+\mu+1)/2}. \qquad (15)
\end{aligned}
$$

From this expression it is seen that for forward scattering the Coulomb excitation function $I_{\lambda-\lambda}$ dominates over all the other functions $I_{\lambda\mu}$ with $\mu \neq -\lambda$.

APPENDIX J

Estimates of Dipole Polarization Effects

In this appendix we study the dipole polarization in order to estimate the polarization corrections to the elastic and inelastic scattering. The calculations are based on the expressions (II.6.8) and the empirical knowledge of the giant dipole resonance.

We consider a nucleus with ground state spin I_0 and evaluate first the matrix element of V_{pol} between two magnetic substates M_0 and M_0'. According to eq. (II.6.8) one finds

$$\langle I_0 M_0 | \, V_{\text{pol}} \, | I_0 M_0' \rangle = -\sum_z \frac{\langle I_0 M_0 | \, V_{\text{E1}}(t) \, | I_z M_z \rangle \langle I_z M_z | \, V_{\text{E1}}(t) \, | I_0 M_0' \rangle}{E_z - E_0} \tag{1}$$

where the dipole field $V_{\text{E1}}(t)$ is given by

$$V_{\text{E1}}(t) = -\tfrac{4}{3}\pi \, |\boldsymbol{E}| \sum_\mu (-1)^\mu \, Y_{1-\mu}(\boldsymbol{E}) \, \mathscr{M}(\text{E1}, \mu), \tag{2}$$

and $\boldsymbol{E}$ denotes the electric field strength at the nucleus, i.e.

$$|\boldsymbol{E}| = Z_1 e / r^2. \tag{3}$$

Evaluating eq. (1) one finds

$$\begin{aligned} \langle I_0 M_0 | \, V_{\text{pol}} \, | I_0 M_0' \rangle = & -\tfrac{4}{9}\pi \, |\boldsymbol{E}|^2 \sum_z \frac{B(\text{E1}, I_0 \to I_z)}{E_z - E_0} \\ & \times \left\{ \delta_{M_0 M_0'} + (-1)^{I_z - M_0'} (2I_0 + 1) \begin{pmatrix} I_0 & I_0 & 2 \\ -M_0 & M_0' & -\kappa \end{pmatrix} \right. \\ & \left. \times \begin{Bmatrix} I_0 & I_0 & 2 \\ 1 & 1 & I_z \end{Bmatrix} \sqrt{24\pi} \, Y_{2\kappa}(\boldsymbol{E}) \right\}. \end{aligned} \tag{4}$$

The second term in (4) which depends on the direction of the field $\boldsymbol{E}$ with respect to the nuclear spin vanishes identically for ground state spin $I_0 = 0$

and $I_0 = \frac{1}{2}$. It also vanishes if the dipole mode is weakly coupled to the nuclear spin. In the following we shall neglect this term and write

$$\langle I_0 M_0 | V_{\text{pol}} | I_0 M_0' \rangle = -\delta_{M_0 M_0'} |\boldsymbol{E}|^2 P, \tag{5}$$

where P is the nuclear polarizability

$$P = \tfrac{4}{9}\pi \sum_z \frac{B(\text{E1}, 0 \to z)}{E_z - E_0}. \tag{6}$$

This polarizability can be estimated from the minus-two moment σ_{-2} of the photo-nuclear absorption cross section since [LEV 60]

$$P = \frac{\hbar c}{4\pi^2} \sigma_{-2}. \tag{7}$$

Inserting the empirical result

$$\sigma_{-2} = 3.5 A^{5/3}\ \mu\text{b/MeV} \tag{8}$$

one finds

$$P = 1.7 \times 10^{-3} A^{5/3}\ \text{fm}^3. \tag{9}$$

The polarizability may instead be estimated by writing eq. (6) in the form

$$P = \tfrac{4}{9}\pi \sum_z \frac{B(\text{E1}, 0 \to z)(E_z - E_0)}{(E_z - E_0)^2}. \tag{10}$$

The energy in the denominator may approximately be taken outside the summation and estimated through the known value of the energy of the giant dipole resonance. The rest can then be estimated by the dipole sum rule. Such estimates are in fair agreement with eq. (9).

For a deformed nucleus the giant dipole resonance is known to split, the energy splitting being inversely proportional to the relative change of the nuclear radius of the principal axes. It is seen from eq. (10) that this indicates that also the polarizability P_{ii} in the direction of the ith principal axes is different for $i = 1$, 2 and 3. In fact, one expects from (10)

$$P_{ii} = P_0 R_i^2 / R_0^2, \tag{11}$$

where P_0 is the polarizability of a spherical nucleus of the radius R_0, while R_i is the length of the ith axis.

In such a situation the polarizability is a tensor of second rank and the dipole moment which is induced by an external homogeneous electric field is only in the direction of the field if $\boldsymbol{E}$ is parallel to one of the three principal axes.

The situation is indicated in fig. 1 from which it is evident that such a polarizability may give rise to excitation of rotational states.

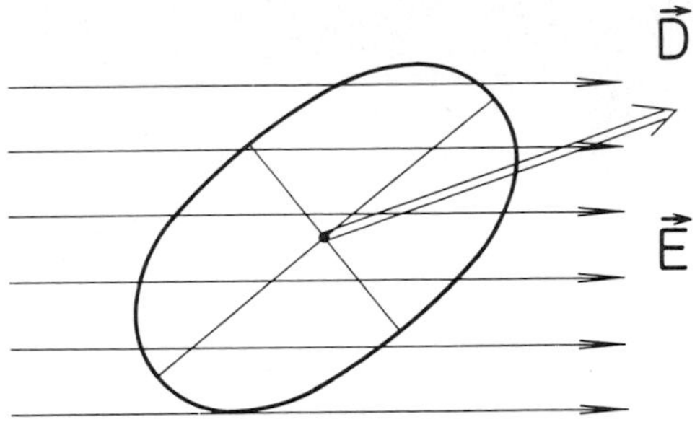

Fig. 1. A deformed nucleus in an external homogeneous electric field $\boldsymbol{E}$. The induced dipole moment $\boldsymbol{D}$ is not in the direction of the external field. Its direction depends on the orientation of the principal axes, which are indicated. The torque which is produced by the interaction between $\boldsymbol{E}$ and $\boldsymbol{D}$ may set the nucleus into rotation.

One may estimate the non-diagonal matrix elements of V_{pol} between nuclear states which are described by nuclear quadrupole deformations. We thus assume that the nuclear surface can be described by (see chapter VII)

$$R(\theta, \phi) = R_0\left[1 + \sum_{\mu} \alpha_{2\mu} Y^*_{2\mu}(\theta, \phi)\right] \tag{12}$$

and consider that the deformation parameters $\alpha_{2\mu}$ only influence the giant dipole resonance in an adiabatic fashion. Since the component of the induced dipole moment D_E in the direction of the external field θ, ϕ is given by

$$D_E = EP_0 R^2(\theta, \phi)/R_0^2 . \tag{13}$$

The polarization potential is given by

$$\begin{aligned} V_{\mathrm{pol}} &= -D_E E \\ &= -P_0 \boldsymbol{E}^2\left[1 + 2\sum_{\mu} \alpha_{2\mu} Y^*_{2\mu}(\boldsymbol{E})\right] \end{aligned} \tag{14}$$

to lowest order in $\alpha_{2\mu}$. Assuming the connection (see § VII.1)

$$\mathscr{M}(\mathrm{E2}, \mu) = \frac{3}{4\pi} Z_2 e R_0^2 \alpha_{2\mu} \tag{15}$$

between the electric quadrupole moment and the deformation parameter, we may write

$$V_{\mathrm{pol}} = -P_0 \frac{Z_1^2 e^2}{r^4}\left[1 + \frac{8\pi}{3Z_2 e} \sum_{\mu} \frac{1}{R_0^2} \mathscr{M}(\mathrm{E2}, \mu) Y^*_{2\mu}(\boldsymbol{E})\right]. \tag{16}$$

It is expected that eq. (16) gives an estimate of the non-diagonal matrix elements between nuclear states which can be described in terms of collective surface deformation, whether they are of vibrational or rotational nature. The total quadrupole interaction $V_{E2}(t) + V_{pol}(t)$ can thus be written as

$$V_{E2}(t) + V_{pol}(t) = Z_1 e \frac{4\pi}{5} \sum_{\mu} (-1)^{\mu} \frac{1}{r^3} \mathscr{M}(E2, -\mu) Y_{2\mu}(\theta, \phi) \left[1 - z(1, 1, 2) \frac{a}{r} \right], \tag{17}$$

with

$$z(1, 1, 2) = \frac{10 Z_1 P_0}{3 Z_2 R_0^2 a} \sim 0.5 \times 10^{-2} \frac{E_{MeV} A_2}{Z_2^2 (1 + A_1/A_2)}. \tag{18}$$

Alphabetical List of Symbols

Below we list alphabetically the symbols occurring in the text. First are listed symbols based on Latin letters, then on Greek letters, in order of increasing complexity. The reference to the first occurrence in the text is given after a short definition of the symbol. Symbols not based on letters are given at the end.

For the mathematical functions not given in the list, we use the notation of Handbook of Mathematical Functions, National Bureau of Standards, Applied Mathematics Series 55 (eds. M. Abramowitz and I. Stegun), Washington D.C., 1964, page 1044 [ABR 64].

a, A

a	half the distance of closest approach in a head-on collision in Coulomb field $(=b/2)$	(I.3)
a_n	excitation amplitude to state n	(I.8)
a_{if}	symmetrized distance of closest approach	(IV.7.2)
$a_{\lambda\mu}$, $a^{\dagger}_{\lambda\mu}$	annihilation and creation operator for phonon	(VII.1.3)
$a^{\pm}(r)$	quantal amplitudes	(IX.3.30)
$a_n(t)$	amplitude on state n as function of time	(II.3.4)
$a_n^{(0)}(t)$	zeroth-order amplitude	(II.4.27)
$a_n^{(1)}(t)$	first-order amplitude	(II.4.27)
$a_{n,k}(t) = a_{k\to n}(t)$	amplitude on the state n with initial condition $a_n(-\infty) = \delta_{nk}$	(II.4.8)
$\bar{a}_{l,n}(t)$	backwards solution with initial condition $a_l(+\infty) = \delta_{ln}$	(II.4.34)
$a^{\text{sim}}_{0\to n}(t)$	simultaneous excitation amplitude	(VI.7.1)
$a^{\text{mut}}_{0\to n}(t)$	mutual excitation amplitude	(VI.7.4)
$a_k^{\text{E}\lambda}(\xi)$	normalized, integrated particle parameter	(IV.4.27)

$a(\vartheta, s, \xi)$	coefficient of E4 excitation	(V.5.14)
$a_{k0}^{E\lambda}(\vartheta, \xi)$	normalized particle parameter for ring counter	(IV.4.25)
$a_{I_f M_f, I_0 M_0}$	excitation amplitude of state $I_f M_f$	(II.8.16)
A	Lenz vector	(II.9.5)
$A_{l\nu}$	real part of reduced amplitude	(V.4.1)
$A_{RM}(\vartheta, q)$	amplitude for transfer of rotational angular momentum	(VIII.2.13)

b, B

b	distance of closest approach in head-on collision	(I.1)
$b(\vartheta)$	distance of closest approach for deflection angle ϑ	(I.12)
$b_{\kappa\kappa}^{\lambda\lambda'}$	particle parameter for mixed first-order excitation	(IV.4.23)
$b_k^{\lambda}(\xi)$	particle parameter for λ-pole first-order excitation	(IV.4.28)
B_k	attenuation coefficient for finite counter	(X.6.16)
$B_{l\nu}$	imaginary part of reduced amplitude	(V.4.1)
$B(\pi\lambda, I_0 \rightarrow I_f)$	reduced transition probability for λ-pole γ-transition	(IV.4.4)

c, C

c	velocity of light	(II.1.2)
$c_n(t)$	excitation amplitude in adiabatic representation	(II.4.14)
$c(\lambda_1, \lambda_2)$	coefficient in multipole expansion	(II.1.8)
$c_{E\lambda}(\xi)$	quantal correction coefficient for total cross section	(IX.4.4)
$c(\vartheta, s, \xi)$	coefficient of interference between E2 and E2-E2 excitation	(V.5.19)
C_λ	restoring force in vibrational model	(VII.1.1)

d, D

$d_k^{E\lambda}(\xi)$	quantal correction coefficient for particle parameter	(IX.4.5)
$d_{mm'}^{j}(\beta)$	rotation matrix for $\alpha = \gamma = 0$	(D.12)

$d(\vartheta, s, \xi)$	coefficient of interference of E4 and E2-E2 excitation	(V.5.13)
D_λ	kinetic energy parameter in vibrational model	(VII.1.1)
D_E	induced dipole moment	(J.13)
$D_k(\vartheta, \xi)$	reorientation function for angular distribution	(V.6.9)
$D^j_{mm'}(\alpha, \beta, \gamma)$	rotation matrix (D-function)	(D.9)
$D^j_{mm'}(\mathrm{S} \to \mathrm{S}')$	rotation matrix for rotation $\mathrm{S} \to \mathrm{S}'$	(D.27)

differentials

$\mathrm{d}\tau$	volume element	(II.1.2)
$\mathrm{d}\omega_\mathrm{P}$	laboratory solid angle for projectile detection	(X.1.15)
$\mathrm{d}\omega_\mathrm{T}$	laboratory solid angle for target detection	(X.1.16)
$\mathrm{d}\Omega_\mathrm{T}$	centre-of-mass solid angle for target detection	(X.1.16)
$\mathrm{d}\Omega_\mathrm{P}$	centre-of-mass solid angle for projectile detection	(X.1.15)
$\mathrm{d}\sigma_{\mathrm{E}\lambda}$	differential cross section for electric λ-pole excitation	(IV.4.2)
$\mathrm{d}\sigma_{\mathrm{M}\lambda}$	differential cross section for Mλ-excitation	(IV.5.18)
$\mathrm{d}P_\gamma(\boldsymbol{k})$	probability of observing γ-quantum in direction $\boldsymbol{k}$	(III.3.2)
$\mathrm{d}f_{\mathrm{E}\lambda}(\vartheta, \xi)$	differential cross section function for Eλ excitation in first-order theory	(IV.4.5)
$\mathrm{d}f_{\mathrm{M}\lambda}(\vartheta, \xi)$	differential magnetic cross section function in first-order theory	(IV.5.19)
$\mathrm{d}N/\mathrm{d}E$	density of final states	(F.29)
$\mathrm{d}c_{\mathrm{E}\lambda}/\mathrm{d}\Omega$	quantal correction coefficient for differential cross section	(IX.4.3)
$(\mathrm{d}\sigma/\mathrm{d}\Omega)_n$	excitation cross section to state n	(I.5)
$(\mathrm{d}\sigma/\mathrm{d}\Omega)_\mathrm{R}$	Rutherford cross section	(I.3)

e, *E*

e	elementary charge ($e > 0$)	(I.1)
$e(\vartheta, \xi)$	quantal differential coefficient for reorientation effect	(IX.4.7)
E	energy in centre-of-mass system	(II.9.1)

E_P	bombarding energy	(X.1.1)
E_c	threshold energy in channel c	(II.7.1)
$\mathbf{E}$	electric field strength	(J.3)
E_n	energy of state n	(I.9)
$E_n^{(1)}$	energy of state n in nucleus 1	(C.4)
$\tilde{E}'$	effective final centre-of-mass energy	(X.1.6)
E'_T	final target energy in laboratory system	(X.1.20)
E'_P	final projectile energy in laboratory system	(X.1.19)
$E_n(t)$	eigenvalue of $H(t)$	(II.4.13)
$E_{(\lambda'\lambda'')\lambda}(\vartheta, \xi)$	polarization function	(V.7.4)

f, F

$f_{E\lambda}(\xi)$	total cross section function	(IV.4.10)
$f_{M\lambda}(\xi)$	total magnetic cross section function	(IV.5.20)
$f_n(\vartheta, \varphi)$	scattering amplitude	(IX.1.5)
$F_n(\omega)$	complex level shift	(II.7.12)
$F^c_{nm}(\omega)$	complex level shift function	(II.7.9)
$F_l(kr)$	regular Coulomb wave function	(II.7.22)
$F_k(LL'I_2I_1)$	F coefficient for γ-distribution	(G.11)
$F_k^{k_2k_1}(LL'I_2I_1)$	F coefficient for triple correlation	(G.21)

g, G

$g_{(lI_n)J}(r)$	radial wave function	(IX.1.9)
G	second-order term in K	(II.3.28)
$G(r, r')$	radial Green function	(IX.2.1)
$G_l(kr)$	irregular Coulomb wave function	(II.7.22)
$G_{kk'}^{\mu\mu'}$	perturbation coefficient for perturbed angular correlations	(III.5.6)
$G_{\mu\mu'}^{\lambda\lambda'}(\vartheta, \xi, \xi')$	imaginary part of second-order orbital integral	(V.1.13)
$G_{(\lambda\lambda')k\kappa}(\vartheta, \xi, \xi')$	imaginary part of second-order orbital integral tensors	(V.1.14)

h, H

h	helicity	(F.17)
$\hbar$	Planck's constant divided by 2π	(I.1)
h_κ	coupling operator	(VIII.1.7)

$h_l^{\pm}(kr)$	outgoing and incoming Coulomb wave function	(IX.2.2)
H	total Hamiltonian	(II.1.1)
H	third-order term in K	(II.3.29)
H_{o}	intrinsic Hamiltonian of free nucleus	(I.6)
H_{i}	intrinsic Hamiltonian	(VIII.1.1)
H_{r}	rotational Hamiltonian	(VIII.1.1)
H_{c}	coupling Hamiltonian	(VIII.1.1)
H_{γ}	interaction with radiation field	(G.2)
$H_{\mathrm{o}}(1)$	intrinsic Hamiltonian of nucleus 1	(C.2)
$H_{\mathrm{o}}'(1)$	Hamiltonian of free nucleus 1	(II.1.1)
$H(t)$	total Hamiltonian as function of time	(II.4.2)
$H_k(I_2I_1)$	coefficient for unobserved γ-transitions	(G.24)
$H_{\mu_1\mu_2\mu_3}^{\lambda_1\lambda_2\lambda_3}(\vartheta, \xi_1, \xi_2, \xi_3)$	principal part integral for third-order perturbation theory	(V.8.4)

i, *I*

$\boldsymbol{I}$	total angular momentum operator	(II.8.1)
$I_{\pm}$	$I_1 \pm \mathrm{i}I_2$ in intrinsic system	(VIII.1.7)
$\mathscr{I}$	moment of inertia	(VIII.1.2)
$I_{\lambda\mu}(\vartheta, \xi)$	Coulomb excitation function	(H.1)
$I_{l_0 l}^{\lambda}(\eta_0, \eta)$	radial matrix element	(IX.2.11)

j, *J*

$\boldsymbol{j}(\boldsymbol{r})$	current density	(II.1.2)
J	total angular momentum	(IX.1.8)

k, *K*

k	wave number	(II.7.22)
$\boldsymbol{k}$	wavevector	(G.1)
$\boldsymbol{k}_n$	wavevector for channel n	(IX.1.6)
$k_0(\theta, \phi)$	action integral for rotational motion	(VIII.2.8)
$k_{\lambda}(\theta, \phi)$	action integral for λ-pole intrinsic interaction for fixed orientation	(VIII.4.3)
K	projection of intrinsic angular momentum on symmetry axis	(VIII.1.4)
K	K-matrix	(II.3.18)
K_{λ}	reduced oscillator matrix element	(VII.1.12)

$K(\vartheta, \xi)$	reorientation function	(V.6.6)
$K_n(t' - t)$	kernel for decaying states	(II.7.26)

l, L

$\boldsymbol{l}$	orbital angular momentum	(II.9.3)
L	multipolarity of γ-radiation	(G.2)
$\boldsymbol{L}$	angular momentum operator	(II.1.5)

m, M

m_0	reduced mass	(I.4)
m_{P}	projectile mass	(X.1.2)
m_{T}	target mass	(X.1.2)
$\mathscr{M}(\mathrm{E}\lambda, \mu)$	electric multipole operator	(II.1.3)
$\mathscr{M}(\mathrm{M}\lambda, \mu)$	magnetic multipole operator	(II.1.4)

n, N

n	number of atoms per unit volume	(X.5.1)
N	principal quantum number	(VII.1.10)
N	total number of states	(VI.1.15)

p, P

$\boldsymbol{p}$	relative momentum	(II.9.4)
$\boldsymbol{p}_0$	average momentum in wavepacket	(B.11)
p_φ	generalized momentum conjugate to orientation angle φ	(VIII.6.1)
p_ϑ	generalized momentum conjugate to orientation angle ϑ	(VIII.6.1)
P	parity operator	(II.8.2)
P	nuclear polarizability	(J.6)
$\boldsymbol{P}$	vector polarization	(III.2.15)
P_n	excitation probability to state n	(I.5)
P^c	total probability of particle emission in channel c	(II.7.17)
$P^{(1)}$	first-order excitation probability	(V.3.2)
$P^{(12)}$	interference between first- and second-order probability	(V.3.3)
$P^{(2)}$	second-order excitation probability	(V.3.4)
$P_\perp$	parity reflection in plane of orbit	(II.8.6)
P_{ij}	tensor polarization	(III.2.19)

P_{ii}	nuclear polarizability along *i*th principal axis	(J.11)
$P^c(E)$	probability of particle emission at energy E in channel c	(II.7.16)
$P_n(t)$	probability for state n as a function of time	(II.4.11)
$P_{k\kappa}(I)$	polarization tensor	(F.41)
$P_{n\to m}(\vartheta, \xi)$	excitation probability of state m if nucleus initially in state n	(I.25)
$P_{(\lambda\lambda')k\kappa}(\vartheta, \xi)$	orbital integrals for polarization effects	(V.2.11)

q, *Q*

q	reorientation coefficient	(V.6.4)
q	strength parameter for rotational excitation	(VIII.2.9)
Q	quadrupole moment of nucleus	(I.20)
Q	phaseshift operator	(II.3.19)
$\boldsymbol{Q}$	vector describing polarization measurement of γ quantum	(F.34)
Q_0	intrinsic quadrupole moment	(VIII.1.12)
Q_μ	μ component of phaseshift	(VI.5.2)
$Q(E)$	penetration factor	(II.7.21)
$Q_{\lambda\mu}(\varepsilon, w)$	normalized collision function	(II.9.18)
$Q^{\mathrm{eff}}_{\lambda\mu}$	normalized collision functions including polarization	(VI.1.11)
$Q_{\mathrm{M}\lambda\mu}(\varepsilon, w)$	collision function for magnetic excitations	(II.9.28)

r, *R*

$\boldsymbol{r}$	relative position vector from 2 to 1	(I.6)
$\hat{\boldsymbol{r}}$	unit vector	(II.1.3)
$\dot{\boldsymbol{r}}$	$\mathrm{d}\boldsymbol{r}/\mathrm{d}t$	(II.1.9)
$r^J_{lI_n, l_0I_0}$	R-matrix in channel spin representation	(IX.1.12)
R	first-order term in K	(II.3.27)
R	nuclear radius	(II.7.22)
$\boldsymbol{R}$	rotational angular momentum in units of $\hbar$	(VIII.1.2)
$\boldsymbol{R}$	average position in wavepacket	(B.11)
R_0	charge radius	(VII.1.15)

$R_{\perp}$	rotation of 180° around normal to scattering plane	(II.8.6)
$R_{\parallel}$	rotation of 180° around symmetry axis	(II.8.7)
$R(\vartheta)$	relative excitation amplitude for different angles	(I.26)
$\boldsymbol{R}_s(t)$	classical relative position in channel s	(C.6)
$R_{\hat{\omega}}(\omega)$	rotation operator of state vector	(D.1)
$R(\alpha, \beta, \gamma)$	rotation operator specified by Eulerian angles	(D.7)
$R_{\lambda\mu}(\vartheta, \xi)$	normalized orbital integrals	(IV.2.2)
$R_{\mathrm{M}\lambda\mu}(\vartheta, \xi)$	normalized orbital integrals for magnetic excitation	(IV.5.7)
$R_{\lambda}^2(\vartheta, \xi)$	relative probabilities for electric excitations	(IV.3.3)
$R_{\mathrm{M}\lambda}^2(\vartheta, \xi)$	relative probabilities for magnetic excitations	(IV.5.12)
$R_{\mu\mu'}^{\lambda\lambda'}(\vartheta, \xi, \xi')$	real part of second-order orbital integral	Table V.1
$R_{(\lambda\lambda')k\kappa}(\vartheta, \xi, \xi')$	real part of second-order orbital integral tensors	(V.1.12)

s, S

s	energy ratio in double excitation	(V.5.6)
s	spin in units of $\hbar$	(F.17)
S	coordinate systems	(D.5)
S	classical S-matrix	(II.3.17)
$\boldsymbol{s}_i$	spin of particle i	(D.1)
$S(E)$	stopping cross section	(X.5.1)
$S_{\mathrm{E}\lambda\mu}$	orbital integral for electric excitations	(IV.1.4)
$S_{\mathrm{M}\lambda\mu}$	orbital integrals for magnetic excitations	(IV.5.2)
$\bar{S}_{\mathrm{M}\lambda\mu}(t)$	collision function for magnetic λ-pole excitation	(II.2.18)
$\bar{S}_{\mathrm{E}\lambda\mu}(t)$	collision function for electric λ-pole excitation	(II.2.17)

t, T

t	time	(I.6)
t	target thickness	(X.5.8)
T	time reversal operator	(II.8.4)

T	quantal T-matrix	(III.1.4)
$\mathcal{T}$	time ordering operator	(II.3.11)
T_{rel}	relative kinetic energy	(IX.1.1)

u, *U*

u_{P}	initial projectile centre-of-mass velocity	(X.1.2)
u_{T}	initial target centre-of-mass velocity	(X.1.3)
u'_{P}	final projectile centre-of-mass velocity	(X.1.4)
u'_{T}	final target centre-of-mass velocity	(X.1.5)
U	unitary matrix diagonalizing Q	(VI.2.3)
$U(t_2, t_1)$	propagator in interaction representation	(II.3.6)

v, *V*

v	relative velocity at large distances	(I.1)
v	symmetrized velocity	(VI.8.9)
v_{i}	initial relative velocity at large distances	(IV.7.1)
v_{f}	final relative velocity at large distances	(IV.7.1)
v_n	relative velocity at large distances when nucleus in state n	(VI.8.8)
$\boldsymbol{v}_s(t)$	classical relative velocity in channel s	(C.7)
V	normalization volume	(G.2)
$V(t)$	interaction energy	(II.2.15)
$\tilde{V}(t)$	interaction energy in interaction representation	(II.3.3)
$V_{\text{pol}}(t)$	polarization potential	(II.6.8)
$V_{\text{E}\lambda}(t)$	electric λ-pole interaction	(J.2)
$V(1, \boldsymbol{r})$	interaction of charge of 2 with multipoles of 1	(II.1.10)
$V_{\text{E}}(2, \boldsymbol{r})$	interaction of electric multipoles of 2 with charge of 1	(II.1.12)
$V_{\text{M}}(2, \boldsymbol{r})$	interaction of magnetic multipoles of 2 with charge of 1	(II.1.13)
$\tilde{V}_{\text{XY}}(t)$	interaction operator between groups X and Y	(VI.6.2)
$\hat{V}_{n,m}(x)$	Fourier transformed interaction matrix element	(II.5.5)
$V_{\text{o}}(\theta, \phi, t)$	interaction with rotational motion	(VIII.2.1)
$V_{\text{i}}(\theta, \phi, t)$	interaction with intrinsic motion for fixed orientation	(VIII.2.1)
$V^J_{lI_n, l'I_m}(r)$	coupling potential	(IX.1.11)

w, *W*

w	parameter in hyperbolic orbit	(II.9.10)
w'_{P}	final laboratory velocity of projectile	(X.1.10)
w'_{T}	final laboratory velocity of target	(X.1.11)
W	coupling to continuum	(II.7.3)
W^{C}	normalized γ-distribution (in coordinate system C)	(III.3.6)
$W(1, 2)$	mutual electromagnetic interaction	(II.1.1)
$W_{\mathrm{E}}(1, 2)$	electric multipole-multipole interaction	(II.1.6)
$W_{\mathrm{M}}(1, 2)$	magnetic multipole-multipole interaction	(II.1.6)
$W_{\mathrm{EM}}(1, 2)$	electric-magnetic multipole-multipole interaction	(II.1.6)
$W_j^c(E')$	matrix element of W	(II.7.3)
$W^{cc'}(E', E'')$	matrix element of W	(II.7.3)

x, *X*

x	parameter for E4 excitation	(V.5.10)
$X_s(\boldsymbol{r}, t)$	wavepacket of relative motion	(C.5)

y, *Y*

y	parameter for indirect excitation	(V.5.18)
Y_{pk}	yield of process p on nucleus k	(X.5.5)
$Y_{\lambda\mu}(\hat{\boldsymbol{r}})$	spherical harmonics (Condon and Shortley phases)	(II.1.3)

z, *Z*

$z(\lambda'\lambda''\lambda)$	polarization coefficient	(V.7.5)
Z_1	charge number of nucleus 1	(I.1)
Z_2	charge number of nucleus 2	(I.1)

α

$\alpha_{l\kappa\kappa}$	second-order integrals old notation	(V.2.4)
$\alpha_{\lambda\mu}$	deformation parameter in vibrational model	(VII.1.1)
$\alpha_n(\omega)$	Fourier transformed amplitude	(II.5.7)

β

β	β-deformation amplitude	(VII.4.1)
β_k	statistical tensor for ring counter	(III.3.10)
$\beta_{k\kappa}$	second-order integrals old notation	(V.2.4)
$\beta_n(\omega)$	Fourier transformed amplitude	(II.5.2)
$\beta^c_{E'}(\omega)$	Fourier components in continuum	(II.7.4)

γ, Γ

γ	γ-deformation amplitude	(VII.4.1)
γ_i	total decay probability of state i	(F.28)
$\gamma_{i\to f}$	partial decay probability	(F.30)
$\Gamma_n(\omega)$	level width	(II.7.14)
$\Gamma^c_n(\omega)$	partial level width	(II.7.16)

δ, Δ

δ_{0n}	Kronecker symbol	(II.4.6)
$\delta_{\pi L}$	transition amplitude for γ radiation of multipolarity πL	(G.2)

differentials

$\delta\vartheta$	uncertainty in deflection angle	(B.2)
Δl	angular momentum transfer	(II.2.9)
ΔE_n	excitation energy $0 \to n$	(I.11)
Δa_n	correction to excitation amplitude	(VI.4.4)
$\Delta^{(1)} a_n$	first-order correction to $\chi(\vartheta)$ approximation	(VI.5.5)
$\Delta^{(2)} a_n$	second-order correction to $\chi(\vartheta)$ approximation	(VI.5.6)

ε

ε	excentricity of hyperbola	(II.9.9)
ε	efficiency operator	(F.11–14)
ε_k	mixing coefficients	(VIII.1.10)
$\varepsilon(t' - t)$	step function	(II.3.15)

ζ

ζ_1	intrinsic coordinates in nucleus 1	(C.3)
$\zeta^{\lambda\mu}_{mn}$	geometrical coefficient	(V.8.8)

η

η	Coulomb parameter	(I.1)
$\bar{\eta}$	Average Coulomb parameter	(IX.3.47)
η_{cl}	classical action	(II.9.6)

ϑ, θ

ϑ	deflection angle in CM system	(I.3)
θ	polar coordinate	(II.1.7)
ϑ_{P}	angle between $\boldsymbol{v}_{\text{i}}$ and $\boldsymbol{v}'_{\text{P}}$	(X.1.8)
ϑ_{T}	angle between $\boldsymbol{v}_{\text{i}}$ and $\boldsymbol{v}'_{\text{T}}$	(X.1.9)
θ_{P}	angle between $\boldsymbol{v}_{\text{i}}$ and $\boldsymbol{w}'_{\text{P}}$	(X.1.8)
θ_{T}	angle between $\boldsymbol{v}_{\text{i}}$ and $\boldsymbol{w}'_{\text{T}}$	(X.1.9)
$\theta(t - t')$	step function	(II.3.15)

λ, Λ

λ	multipole order	(I.19)
$\lambdabar$	wave length in relative motion at large distances	(I.1)
λ_n	eigenvalues of Q	(VI.2.3)
Λ	number of channels	(II.7.1)

μ

μ	z-component of λ	(II.1.7)
μ_m	eigenvalues of ρ	(VI.3.7)

ν

ν_m	eigenvalues of ρR	(VI.3.4)
ν_i	relative atomic abundance of substance i	(X.5.3)

ξ

$\xi_{0 \to n} = \xi_{n,0}$	adiabaticity parameter	(I.15)
$\xi(\vartheta)$	adiabaticity parameter for deflection angle ϑ	(I.14)
$\bar{\xi}$, $\tilde{\xi}$	adiabaticity parameters for second-order processes	(V.2.7–8)

π, Π

$\pi_I(\vartheta, s, \xi)$	reduced double excitation probability	(V.5.3)
$\Pi_{\text{E}\lambda}(\vartheta, \xi)$	polarization function for Eλ excitation	(IV.4.19)

ρ

ρ	impact parameter	(II.9.7)
ρ	density operator	(F.2)

ρ_σ	density operator normalized to cross section	(F.27)
$\rho_{pq}^{\lambda\mu}$	geometrical coefficient used in diagonalization	(VI.3.3)
$\rho(\boldsymbol{r})$	charge density	(II.1.2)
$\rho_{k\kappa}(I_1, I_2)$	statistical tensor	(F.37)

σ

σ	width of wavepacket	(C.9)
σ_0	Coulomb phaseshift	(III.1.6)
σ_x	uncertainty in coordinate x	(B.11)
$\boldsymbol{\sigma}$	Pauli matrices	(F.34)
σ_{-2}	-2 moment of photonuclear cross section	(J.7)
$\sigma_{\mathrm{E}\lambda}$	total cross section for electric λ-pole excitation	(IV.4.1)
$\sigma_{\mathrm{M}\lambda}$	total cross section for Mλ excitation	(IV.5.20)
σ_{p_x}	uncertainty in momentum p_x	(B.14)
$\sigma_l(\eta)$	Coulomb phase shift	(IX.1.14)

τ

τ	collision time	(I.13)
τ	parameter for centre-of-mass transfer	(X.1.8)
τ	circular polarization quantum number for γ-quantum	(G.1)
$\tilde{\tau}$	parameter for centre-of-mass transfer	(X.1.9)
$\tau_n(\omega)$	classical T-matrix	(II.5.10)

φ, ϕ, Φ

φ	azimuthal scattering angle	(III.2.7)
ϕ	polar coordinate	(II.1.7)
ϕ_0	incoming flux	(III.2.7)
$\varphi_{I_n l}(r)$	asymptotic phase	(IX.1.13)
$\boldsymbol{\Phi}_{l\lambda\mu}$	vector spherical harmonic	(II.1.9)

χ

$\chi^{(\lambda)}$	strength parameter for multipole order λ	(I.19)
χ_{eff}	effective χ-parameter	(VI.3.10)
χ_{int}	intrinsic χ-parameter	(VII.4.9–10)
$\chi_{n\to m}$	strength parameter for backward scattering	(I.27)

$\chi_{n\to m}(\vartheta)$	strength parameter for deflection angle ϑ	(I.18)
$\chi(\boldsymbol{r})$	shape of wavepacket	(C.5)
$\chi_{n,0}^{(\lambda)} = \chi_{0\to n}^{(\lambda)}$	strength parameter for λ-pole interaction	(IV.3.6)
$\chi_{0\to n}^{(M\lambda)}$	magnetic strength parameter	(IV.5.5)

ψ, Ψ

$\psi_n^{(1)}(\zeta_1)$	intrinsic eigenstate n of nucleus 1	(C.4)
$\psi_{jm}(\boldsymbol{r})$	wave function of total angular momentum j, m	(D.10)
$\Psi(t)$	total wave function including relative motion	(C.1)

ω

ω	frequency	(II.5.1)
ω^+	$\omega + \mathrm{i}\varepsilon$	(II.5.8)
ω_{L}	Larmor precession frequency	(III.5.9)
ω_λ	oscillator frequency	(VII.1.2)

Symbols

kets

$\|0\rangle$	ground state	(I.7)
$\|n\rangle$	free nuclear eigenstate	(I.8)
$\|z\rangle$	intermediate state	(V.1.4)
$\|\Phi\rangle$	state vector in intcraction representation	(II.3.1)
$\|\phi_m\rangle$	eigenstates of $H_0 - W$	(II.7.1)
$\|IM\rangle$	eigenstate of $\boldsymbol{I}^2$, I_z	(II.8.3)
$\|n(t)\rangle$	eigenstate of $H(t)$	(II.4.13)
$\|\chi_{E'}^{c}\rangle$	continuum eigenstates of $H_0 - W$	(II.7.1)
$\|\Psi(\boldsymbol{k}_0)\rangle$	scattering state	(IX.1.5)
$\|\psi(t)\rangle$	time-dependent state vector	(I.6)
$\|\hat{\psi}(\omega)\rangle$	Fourier transformed state vector	(II.5.1)
$\|\psi_{\mathrm{int}}(t)\rangle$	intrinsic state vector for both nuclei	(II.2.13)
$\|IKMN\rangle$	pure rotational state	(VIII.1.6)
$\|IKMN\rangle_m$	mixed rotational state	(VIII.1.9)
$\|(lI_n)JN\rangle$	channel spin wave function	(IX.1.8)
$\langle I\\|\mathscr{M}(\lambda)\\|I'\rangle$	reduced matrix element	(IV.1.8)

Mathematical symbols

Tr	trace of matrix	(F.3)
arg	argument of complex number	(III.1.6)

det	determinant	(II.7.11)
$\binom{n}{m}$	binomial coefficient	(D.15)
† , h.c.	Hermitian conjugate	(II.8.14)
*	complex conjugate	(II.3.14)

Geometrical coefficients

$\begin{pmatrix} j_1 & j_2 & j_3 \\ m_1 & m_2 & m_3 \end{pmatrix}$	3-j symbol	(II.1.7)
$\begin{Bmatrix} a & b & c \\ d & e & f \end{Bmatrix}$	6-j symbol	(G.11)
$\begin{Bmatrix} a & b & c \\ d & e & f \\ g & i & j \end{Bmatrix}$	9-j symbol	(G.6)

References

ABR 64 Abramowitz, M. and I. A. Stegun, eds., 1964, *Handbook of Mathematical Functions* (National Bureau of Standards, Washington D.C.).

ALD 54 Alder, K. and A. Winther, 1954, Phys. Rev. **96**, 237.

ALD 55 Alder, K. and A. Winther, 1955, Mat. Fys. Medd. Dan. Vid. Selsk **29**, Nos. 18 and 19.

ALD 56 Alder, K., A. Bohr, T. Huus, B. Mottelson and A. Winther, 1956, Rev. Mod. Phys. **28**, 432, reprinted in [ALD 66] p. 77.

ALD 56a Alder, K. and A. Winther, 1956, Mat. Fys. Medd. Dan. Vid. Selsk. **31**, No. 1.

ALD 60 Alder, K. and A. Winther, 1960, Mat. Fys. Medd. Dan. Vid. Selsk. **32**, No. 8, reprinted in [ALD 66] p. 209.

ALD 62 Alder, K. and A. Winther, 1962, Nucl. Phys. **37**, 194.

ALD 64 Alder, K. and R. M. Steffen, 1964, Ann. Rev. Nucl. Sci. **14**, 403.

ALD 66 Alder, K. and A. Winther, 1966, *Coulomb Excitation* (Academic Press, New York).

ALD 69 Alder, K. and H. K. A. Pauli, 1969, Nucl. Phys. **A128**, 193.

ALD 69a Alder, K. and A. Winther, 1969, Nucl. Phys. **A132**, 1.

ALD 72 Alder, K., F. Roesel and R. Morf, 1972, Nucl. Phys. **A186**, 449.

BEN 56 Benedict, F. D., P. B. Daitch and G. Breit, 1956, Phys. Rev. **101**, 171.

BEY 69 Beyer, K. and A. Winther, 1969, Phys. Lett. **30B**, 296.

BIA 69 Bialynicki-Birula, I., B. Mielnik and J. Plebonski, 1969, Ann. of Phys. **51**, 187.

BIE 55 Biedenharn, L. C., J. L. McHale and R. M. Thaler, 1955, Phys. Rev. **100**, 376.

BIE 65 Biedenharn, L. C. and P. J. Brussaard, 1965, *Coulomb Excitation* (Clarendon Press, Oxford).

BLA 52 Blatt, J. M. and L. C. Biedenharn, 1952, Rev. Mod. Phys. **24**, 258.

BOE 65 De Boer, J., R. G. Stokstad, G. D. Symons and A. Winther, 1965, Phys. Rev. Lett. **14**, 564.

BOE 66 Boehm, F., G. B. Hagemann and A. Winther, 1966, Phys. Lett. **21**, 217.

BOE 68 De Boer, J. and J. Eichler, 1968, *Advances in Nuclear Physics*, Vol. 1, eds. M. Baranger and E. Vogt, p. 1.

BOH 13 Bohr, Niels, 1913, Phil. Mag. (6) **25**, 10.

BOH 48 Bohr, Niels, 1948, Mat. Fys. Medd. Dan. Vid. Selsk. **18**, No. 8.

BOH 69 Bohr, A. and B. Mottelson, 1969, *Nuclear Structure*, Vol. I (Benjamin Press, New York).

BOH 75 Bohr, A. and B. Mottelson, 1975, *Nuclear Structure*, Vol. II (Addison-Wesley Publ. Co. Inc., Reading, Mass.).

BRE 54 Breit, G. and P. B. Daitch, 1954, Phys. Rev. **96**, 1447.
BRE 55 Breit, G. and P. B. Daitch, 1955, Proc. Nat. Acad. Sci. **41**, 653.
BRE 56 Breit, G., R. L. Gluckstern and J. E. Russell, 1956, Phys. Rev. **103**, 727, reprinted in [ALD 66] p. 189.
BRE 59 Breit, G. and R. L. Gluckstern, 1959, *Encyclopedia of Physics*, Vol. 41/1 (Springer Verlag, Berlin) p. 496.
BRE 60 Breit, G. and R. L. Gluckstern, 1960, Nucl. Phys. **20**, 188.
BRO 72 Broglia, R. A. and A. Winther, 1972, Phys. Reports **4C**, 153.
BRU 62 Brussaard, P. J., T. A. Griffy and L. C. Biedenharn, 1962, Ann. d. Phys. (7) **10**, 47.
BRU 64 Brussaard, P. J., T. A. Griffy and L. C. Biedenharn, 1964, Ann. d. Phys. (7) **13**, 208.
CON 51 Condon, E. U. and G. H. Shortley, 1951, *The Theory of Atomic Spectra* (Cambridge University Press, Cambridge).
DAV 72 Davies, J., 1972, Lecture presented at Trieste, March 1971 (IAEA, Wien).
DEV 57 Devons, S. and L. J. B. Goldfarb, 1957, *Encyclopedia of Physics*, Vol. 42, ed. S. Flügge (Springer Verlag, Berlin) p. 362.
DOU 62 Douglas, A. C., 1962, Double Excitation in Even Nuclei. Tables of Cross Section Functions, AWRE Report No. NR/P-2/62, Aldermaston.
EDM 57 Edmonds, A. R., 1957, *Angular Momentum in Quantum Mechanics* (Princeton University Press, Princeton).
FOL 54 Foldy, L. and E. Eriksen, 1954, Phys. Rev. **95**, 1048.
FRA 65 Frauenfelder, H. and R. M. Steffen, 1965, in: *Alpha-, Beta and Gamma Ray Spectroscopy*, Vol. 2, ed. K. Siegbahn (North-Holland Publ. Co., Amsterdam) p. 997.
FRA 65a Franz, W., 1965, Z. f. Physik **184**, 85.
GAB 70 Gabriel, H. and J. Bosse, 1970, in: *Proc. Intern. Conf. on Angular Correlations in Nuclear Disintegration*, eds. H. van Kruyten and B. van Nooijen (Rotterdam University Press, Groningen 1971) p. 394.
GLU 60 Gluckstern, R. L. and G. Breit, 1960, *Proc. Second Conf. on Reactions Between Complex Nuclei* (John Wiley and Sons Inc., New York) p. 77.
GOL 64 Goldberger, M. L. and K. M. Watson, 1964, *Collision Theory* (John Wiley and Sons Inc., New York).
GOM 49 Gombàs, P., 1949, *Die statistische Theorie des Atoms und ihre Anwendungen* (Springer, Wien).
GRO 66 Grodzins, L., R. Borchers and G. B. Hagemann, 1966, Phys. Lett. **21**, 214.
HAM 74 Hamilton, W. D., ed., 1974, *The Electromagnetic Interaction in Nuclear Physics* (North-Holland Publ. Co., Amsterdam).
HAN 72 Hansteen, J. M. and O. P. Mosebekk, 1972, Nucl. Phys. **A201**, 541.
HEY 56 Heydenburg, N. P. and G. M. Temmer, 1956, Ann. Rev. Nucl. Sci. **6**, 77.
HIR 54 Hirschfelder, J. O., C. F. Curtiss and R. B. Bird, 1954, *Molecular Theory of Gases and Liquids* (Wiley and Son Inc., New York).
HUB 58 Huby, R., 1958, *Reports on Progress in Physics*, Vol. XXI (The Physical Society, London) p. 59.
HUU 53 Huus, T. and C. Zupancic, 1953, Mat. Fys. Medd. Dan. Vid. Selsk. **28**, No. 1.
LAN 62 Landau, L. D. and E. M. Lifshitz, 1962, *The Classical Theory of Fields* (Pergamon Press, Oxford).
LAZ 55 Lazarus, J. P. and S. Sack, 1955, Phys. Rev. **100**, 370.
LEV 60 Levinger, J. S., 1960, *Nuclear Photo-Disintegration* (Oxford University Press, Oxford).
LÜT 64 Lütken, H. and A. Winther, 1964, Mat. Fys. Skr. Dan. Vid. Selsk. **2**, No. 6.

MAH 69 MAHAUX, C. and H. A. WEIDENMÜLLER, 1969, *Shell Model Approach to Nuclear Reactions* (North-Holland Publ. Co., Amsterdam).

MAS 65 MASSO, J. F. and D. L. LIN, 1965, Phys. Rev. **B140**, 1182.

MCC 53 MCCLELLAND, C. I . and C. GOODMAN, 1953, Phys. Rev. **91**, 760.

MES 64 MESSIAH, A., 1964, *Quantum Mechanics* (North-Holland Publ. Co., Amsterdam).

MIC 74 MICHELIN, Guide France, 1974.

MOT 49 MOTT, N. F. and H. S. MASSEY, 1949, *The Theory of Atomic Collision* (Oxford University Press, Oxford).

MOT 52 MOTTELSON, B., 1952, *Intern. Phys. Conf.*, Copenhagen, reprinted in [ALD 66] p. 11.

NAK 66 NAKAMURA, K., 1966, Phys. Rev. **152**, 955.

ROB 61 ROBINSON, D. W., 1961, Nucl. Phys. **25**, 459.

ROB 63 ROBINSON, D. W., 1963, Helv. Phys. Acta **36**, 140.

ROM 65 ROMAN, P., 1965, *Advanced Quantum Theory* (Addison-Wesley Publ. Co. Inc.).

ROS 67 ROSE, H. J. and D. M. BRINK, 1967, Rev. Mod. Phys. **39**, 306.

ROS 73 ROESEL, F., K. ALDER and U. SMILANSKY, 1973, Ann. of Phys. **78**, 518.

SIM 65 SIMONIUS, M., W. BIERTER, A. ZWICKY and K. ALDER, 1965, Helv. Phys. Acta **38**, 669.

SMI 68 SMILANSKY, U., 1968, Nucl. Phys. **A112**, 185.

SMI 68a SMILANSKY, U., 1968, Nucl. Phys. **A118**, 529.

SOM 39 SOMMERFELD, A., 1939, *Atombau und Spektrallinien* (F. Vieweg und Sohn, Braunschweig) pp. 495 ff.

SPE 70 SPETH, E., K. O. PFEIFFER and K. BETHGE, 1970, Phys. Rev. Lett. **24**, 1493.

STE 59 STEPHENS, F. S., R. M. DIAMOND and I. PERLMAN, 1959, Phys. Rev. Lett. **3**, 435, reprinted in [ALD 66] p. 205.

STE 63 STELSON, P. H. and F. K. MCGOWAN, 1963, Ann. Rev. Nucl. Sci. **13**, 163.

TAY 69 TAYLOR, B. N., W. H. PARKER and D. N. LANGENBERG, 1969, Rev. Mod. Phys. **41**, 374.

TER 52 TER-MARTIROSYAN, K. A., 1952, Zh. Eksperim. i. Teor. Fiz. **22**, 284; English transl. in [ALD 66] p. 15; German transl. in: Fortschr. der Physik, Bd. 1.

TRA 70 TRAUTMANN, D. and K. ALDER, 1970, Helv. Phys. Acta **43**, 363.

UEB 71 UEBERALL, H., 1971, *Electron Scattering from Complex Nuclei* (Academic Press, New York and London).

WEI 71 WEIDENMÜLLER, H. and A. WINTHER, 1971, Ann. of Phys. **66**, 218.

WIN 65 WINTHER, A. and J. DE BOER, 1965, A Computer Program for Multiple Coulomb Excitation, California Institute of Technology, Technical Report, reprinted in [ALD 66] p. 303.

Subject Index